Linear Algebra

Linear Algebra

Prof. Mohd. Zubair Khan (Retd.)
Department of Mathematics
Aligarh Muslim University
Aligarh (UP)

I.K. International Pvt. Ltd.
NEW DELHI

Published by
I.K. International Pvt. Ltd.
4435–36/7, Ansari Road, Daryaganj
New Delhi-110 002 (India)
E-mail: info@ikinternational.com
Website: www.ikbooks.com

ISBN 978-93-90620-11-1

Published by Krishan Makhijani for I.K. International Pvt. Ltd., 4435–36/7, Ansari Road, Daryaganj, New Delhi-110 002.

Preface

Linear algebra is basically the study of linear sets of equations and their transformation properties. Matrices, determinants and elementary abstract algebra are useful tools of linear algebra. While teaching this subject, I came across three types of books, viz., books dealing with applications and avoiding the proofs of important results, advanced level books that are excellent but very often the students and even some teachers are not ready for them, and texts of elementary nature. Apart from these aspects, the books depend on other other literature for different concepts.

This book contains nine chapters. Chapter 1 discusses different types of matrices along with their properties, determinants and different methods of solutions of linear equations with sufficient examples. Chapter 2 deals with sets, groups, rings and fields with their elementary properties along with a good number of examples and counter-examples. In Chapter 3, vector spaces, linear dependence and independence along with examples are presented. Chapter 4 covers linear transformations and related example-oriented topics. In Chapter 5, linear transformations are described along with the matrix representation of the sum and composition of linear transformations. The concepts are further illustrated with solutions of many problems. Chapter 6 presents dual spaces with many examples to find the dual basis. Chapter 7 deals with eigenvalues, eigenvectors and diagonalisation problems. Chapter 8 deals with inner products and many solved examples of related topics while Chapter 9 describes bilinear functional and quadratic forms with examples.

The present book is self-explanatory. All the concepts and topics of matrices, sets and elementary abstract algebra needed for subsequent use are included. This book also contains more examples and counter-examples of the concepts and topics used in the text. More stress is laid on understanding of the subject and, therefore, the approach is not too advanced for the reader and motivations are clearly mentioned. I feel that after going through the book, the reader will not have any difficulty in understanding the subject.

I am very thankful to all my colleagues and friends for inspiring me to write this book.

Prof. Mohd. Zubair Khan

Preface

[illegible]

Contents

1

Matrices and Linear Equations

In this chapter we shall mostly discuss the working rules instead of dealing with theory on above topics. We shall also deal with different types of matrices and operations on matrices and methods of solving of linear equations with examples.

1 MATRICES

A collection of *mn* numbers (scalars, elements of a field) arranged in a rectangular array of the form

$$A = \begin{bmatrix} a_{11} & a_{12} & \cdots & a_{1n} \\ a_{21} & a_{22} & \cdots & a_{2n} \\ \vdots & \vdots & \vdots & \vdots \\ a_{m1} & a_{m2} & \cdots & a_{mn} \end{bmatrix}$$

is called a *matrix* of order $m \times n$ and real m by n and a_{ij} are known to be entries of this matrix. The m horizontal array $[a_{i1}, a_{i2}, \ldots a_{in}]$ are known as rows and n perpendicular (vertical) arrays

$$\begin{bmatrix} a_{1j} \\ a_{2j} \\ \vdots \\ a_{mj} \end{bmatrix}$$

are known as columns. In short we write $A = (a_{ij})$ for $i = 1, 2, \; = 1, 2, \ldots, m; j = 1, 2, \ldots, n$ and a_{ij} is i^{th} row and j^{th} column entry. We also denote by R_i the ith row and C_i and j^{th} column. Thus

$$A = \begin{bmatrix} R_1 \\ R_2 \\ \vdots \\ R_m \end{bmatrix} = [C_1 \quad C_2 \quad \cdots \quad C_n]$$

If A is of order $m \times n$, then we also say that A is of size m by n. If $m = n$, then A is called a *square matrix.* If A is a square matrix, then the entries

$a_{11}, a_{22}, \ldots, a_{mn}$ are the entries of main diagonal. If $a_{11} = a_{22} = \ldots = a_{nn}$ and $a_{ij} = 0$ for $i \neq j$, then A is called a *scalar matrix*. If some $a_{ii} \neq a_{jj}$ and the rest entries $a_{ij} \neq 0$ for $i \neq j$, then it is called a *diagonal matrix*. If A is a square matrix and $a_{ii} = 1$ for all $i = 1, 2, \ldots, n$ and $a_{ij} = 0$ for $i \neq j$, then A is called *identity matrix* of order (size) n. If

$$A = \begin{bmatrix} 1 & 2 & 0 \\ -1 & 3 & 4 \end{bmatrix}, B = \begin{bmatrix} 1 & -1 \\ 2 & 3 \\ 0 & 4 \end{bmatrix} \text{ and } C = [5]$$

the A is 2×3, B is 3×2 and C is 1×1, i.e., every scalar can be considered to be matrix of order 1×1.

If $A = [2 \quad 4 \quad 0 \quad -3]$, $B = \begin{bmatrix} 2 \\ 4 \\ 0 \\ -3 \end{bmatrix}$, $C = \begin{bmatrix} -4 & 0 \\ 0 & 2 \end{bmatrix}$, $D = \begin{bmatrix} 1 & 0 \\ 0 & 1 \end{bmatrix}$ and

$E = \begin{bmatrix} 2 & 0 \\ 0 & 2 \end{bmatrix}$, then A is a row matrix (row vector), B is column matrix (column vector), C is a diagonal matrix, D is identity matrix of size 2, E is a scalar matrix and $-4, 2; 1, 1; 2, 2$ are the main diagonal entries of C, D and E, respectively. If $A = (a_{ij})$ such that $a_{ij} = 0$ for all i and j, then A is called *zero matrix*. A square matrix A is called *upper triangular* if $a_{ij} = 0$ for $i > j$, i.e., all the entries below main diagonal are zero. Similarly, a square matrix A is called *lower triangular* if $a_{ij} = 0$ for $i < j$, i.e., all the entries above the main diagonal are zero. For example

$$\begin{bmatrix} 1 & 2 & 0 \\ 0 & 0 & 0 \\ 0 & 0 & -3 \end{bmatrix} \text{ and } \begin{bmatrix} 2 & 0 & 0 \\ 0 & 1 & 0 \\ -1 & 5 & 0 \end{bmatrix}$$

are upper and lower triangulars, respectively. A square matrix is called triangular if it is either lower triangular or upper triangular. A matrix B, obtained after interchanging the ith row (column) into the respectively i^{th} column (row) of a matrix A for every i, is called *transpose* of A and is denoted by A', i.e., $B = A'$. For example, if

$$A = \begin{bmatrix} 1 & 2 & -3 \\ 0 & -1 & 4 \end{bmatrix}, \text{ then } A' = \begin{bmatrix} 1 & 0 \\ 2 & -1 \\ -3 & 4 \end{bmatrix}$$

Thus, if A of order m by n, then A^t is of order n by m.

1.1 Operations on Matrices

If $A = (a_{ij})$ and $B = (b_{ij})$ are matrices of same order $m \times n$, then $A + B$ is defined as $(a_{ij} + b_{ij})$, i.e., the sum of corresponding entries of A and B. Thus, $A + B$ is also a matrix of same order m by n. For example, if

$$A = \begin{bmatrix} 1 & 2 & 3 \\ 0 & 5 & 4 \end{bmatrix} \text{ and } B = \begin{bmatrix} -1 & 3 & -4 \\ 2 & 1 & -2 \end{bmatrix}, \text{ then } A + B = \begin{bmatrix} 0 & 5 & -1 \\ 2 & 6 & 2 \end{bmatrix}$$

It is trivial to see that $A + B = B + A$.

If $A = (a_{ij})$, then for any scalar α, we define $\alpha A = (\alpha a_{ij})$ which is obtained by multiplying each entry by α. For example, if

$$A = \begin{bmatrix} 2 & 3 & 0 \\ 1 & -2 & 1 \end{bmatrix}, \text{ then } (-2)A = \begin{bmatrix} -4 & -6 & 0 \\ -2 & 4 & -2 \end{bmatrix}.$$

If $A = (a_{ij})$, $B = (b_{ij})$ are m by n, then $A = B$ iff $a_{ij} = b_{ij}$, i.e., the corresponding entries of A and B are equal. For example, if

$$A = \begin{bmatrix} 1 & 2 & -1 \\ 3 & 0 & -2 \\ 4 & 5 & 7 \end{bmatrix} \text{ and } B = \begin{bmatrix} 1 & x & -1 \\ y & 0 & -2 \\ 4 & z & 7 \end{bmatrix}$$

then $A = B$ iff $x = 2$, $y = 3$, $z = 5$. Now, it is trivial to see that for any matrix A, $A + (-1)A = A - A = O$ (zero matrix). If we take

$$x = \begin{bmatrix} a_{i1} & a_{i2} & \dots & a_{in} \end{bmatrix} \text{ and } y = \begin{bmatrix} b_{1j} \\ b_{2j} \\ \vdots \\ b_{nj} \end{bmatrix},$$

then we define

$$x \cdot y = a_{i1}b_{1j} + a_{i2}b_{2j} + \dots + a_{in}b_{nj}$$

and sometimes known as dot product of x and y. Here we see that x and y are $1 \times n$ and $n \times 1$ matrices, respectively. In $x \cdot y$, x is known as *prefactor* and y as the *postfactor* and we see that multiplication of x and y is only possible if the number of columns of x is equal to the number of rows of y.

Now, suppose $A = (a_{ij}) = \begin{bmatrix} R_1 \\ R_2 \\ \vdots \\ R_m \end{bmatrix}$ is a $m \times n$ matrix and $B = (b_{ij}) =$

$[C_1 \quad C_2 \quad \ldots \quad C_p]$ is a $n \times p$ matrix, then we define

$$AB = \begin{bmatrix} R_1C_1 & R_1C_2 & \ldots & R_1C_p \\ R_2C_1 & R_2C_2 & \ldots & R_2C_p \\ \vdots & \vdots & \vdots & \vdots \\ R_mC_1 & R_mC_2 & \ldots & R_mC_p \end{bmatrix}$$

where R_i is i^{th} row of A and C_j is the jth column of B. Now, $(i, j)^{\text{th}}$ entry of AB is

$$R_iC_j = [a_{i1} \quad a_{i2} \quad \ldots \quad a_{in}] \begin{bmatrix} b_{1j} \\ b_{2j} \\ \vdots \\ b_{nj} \end{bmatrix}$$

$$= a_{i1}b_{1j} + a_{i2}b_{2j} + \ldots + a_{in}b_{nj}$$

$$= \sum_{k=1}^{n} a_{ik}b_{kj}$$

Thus if $AB = (c_{ij})$, then $c_{ij} = \sum_{k=1}^{n} a_{ik}b_{kj}$ and AB is of order $m \times p$. Therefore to get the ith row entry of AB for any i, we start with i^{th} row vector R_i of prefactor A and multiply with column vectors $C_1, C_2, \ldots, C_p$ of the postfactor B to get first, second, third entry, and so on... in i^{th} row.

For example, if

$$A = \begin{bmatrix} 1 & 2 & 0 \\ 0 & -1 & 3 \end{bmatrix} \text{ and } B = \begin{bmatrix} 2 & 0 \\ -1 & 1 \\ 3 & 4 \end{bmatrix}$$

then A is of order 2×3 and B is of order 3×2. So, both AB and BA are possible. Now

$$AB = \begin{bmatrix} 1\cdot 2 + 2\cdot(-1) + 0\cdot 3 & 1\cdot 0 + 2\cdot 1 + 0\cdot 4 \\ 0\cdot 2 + (-1)\cdot(-1) + 3\cdot 3 & 0\cdot 0 + (-1)\cdot 1 + 3\cdot 4 \end{bmatrix} = \begin{bmatrix} 0 & 2 \\ 10 & 11 \end{bmatrix}$$

Similarly

$$BA = \begin{bmatrix} 2 & 4 & 0 \\ -1 & -3 & 3 \\ 3 & 2 & 12 \end{bmatrix}$$

Thus, by definition, if A is of order $m \times n$, then $A^2 = A \cdot A$ is not possible. But if A is square matrix, then $A^2, A^3, \ldots$ are possible.

Let $A = \begin{bmatrix} 0 & 1 \\ 1 & 0 \end{bmatrix}$ and $B = \begin{bmatrix} 1 & 1 \\ 0 & 0 \end{bmatrix}$, then $AB = \begin{bmatrix} 0 & 0 \\ 1 & 1 \end{bmatrix}$, $BA = \begin{bmatrix} 1 & 1 \\ 0 & 0 \end{bmatrix}$

and we see that $AB \neq BA$. So, in general, we can say that if AB and BA are permissible, then AB need not be the same as BA.

If $A = \begin{bmatrix} 0 & 1 \\ 0 & 0 \end{bmatrix}$, then $A^2 = \begin{bmatrix} 0 & 0 \\ 0 & 0 \end{bmatrix}$ but $A \neq \begin{bmatrix} 0 & 0 \\ 0 & 0 \end{bmatrix}$. Now onwards zero matrix will be simply denoted by O. Let $B = \begin{bmatrix} 1 & 0 \\ 0 & 0 \end{bmatrix}$, then

$$AB = \begin{bmatrix} 0 & 0 \\ 0 & 0 \end{bmatrix}$$

But neither $A = O$ nor $B = O$, where $A = \begin{bmatrix} 0 & 1 \\ 0 & 0 \end{bmatrix}$, here

$$BA = \begin{bmatrix} 0 & 1 \\ 0 & 0 \end{bmatrix} = A \neq O$$

If

$$A = \text{diag }(0, 2, 0, 1) = \begin{bmatrix} 0 & 0 & 0 & 0 \\ 0 & 2 & 0 & 0 \\ 0 & 0 & 0 & 0 \\ 0 & 0 & 0 & 1 \end{bmatrix}$$

and $B = \text{diag }(3, 0, 4, 0)$, then $AB = BA = O$, where neither $A = O$ nor $B = O$.

Now we state the following result and omit the proof which may be seen in any book on matrices, however, we give a sketchy proof.

Theorem 1.1.1. If A, B and C are matrices of suitable orders,then

(*i*) $A(BC) = (AB)C$

(*ii*) $A(B + C) = AB + AC$

(*iii*) $A + (-A) = O$

(*iv*) $(A + B) + C = A + (B + C)$

(*v*) $A + O = A$

Proof. Let $A = (a_{ij})$, $B = (b_{jk})$ and $C = (c_{kl})$ be matrices of orders $m \times n$, $n \times r$ and $r \times s$, respectively.

$$(i, k)^{\text{th}} \text{ element of } AB = \sum_{j=1}^{n} a_{ij}b_{jk} = d_{ik} \text{ (say)}$$

$$(i, l)^{\text{th}} \text{ element of } (AB)C = \sum_{k=1}^{r} d_{ik}c_{kl} = \sum_{k=1}^{r}\sum_{j=1}^{n} a_{ij}b_{jk}c_{kl}$$

$$= \sum_{j=1}^{n} a_{ij} \sum_{k=1}^{r} b_{jk}c_{kl}$$

$$= (i, l)^{\text{th}} \text{ element of } A(BC)$$

$$\Rightarrow \qquad (AB)C = A(BC)$$

Now, we prove that $A(B + C) = AB + AC$.

Let $A = (a_{ij})$, $B = (b_{jk})$ and $C = (c_{jk})$ be matrices of orders $m \times n$, $n \times r$ and $n \times r$, respectively. Then

$$(i, k)^{\text{th}} \text{ element of } A(B + C) = \sum_{j=1}^{n} a_{ij}(b_{jk} + c_{jk})$$

$$= \sum_{j=1}^{n} a_{ij}b_{jk} + \sum_{j=1}^{n} a_{ij}c_{jk}$$

$$= (i, k)^{\text{th}} \text{ element of } (AB + AC)$$

$$\Rightarrow \qquad A(B + C) = (AB + AC).$$

Similarly, for the matrices of suitable size we can show that $(A + B)C = AC + BC$.

Thus, the associative and distributive laws are satisfied. The proofs of the rest are trivial. Now, for any square matrix A, we define $A^3 = (A \cdot A)\, A$, and so on. If m is positive we defined $A^m = A^{m-1} \cdot A$. We also denote $A^0 = 1$. Trivially $A^m \cdot A^n = A^{m+n}$ and $(A^m)^n = A^{mn}$, where m and n are positive integers.

Now, giving similar arguments, we can prove the following.

Theorem 1.1.2. For any matrix A, $(A^t)^t = A$ and $(\alpha A)^t = \alpha A^t$.

Theorem 1.2.3. For matrices A and B of suitable orders

$$(AB)^t = B^tA^t.$$

Proof. Let $A = (a_{ij})$ and $B = (b_{jk})$ be of orders $m \times n$ and $n \times r$, respectively. Then AB is of order $m \times r$. Hence $(AB)^t$ is of order $r \times m$. Also B^t and A^t are matrices of orders $r \times n$ and $n \times m$. So, B^tA^t is of order $r \times m$. Now,

$(i, k)^{\text{th}}$ element of $AB = \sum_{j=1}^{n} a_{ij}b_{jk} = d_{ik}$ (say).

Then $(AB)^t = (d_{ki})$, where $d_{ki} = d_{ik}$

$\Rightarrow (k, i)^{\text{th}}$ element of $(AB)^t = d_{ki} = d_{ik} = \sum_{j=1}^{n} a_{ij}b_{jk}$.

Now $B^t = (b_{kj})$, where $b_{kj} = b_{jk}$ and $A^t = (a_{ji})$, here $a_{ji} = a_{ij}$. Thus,

$$(k, i)^{\text{th}} \text{ element of } B^t A^t = \sum_{k=1}^{n} b'_{kj}a'_{ji}$$

$$= \sum_{j=1}^{n} b_{jk}a_{ij} = \sum_{j=1}^{n} a_{ij}b_{jk} = d'_{ki}$$

$$\Rightarrow \qquad (AB)^t = B^tA^t$$

Theorem 1.1.4. If A and B are matrices of orders $m \times n$, then

$$(A + B)^t = A^t + B^t.$$

Proof. Since it is routine work, it is left as an exercise.

Now, we define and discuss various types of matrices. A square matrix $A = (a_{ij})$ is called *symmetric matrix* if $A^t = A$ i.e., $a_{ij} = a_{ji}$ for all i and j. A square matrix $A = (a_{ij})$ is called *skew-symmetric matrix* if $A^t = -A$ i.e., $a_{ij} = -a_{ji}$ for all i and j.

In skew-symmetric matrices diagonal elements are zero. Since

$$a_{ij} = -a_{ji} \text{ for all } i \text{ and } j$$

$$\Rightarrow \qquad a_{ii} = -a_{ii}$$

$$\Rightarrow \qquad (1 + 1)a_{ii} = 0$$

$$\Rightarrow \qquad a_{ii} = 0,$$

For example,

$$A = \begin{bmatrix} 2 & 4 & 3 \\ 4 & 1 & 5 \\ 3 & 5 & 6 \end{bmatrix}, B = \begin{bmatrix} 0 & -4 & -3 \\ 4 & 0 & -5 \\ 3 & 5 & 0 \end{bmatrix}$$

are symmetric and skew-symmetric matrices, respectively. We can construct infinite number of symmetric and skew-symmetric matrices as we can see from the above example. If we take

$$C = \begin{bmatrix} 0 & 4 & 3 \\ -4 & 0 & 5 \\ -3 & 5 & 0 \end{bmatrix}$$

then C is also skew-symmetric.

Let A be any square matrix. We can express A as

$$A = P + Q, \text{ where } P = \frac{A + A^t}{2} \text{ and } Q = \frac{A - A^t}{2}$$

Now, we have

$$P^t = \left(\frac{A + A^t}{2}\right)^t = \frac{A + A^t}{2} = P$$

and

$$Q^t = \left(\frac{A - A^t}{2}\right)^t = \frac{A^t - A}{2}$$

$$= \left(\frac{A - A^t}{2}\right) = -Q$$

Thus, $P = \frac{A + A^t}{2}$ is symmetric and $Q = \frac{A - A^t}{2}$ is skew-symmetric.

Hence, every square matrix can be uniquely expressed as the sum of its symmetric and skew-symmetric parts. In the above example

$$A = \begin{bmatrix} 2 & 4 & 3 \\ 4 & 1 & 5 \\ 3 & 5 & 6 \end{bmatrix}$$

since A itself is symmetric. Therefore, its skew-symmetric part is 0 (zero matrix).

Now, let

$$A = \begin{bmatrix} 1 & -1 & 2 \\ 2 & -2 & 3 \\ 0 & 1 & 4 \end{bmatrix}, \text{ then } A' = \begin{bmatrix} 1 & 2 & 0 \\ -1 & -2 & 1 \\ 2 & 3 & 4 \end{bmatrix}$$

$$\Rightarrow \qquad \frac{A + A^t}{2} = \frac{1}{2}\begin{bmatrix} 2 & 1 & 2 \\ 1 & -4 & 4 \\ 2 & 4 & 8 \end{bmatrix}$$

and

$$\frac{A - A^t}{2} = \begin{bmatrix} 0 & -3 & 2 \\ 3 & 0 & 2 \\ -2 & -2 & 0 \end{bmatrix}$$

are symmetric and skew-symmetric parts of A, respectively.

From Theorem 1.1.4, it is trivial to see that the sum of symmetric (skew-symmetric) matrices is symmetric (skew-symmetric). Also it follows from Theorem 1.1.3 that if A and B are square matrices of same orders then AB need not be symmetric (skew-symmetric) though both, A and B, are symmetric (skew-symmetric). But if $AB = BA$, A and B are symmetric, then AB is symmetric. Also $(AA^t)^t = AA^t \Rightarrow AA^t$ is symmetric and $(A^tA)^t = A^tA \Rightarrow A^tA$ is also symmetric. Let $A = (a_{ij})$ be a square matrix where a_{ij}'s are either real or complex numbers, then A is called *Hermitian* if $(\overline{A})^t = A$ i.e., $\overline{a}_{ij} = a_{ji}$ for all i and j, where $\overline{a}_{ij}$ is complex conjugate of a_{ij}. As we know that if a_{ij} is real then $\overline{a}_{ij} = a_{ij}$. This implies that in case of real entities of the matrices, Hermitian and symmetric matrices are same. Similarly, $A = (a_{ij})$ is called *skew-Hermitian* if $(\overline{A})^t = -A$ i.e., $\overline{a}_{ij} = -aji$ which, in other words implies that $\overline{a}_{ii} + a_{ii} = 0$. This implies that diagonal entries are either zero or purely imaginary. In case of real entries skew-symmetric and skew-Hermitian matrices are same. Let

$$A = \begin{bmatrix} 5 & 3+4i \\ 3-4i & 6 \end{bmatrix}$$

then
$$(\overline{A})^t = \begin{bmatrix} 5 & 3+4i \\ 3-4i & 6 \end{bmatrix} = A$$

$\Rightarrow A$ is Hermitian. In case of Hermitian matrices we know that $\overline{a}_{ii} = a_{ii}$.

$\Rightarrow$ The diagonal entries of Hermitian matrices will be always real numbers.

Let

$$A = \begin{bmatrix} 5 & 2+3i & 3-2i \\ 2-3i & 2 & 1+i \\ 3+2i & 1-i & 7 \end{bmatrix}$$

then $(\overline{A})^t = A$ and one can see the method of getting Hermitian matrices. Now, let

$$A = \begin{bmatrix} 0 & 2+i \\ -2+i & 0 \end{bmatrix} \text{ and } B = \begin{bmatrix} 2i & 4+i \\ -4+i & 3i \end{bmatrix}$$

then A and B are skew-Hermitian. Similarly, we can see that

$$A = \begin{bmatrix} i & 2+3i & 3-2i \\ -2+3i & 0 & 1+i \\ -3-2i & -1+i & 2i \end{bmatrix}$$

is skew-Hermitian.

If we take

$$B = \begin{bmatrix} 1 & 1+i \\ -1+i & 2 \end{bmatrix} \text{ then } (\overline{B})^t = \begin{bmatrix} 1 & -1-i \\ 1-i & 2 \end{bmatrix} \neq B \text{ or } -B.$$

Thus, B is neither Hermitian nor skew-Hermitian.

Now, we state a result whose proof can be well adopted on similar lines as in Theorems 1.1.2., 1.1.3 and 1.1.4.

Theorem 1.1.5. If A and B are matrices of suitable orders, then

$$\overline{((\overline{A}))^t}^t = A, \ (\overline{A+B})^t = (\overline{A})^t + (\overline{B})^t,$$

$$(\overline{\alpha A})^t = \overline{\alpha}\,(\overline{A})^t \text{ and } (\overline{AB})^t = (\overline{B})^t\,(\overline{A})^t.$$

It may be noted that $(\overline{A})^t = (\overline{A^t})$.

A square matrix A is called *non-singular* if there exists a matrix B such that $AB = BA = I$ and B is called inverse of A and is denoted by A^{-1}. Inverse of a square matrix, if it exists, is always unique. For if $AB = BA = I$ and $CA = AC = I$ for some matrices B and C, then

$B = IB = (CA)B = C(AB) = CI = C.$

Thus, if A^{-1} exists, it is always unique.

The method and condition to find A^{-1} will be dealt with in the next sections.

However, for 2×2 matrix $A = \begin{bmatrix} 1 & 3 \\ 2 & 5 \end{bmatrix}$ we suppose that $B = \begin{bmatrix} a & b \\ c & d \end{bmatrix}$ be such that

$$\begin{bmatrix} 1 & 3 \\ 2 & 5 \end{bmatrix}\begin{bmatrix} a & b \\ c & d \end{bmatrix} = \begin{bmatrix} 1 & 0 \\ 0 & 1 \end{bmatrix}$$

then

$$a + 3c = 1, \ b + 3d = 0$$

$$2a + 5c = 0, \ 2b + 5d = 1.$$

Now, solving these equations we get $a = -5$, $b = 3$, $c = 2$ and $d = -1$.

Thus

$$A^{-1} = \begin{bmatrix} -5 & 3 \\ 2 & -1 \end{bmatrix}.$$

Theorem 1.1.6. If A and B are non-singular matrices, then

$$(A^{-1})^{-1} = A, \ (A^{-1})^t = (A^t)^{-1}$$

and $(AB)^{-1} = B^{-1}A^{-1}$.

Proof. Since

$$A^{-1}A = AA^{-1} = I$$

$\Rightarrow$ $$A = (A^{-1})^{-1}$$

Also

$$AA^{-1} = I$$

$\Rightarrow$ $$(AA^{-1})^t = (A^{-1})^t A^t = I^t = I$$

$\Rightarrow$ $$(A^t)^{-1} = (A^{-1})^t.$$

Since

$$(B^{-1}A^{-1})(AB) = B^{-1}(A^{-1}A)B$$
$$= B^{-1}(IB) = B^{-1}B = I$$

$\Rightarrow$ $$(AB)^{-1} = B^{-1}A^{-1}$$

Theorem 1.1.7. If A and B are square matrices of order n such that $AB = I$, then $BA = I$.

Proof. Suppose for some matrix C of order n, $CA = I$, then

$$C = CI = C(AB) = (CA)B = IB = B$$

$\Rightarrow$ $$BA = I.$$

Thus, to find the inverse of a matrix it is sufficient to find either left or right inverse. If a matrix is not non-singular, then it is called singular.

A square matrix A is called *idempotent* if for some positive integer k, $A^k = 0$ but $A^{k-1} \neq 0$ and k is called *index of nilpotency*.

A square matrix A is called *orthogonal* if $AA^t = I \Rightarrow A^{-1} = A^t$. So, as done in Theorem 1.1.7, we get $AA^t = I$.

A square matrix A is called *unitary* if $A(\overline{A})^t = I \Rightarrow (\overline{A})^t A = I$ is also valid. However, if the entries of the matrix are real then orthogonal and unitary matrices coincide. Now we give some examples. Let

$$A = \begin{bmatrix} 2 & -1 & 1 \\ -3 & 3 & -2 \\ -4 & 4 & -3 \end{bmatrix}$$

then $A^2 \neq A$. Hence A is not idempotent, but

$$B = \begin{bmatrix} 2 & -1 & 1 \\ -2 & 3 & -2 \\ -4 & 4 & -3 \end{bmatrix}$$

is idempotent.

If any square matrix A is idempotent, then A^t is also idempotent. The matrices

$$\begin{bmatrix} 3 & 9 \\ -1 & -3 \end{bmatrix} \text{ and } \begin{bmatrix} 0 & 1 & 1 \\ 0 & 0 & 1 \\ 0 & 0 & 0 \end{bmatrix}$$

and nilpotent. Any matrix of the type

$$\begin{bmatrix} ab & b^2 \\ -a^2 & -ab \end{bmatrix}$$

is always nilpotent. Further, if A is nilpotent, then A^t is also nilpotent.
If

$$A = \begin{bmatrix} 0 & 1 & -1 \\ 4 & -3 & 4 \\ 3 & -3 & 4 \end{bmatrix} \text{ and } B = \begin{bmatrix} 0 & 1 \\ 1 & 0 \end{bmatrix},$$

then A and B are involuntary. If

$$A = \begin{bmatrix} \cos\theta & 0 & -\sin\theta \\ 0 & 1 & 0 \\ \sin\theta & 0 & \cos\theta \end{bmatrix}$$

then $AA^t = I$. Thus, A is orthogonal. Let

$$B = \begin{bmatrix} \frac{1}{\sqrt{3}} & \frac{1+i}{\sqrt{3}} \\ \frac{1-i}{\sqrt{3}} & -\frac{1}{\sqrt{3}} \end{bmatrix}$$

then $B(\overline{B})^t = I$ and hence B is unitary.
A square matrix A is called periodic if $A^{p+1} = A$, where p, called the period of A, is the least positive integer. Thus, idempotent matrices are of period 1. If A is of period p then A^t is also of period p.

1.2 Elementary Operations

In this section we shall discuss elementary operations on matrices and their consequences which will be used for solving a system of linear equations. The following types of elementary operations are defined on any matrix $A = (a_{ij})$:

(*i*) Interchanging any two rows (columns) of A.

(*ii*) Multiplying any row (column) by a non-zero scalar α.

(*iii*) Adding in any row (column) any multiple of any other row (column).

If these operations are applied on rows only then they are called *row operations* and if such operations are applied on columns only then they are called *column operations.*

Sometimes it is possible to apply both the operations on any matrix.

As indicated earlier, the rows are denoted by $R_1, R_2, \ldots$ and columns by $C_1, C_2, \ldots$. The three types of rows operations are denoted by $R_i \leftrightarrow R_j$, $R_i \to R_i(\alpha)$ and $R_i \to R_i + R_j(\alpha)$ respectively and similarly for column operations.

Let d

$$A = \begin{bmatrix} 1 & 0 & 2 & 4 \\ -1 & 2 & 3 & 0 \\ 3 & 1 & -2 & 0 \end{bmatrix}.$$

If we apply $R_1 \to R_2$, then

$$A \sim \begin{bmatrix} -1 & 2 & 3 & 0 \\ 1 & 0 & 2 & 4 \\ 3 & 1 & -2 & 0 \end{bmatrix}$$

and if we apply $R_2 \leftrightarrow R_2(2)$ on A, then

$$A \sim \begin{bmatrix} 1 & 0 & 2 & 4 \\ -2 & 4 & 6 & 0 \\ 3 & 1 & -2 & 0 \end{bmatrix}$$

If we apply $R_1 \leftrightarrow R_1 + R_2(2)$ on A, then

$$A \sim \begin{bmatrix} -1 & 4 & 8 & 4 \\ -1 & 2 & 3 & 0 \\ 3 & 1 & -2 & 0 \end{bmatrix}.$$

Similarly, we can illustrate for column operations.

Applying single row operation on identity matrix I, the matrix obtained is called elementary matrix. We can notice that applying column operations on any matrix A amounts the same as applying the corresponding row operations on A^t. Therefore, most of the time we only apply row operations.

Now, we denote by E_{ij}, $E_i(\alpha)$ and $E_{ij}(\alpha)$ the elementary matrices obtained after applying $R_i \leftrightarrow R_j$, $R_i \to R_i(\alpha)$ and $R_i \to R_i + R_i(\alpha)$ on identity matrix. We now show that applying elementary row operation on A amounts the same as pre-multiplying A by the corresponding elementary matrix.

Suppose we apply $R_i \leftrightarrow R_j$ on A, where $i < j$, then

$$E_{ij}A = \begin{bmatrix} 1 & 0 & \cdot & & \cdot & \cdot & 0 & \cdots & 0 \\ 0 & 1 & . & & \cdot & \cdot & 0 & \cdots & 0 \\ \vdots & & & & \vdots & & & \vdots & \vdots \\ 0 & \cdots & 1 & \ldots & 0 & \cdot & \cdot & \cdot & 0 \\ \vdots & & & & \vdots & & & \vdots & \vdots \\ 0 & \cdots & 1 & \cdots & 0 & \cdot & \cdot & \cdot & 0 \\ \vdots & & & & \vdots & & & \vdots & \vdots \\ 0 & \cdots & 0 & \cdots & 0 & \cdot & \cdot & \cdot & 1 \end{bmatrix} \begin{bmatrix} R_1 \\ R_2 \\ \vdots \\ R_i \\ \vdots \\ R_j \\ \vdots \\ R_n \end{bmatrix}$$

$$= \begin{bmatrix} R_1 \\ R_2 \\ \vdots \\ R_i \\ \vdots \\ R_j \\ \vdots \\ R_n \end{bmatrix}$$

Also

$$E_i(\alpha)A = \begin{bmatrix} 1 & 0 & \cdot & \cdot & \cdot & & 0 \\ 0 & 1 & \cdot & \cdot & \cdot & & 0 \\ \vdots & & & \vdots & & & \vdots \\ 0 & \cdots & \alpha & \cdots & 0 & \cdots & 0 \\ \vdots & & & \vdots & & & \vdots \\ 0 & 0 & \cdot & \cdot & \cdot & & 0 \\ \vdots & & & \vdots & & & \vdots \\ 0 & 0 & \cdot & \cdot & \cdot & & 1 \end{bmatrix} \begin{bmatrix} R_1 \\ R_2 \\ \vdots \\ R_i \\ \vdots \\ R_j \\ \vdots \\ R_n \end{bmatrix}$$

$$= \begin{bmatrix} R_1 \\ R_2 \\ \vdots \\ \alpha R_i \\ \vdots \\ R_j \\ \vdots \\ R_n \end{bmatrix}$$

and $$E_{ij}(\alpha)A = \begin{bmatrix} 1 & 0 & & \cdots & & \cdots & & \cdots & & \\ 0 & 1 & & \cdots & & \cdots & & \cdots & & 0 \\ \vdots & & & & & \vdots & & & & \vdots \\ 0 & 0 & \cdots & 1 & 0 & \cdots & 0 & \alpha & \cdots & 0 \\ \vdots & & & & & \vdots & & & & \vdots \\ 0 & \cdots & 0 & \cdots & 1 & 0 & & \cdots & & 0 \\ \vdots & & & & & \vdots & & & & \vdots \\ 0 & 0 & & \cdots & & \cdots & & \cdots & 0 & 1 \end{bmatrix} \begin{bmatrix} R_1 \\ R_2 \\ \vdots \\ R_i \\ \vdots \\ R_j \\ \vdots \\ R_n \end{bmatrix}$$

$$= \begin{bmatrix} R_1 \\ R_2 \\ \vdots \\ R_i + \alpha R_j \\ \vdots \\ R_j \\ \vdots \\ R_n \end{bmatrix}$$

It may be noted that $R_i = R_i(\alpha)$.

Similarly, we can show that applying an elementary column operation on A amounts the same as post-multiplying by the corresponding elementary matrix. We shall denote the elementary matrices by E_{ij}, $E_i(\alpha)$ and $E_{ij}(\alpha)$ corresponding to the three elementary column operations. It is trivial to see that $E_{ij} E_{ij} = I$, $E_i(\alpha) E_i(\alpha^{-1}) = I$ and $E_{ij}(\alpha) E_{ij}(-\alpha) = I$. Hence the elementary matrices are always non-singular and their inverses are also elementary matrices of the same type.

Now, we define:

Definition 1.2.1. A $m \times n$ matrix A is called row echelon matrix or row echelon form if

(*i*) zero rows, if any, occur at the bottom of the matrix.

(*ii*) first non-zero entry in any non-zero row is 1 which is known as leading 1.

(*iii*) the number of zeros before leading 1 in any i^{th} row is less than the number of zeros before leading 1 in $(i + 1)^{\text{th}}$ row.

Definition 1.2.2. A row echelon form matrix is called *row reduced echelon form* if all the entries in the column in which leading 1 occurs are zero.

For example,

$$A = \begin{bmatrix} 0 & 1 & 2 & 3 & 0 \\ 0 & 0 & 1 & -1 & 3 \\ 0 & 0 & 0 & 1 & 5 \end{bmatrix}$$

is row echelon form but

$$B = \begin{bmatrix} 0 & 1 & 2 & 3 & 0 \\ 0 & 1 & 0 & -1 & 3 \\ 0 & 0 & 0 & 1 & 5 \end{bmatrix}$$

is not row echelon form.

The matrix

$$A = \begin{bmatrix} 1 & 0 & 2 & 3 \\ 0 & 1 & 1 & 0 \\ 0 & 0 & 0 & 0 \end{bmatrix}$$

is row reduced echelon form while

$$B = \begin{bmatrix} 1 & 0 & 2 & 3 \\ 0 & 1 & 1 & 0 \\ 0 & 0 & 0 & 1 \end{bmatrix}$$

is not row reduced echelon form though it is row echelon form.

Now, appealing to elementary row operations, it is routine matter to reduce any $m \times n$ matrix A to row reduced echelon form.

Now we illustrate the same by an example.

Let

$$A = \begin{bmatrix} 1 & 3 & -1 & 2 \\ 0 & 11 & -5 & 3 \\ 2 & -5 & 3 & 1 \\ 4 & 1 & 1 & 5 \end{bmatrix}$$

then

$$A \sim \begin{bmatrix} 1 & 3 & -1 & 2 \\ 0 & 11 & -5 & 3 \\ 0 & -11 & 5 & -3 \\ 4 & 1 & 1 & 5 \end{bmatrix} \text{ by } R_3 \to R_3 + R_1(-2)$$

$$\sim \begin{bmatrix} 1 & 3 & -1 & 2 \\ 0 & 11 & -5 & 3 \\ 0 & -11 & 5 & -3 \\ 0 & -11 & 5 & -3 \end{bmatrix} \text{ by } R_4 \to R_4 + R_1(-4)$$

$$\sim \begin{bmatrix} 1 & 3 & -1 & 2 \\ 0 & 11 & -5 & 3 \\ 0 & 0 & 0 & 0 \\ 0 & 0 & 0 & 0 \end{bmatrix} \text{ by } R_3 \to R_3 + R_2(1) \text{ and } R_4 \to$$

$R_4 + R_2(1)$

$$\sim \begin{bmatrix} 1 & 3 & -1 & 2 \\ 0 & 1 & -\frac{5}{11} & \frac{3}{11} \\ 0 & 0 & 0 & 0 \\ 0 & 0 & 0 & 0 \end{bmatrix} \text{ by } R_2 \to R_1\left(-\frac{1}{11}\right)$$

$$\sim \begin{bmatrix} 1 & 0 & \frac{4}{11} & \frac{13}{11} \\ 0 & 1 & -\frac{5}{11} & \frac{3}{11} \\ 0 & 0 & 0 & 0 \\ 0 & 0 & 0 & 0 \end{bmatrix} \text{ by } R_1 \to R_1 + R_2(-3)$$

$$= C$$

which is row reduced echelon form matrix.

Definition 1.2.3. A matrix A of order $m \times n$ is called *row equivalent* to a matrix B of order $m \times n$ if B can be obtained from A after operating a finite number of elementary row operations.

Definition 1.2.4. A matrix A of order $m \times n$ is called column equivalent to a matrix B of order $m \times n$ if B can be obtained from A after operating a finite number of elementary column operations.

We denote $A \underset{c}{\sim} B$ and $A \underset{c}{\sim} B$ for row equivalent and column equivalent, respectively.

As seen earlier that for non-singular matrices A and B, $(AB)^{-1} = B^{-1}A^{-1}$. Similarly, if $A_1, A_2, \ldots, A_n$ are non-singular matrices, then

$$(A_1 A_2 \ldots A_n)^{-1} = A_n^{-1} A_{n-1}^{-1} \ldots A_2^{-1} A_1^{-1}.$$

Hence, we can say that products of non-singular matrices are non-singular. As discussed earlier, operating elementary operations on any matrix A is equivalent to pre-multiplying A by the corresponding elementary matrix and applying elementary column operations are equivalent to post-multiplying by the corresponding elementary matrices. Hence we have the following:

Theorem 1.2.5. If A and B are $m \times n$ matrices, $A \underset{r}{\sim} B$ iff $B = PA$, where P is $m \times m$ non-singular matrix.

Theorem 1.2.6. If A and B are $m \times n$ matrices, then $A \underset{c}{\sim} B$ iff $B = AQ$, where Q is $n \times n$ non-singular matrix.

We can trivially check that '$\underset{r}{\sim}$' and '$\underset{c}{\sim}$' satisfy the following:

(*i*) $A \underset{r}{\sim} A$, $A \underset{c}{\sim} A$.

(*ii*) If $A \underset{r}{\sim} B$, then $B \underset{r}{\sim} A$. Also if $A \underset{c}{\sim} B$ then $B \underset{c}{\sim} A$.

(*iii*) If $A \underset{r}{\sim} B$ and $B \underset{r}{\sim} D$, then $A \underset{r}{\sim} D$. Also if $A \underset{c}{\sim} B$ and $B \underset{c}{\sim} D$, then $A \underset{c}{\sim} D$.

Hence, instead of saying "A is row equivalent to B" we say that A and B are row equivalent matrices. Similarly, for column equivalent, we say that A and B are column equivalent.

Theorem 1.2.7. A square matrix A of order $n \times n$ is non-singular iff $A \underset{r}{\sim} I$.

Proof. Now, suppose A is non-singular. Then applying a finite number of elementary row operations on A we get A to be reduced to row reduced echelon form matrix B. Thus

$E_t E_{t-1} \dots E_2 E_1 A = B \dots (1.2.1)$

Since right side of (1.2.1) is non-singular, therefore, B cannot have any zero row $\Rightarrow B = I$. Hence,

$$A \underset{r}{\sim} I.$$

Now, suppose that $A \underset{r}{\sim} I$, then there exist elementary matrices E_1, E_2, $\dots$, E_p such that

$$E_p E_{p-1} \dots E_2 E_1 A = I$$

$$\Rightarrow \qquad A = E_1^{-1} E_2^{-1} \dots E_p^{-1}.$$

Since the inverse of an elementary matrix is elementary and elementary matrices are non-singular, so A is non-singular.

Corollary 1.2.8. A square matrix A is non-singular iff A is the product of elementary matrices.

Now we discuss some examples.

Let $A = \begin{bmatrix} 1 & 2 \\ 3 & 5 \end{bmatrix}$. Then

$$A \sim \begin{bmatrix} 1 & 2 \\ 0 & -1 \end{bmatrix} \text{ by applying } R_2 \to R_2 + R_1(-3)$$

$$\sim \begin{bmatrix} 1 & 0 \\ 0 & -1 \end{bmatrix} \text{ by applying } R_1 \to R_1 + R_2(2)$$

$$\sim \begin{bmatrix} 1 & 0 \\ 0 & 1 \end{bmatrix} \text{ by applying } R_2 \to R_2(-1).$$

Now, applying $R_2 \to R_2 + R_1(-3)$ on $\begin{bmatrix} 1 & 0 \\ 0 & 1 \end{bmatrix}$, we get

$$E_{21}(-3) = \begin{bmatrix} 1 & 0 \\ -3 & 1 \end{bmatrix}.$$

Applying $R_1 \to R_1 + R_2(2)$ on $\begin{bmatrix} 1 & 0 \\ 0 & 1 \end{bmatrix}$, we get

$$E_{12}(2) = \begin{bmatrix} 1 & 2 \\ 0 & 1 \end{bmatrix}.$$

Applying $R_2 \to R_2(-1)$ on $\begin{bmatrix} 1 & 0 \\ 0 & 1 \end{bmatrix}$, we get

$$E_2(-1) = \begin{bmatrix} 1 & 0 \\ 0 & -1 \end{bmatrix}$$

$$\begin{aligned} A &= (E_{21}(-3))^{-1}(E_{12}(2))^{-1}(E_2(2))^{-1} \\ &= \begin{bmatrix} 1 & 0 \\ 3 & 1 \end{bmatrix}\begin{bmatrix} 1 & -2 \\ 0 & 1 \end{bmatrix}\begin{bmatrix} 1 & 0 \\ 0 & -1 \end{bmatrix} \\ &= \begin{bmatrix} 1 & -2 \\ 3 & -5 \end{bmatrix}\begin{bmatrix} 1 & 0 \\ 0 & -1 \end{bmatrix} \\ &= \begin{bmatrix} 1 & 2 \\ 3 & 5 \end{bmatrix}. \end{aligned}$$

Also

$$\begin{aligned} A^{-1} &= \begin{bmatrix} 1 & 0 \\ 0 & -1 \end{bmatrix}\begin{bmatrix} 1 & 2 \\ 0 & 1 \end{bmatrix}\begin{bmatrix} 1 & 0 \\ -2 & 1 \end{bmatrix} \\ &= \begin{bmatrix} 1 & 2 \\ 0 & -1 \end{bmatrix}\begin{bmatrix} 1 & 0 \\ -2 & 1 \end{bmatrix} = \begin{bmatrix} -5 & 2 \\ 3 & -1 \end{bmatrix}. \end{aligned}$$

Verification: $\begin{bmatrix} 1 & 2 \\ 3 & 5 \end{bmatrix}\begin{bmatrix} -5 & 2 \\ 3 & -1 \end{bmatrix} = \begin{bmatrix} 1 & 0 \\ 0 & 1 \end{bmatrix}$

Now, we give another example to reduce a matrix to I and side by side find the inverse. Let

$$A = \begin{bmatrix} 1 & 2 & 1 \\ 1 & 3 & 2 \\ 1 & 0 & 1 \end{bmatrix}.$$

Consider

$$\begin{aligned} [A : I] &= \begin{bmatrix} 1 & 2 & 1 & : & 1 & 0 & 0 \\ 1 & 3 & 2 & : & 0 & 1 & 0 \\ 1 & 0 & 1 & : & 0 & 0 & 1 \end{bmatrix} \\ &= \begin{bmatrix} 1 & 2 & 1 & : & 1 & 0 & 0 \\ 0 & 1 & 1 & : & -1 & 1 & 0 \\ 0 & -2 & 0 & : & -1 & 0 & 1 \end{bmatrix} \end{aligned}$$

$$R_3 \to R_2 + R_1(-1)$$
$$R_3 \to R_3 + R_1(-1)$$

$$= \begin{bmatrix} 1 & 2 & 1 & : & 1 & 0 & 0 \\ 0 & -2 & 0 & : & -1 & 0 & 1 \\ 0 & 1 & 1 & : & -1 & 1 & 0 \end{bmatrix} R_2 \leftrightarrow R_3$$

$$= \begin{bmatrix} 1 & 2 & 1 & : & 1 & 0 & 0 \\ 0 & 1 & 0 & : & \frac{1}{2} & 0 & -\frac{1}{2} \\ 0 & 1 & 1 & : & -1 & 1 & 0 \end{bmatrix} R_2 \leftrightarrow R_2\left(-\frac{1}{2}\right)$$

$$= \begin{bmatrix} 1 & 2 & 1 & : & 1 & 0 & 0 \\ 0 & 1 & 0 & : & \frac{1}{2} & 0 & -\frac{1}{2} \\ 0 & 0 & 1 & : & -\frac{3}{2} & 1 & \frac{1}{2} \end{bmatrix} R_3 \leftrightarrow R_3 + R_2(-1)$$

$$= \begin{bmatrix} 1 & 0 & 1 & : & 0 & 0 & 1 \\ 0 & 1 & 0 & : & \frac{1}{2} & 0 & -\frac{1}{2} \\ 0 & 0 & 1 & : & -\frac{3}{2} & 1 & \frac{1}{2} \end{bmatrix} R_1 \leftrightarrow R_1 + R_2(-2)$$

$$= \begin{bmatrix} 1 & 0 & 0 & : & \frac{3}{2} & -1 & \frac{1}{2} \\ 0 & 1 & 0 & : & \frac{1}{2} & 0 & -\frac{1}{2} \\ 0 & 0 & 1 & : & -\frac{3}{2} & 1 & \frac{1}{2} \end{bmatrix} R_1 \leftrightarrow R_1 + R_3(-1)$$

$$\Rightarrow \quad A^{-1} = \begin{bmatrix} \frac{3}{2} & -1 & \frac{1}{2} \\ \frac{1}{2} & 0 & -\frac{1}{2} \\ -\frac{3}{2} & 1 & \frac{1}{2} \end{bmatrix} = \frac{1}{2}\begin{bmatrix} 3 & -2 & 1 \\ 1 & 0 & -1 \\ -3 & 2 & 1 \end{bmatrix}$$

Definition 1.2.9. A square matrix A of order $m \times n$ is called equivalent to a matrix B of order $m \times n$ if B can be obtained from A after applying a finite number of elementary row and column operations.

In view of earlier discussions and results, the following is immediate.

Theorem 1.2.10. A matrix A of order $m \times n$ is equivalent to a matrix B of order $m \times n$ iff $B = PAQ$, where P and Q are non-singular matrices of order $m \times m$ and $n \times n$, respectively.

Let us denote by $A \underset{e}{\sim} B$ to mean that A is equivalent to B. It is trivial to see that $A \underset{e}{\sim} A$ if $A \underset{e}{\sim} B$, then $B \underset{e}{\sim} A$ and if $A \underset{e}{\sim} B$ and $B \underset{e}{\sim} D$, then $A \underset{e}{\sim} D$.

Hence onwards if $A \underset{e}{\sim} B$, then we can say that A and B are equivalent.

Theorem 1.2.11. Every non-zero matrix A of order $m \times n$ is equivalent to $\begin{bmatrix} I_r & O \\ O & O \end{bmatrix}$, where I_r is the identity matrix of order $r \times r$, for some r.

Proof. After applying a finite number of the first type elementary row and column operations on A, we get a matrix

$$B = \begin{bmatrix} \alpha_1 & \alpha_2 & \cdots & \alpha_n \\ a_1 & & & \\ \vdots & & B_1 & \\ a_{m-1} & & & \end{bmatrix}.$$

Now, applying second type of elementary row operation, we get a matrix

$$C = \begin{bmatrix} I & \lambda_2 & \cdots & \lambda_n \\ a_1 & & & \\ \vdots & & B_1 & \\ a_{m-1} & & & \end{bmatrix}.$$

Again, applying third type of elementary row and column operations we get

$$D = \begin{bmatrix} 1 & 0 & \cdots & \lambda_n \\ 0 & & & \\ \vdots & & B_2 & \\ 0 & & & \end{bmatrix}.$$

Continuing similar process on B_2, and so on, we get finally a matrix of the form

$$\begin{bmatrix} I_r & O \\ O & O \end{bmatrix}$$

Note. The form $\begin{bmatrix} I_r & O \\ O & O \end{bmatrix}$ is called normal form of A and r is called rank of A. It may also be noted that rank of A is the number of non-zero rows of row reduced echelon form of A. Here r may be equal to m or n according as $m < n$ or $n < m$. For rank of A we denote $r(A)$.

Now, we discuss some examples,

(1) Let $A = \begin{bmatrix} 2 & 3 \\ 1 & 2 \\ 7 & 3 \\ -2 & 5 \end{bmatrix}$, then

$$A \sim \begin{bmatrix} 2 & 3 \\ 1 & 2 \\ 7 & 3 \\ -2 & 5 \end{bmatrix} \sim \begin{bmatrix} 1 & 2 \\ 0 & -1 \\ 0 & -11 \\ 0 & 9 \end{bmatrix} \sim \begin{bmatrix} 1 & 0 \\ 0 & -1 \\ 0 & 0 \\ 0 & 0 \end{bmatrix} = \begin{bmatrix} I_2 \\ O \end{bmatrix}$$

$$\Rightarrow \qquad r(A) = 2$$

(2) Let $B = \begin{bmatrix} 1 & 2 & 0 & -3 & 1 & 0 \\ 1 & 2 & 1 & -3 & 1 & 2 \\ 1 & 2 & 0 & -3 & 2 & 1 \\ 3 & 6 & 1 & -9 & 4 & 3 \end{bmatrix}$, then

$$B \sim \begin{bmatrix} 1 & 2 & 0 & -3 & 1 & 0 \\ 0 & 0 & 1 & 0 & 0 & 2 \\ 0 & 0 & 0 & 0 & 1 & 1 \\ 0 & 0 & 1 & 0 & 1 & 3 \end{bmatrix} \text{ by } R_2 \to R_2 + R_1(-1)$$

$$R_3 \to R_3 + R_1(-1)$$
$$R_4 \to R_4 + R_1(-3)$$

$$\sim \begin{bmatrix} 1 & 2 & 0 & -3 & 1 & 0 \\ 0 & 0 & 1 & 0 & 0 & 2 \\ 0 & 0 & 0 & 0 & 1 & 1 \\ 0 & 0 & 0 & 0 & 1 & 1 \end{bmatrix} \text{ by } R_4 \to R_4 + R_2(-1)$$

$$\sim \begin{bmatrix} 1 & 2 & 0 & -3 & 1 & 0 \\ 0 & 0 & 1 & 0 & 0 & 2 \\ 0 & 0 & 0 & 0 & 1 & 1 \\ 0 & 0 & 0 & 0 & 0 & 0 \end{bmatrix} \text{by } R_4 \to R_4 + R_3(-1)$$

$$\sim \begin{bmatrix} 1 & 0 & 2 & -3 & 1 & 0 \\ 0 & 1 & 0 & 0 & 0 & 2 \\ 0 & 0 & 0 & 0 & 0 & 1 \\ 0 & 0 & 0 & 0 & 0 & 0 \end{bmatrix} \text{by } C_2 \leftrightarrow C_3$$

$$\sim \begin{bmatrix} 1 & 0 & 1 & -3 & 2 & 0 \\ 0 & 1 & 0 & 0 & 0 & 2 \\ 0 & 0 & 1 & 0 & 0 & 1 \\ 0 & 0 & 0 & 0 & 0 & 0 \end{bmatrix} \text{by } C_3 \leftrightarrow C_5$$

$$\sim \begin{bmatrix} 1 & 0 & 1 & 0 & 0 & 0 \\ 0 & 1 & 0 & 0 & 0 & 2 \\ 0 & 0 & 1 & 0 & 0 & 1 \\ 0 & 0 & 0 & 0 & 0 & 0 \end{bmatrix} \text{by } C_3 \to C_3 + C_1(-1)$$

$$C_4 \to C_4 + C_1(3)$$
$$C_5 \to C_5 + C_1(-2)$$

$$\sim \begin{bmatrix} 1 & 0 & 1 & 0 & 0 & 0 \\ 0 & 1 & 0 & 0 & 0 & 0 \\ 0 & 0 & 1 & 0 & 0 & 0 \\ 0 & 0 & 0 & 0 & 0 & 0 \end{bmatrix} \text{by } C_6 \to C_6 + C_2(-2)$$

and again by $C_6 \to C_6 + C_1(-1)$

$$\sim \begin{bmatrix} I_3 & O \\ O & O \end{bmatrix}$$

$\Rightarrow \quad r(B) = 3.$

Definition 1.2.12. A matrix A of order $n \times n$ is said to be similar to a matrix B of order $n \times n$ if $B = P^{-1}AP$ for some non-singular matrix P of order $n \times n$ and is denoted by $A \sim B$.

It is obvious to see that $A \underset{s}{\sim} A$, if $A \underset{s}{\sim} B$, then $B \underset{s}{\sim} A$ and if $A \underset{s}{\sim} B$ and $B \underset{s}{\sim} D$, then $A \underset{s}{\sim} D$, so to mean $A \underset{s}{\sim} B$ we say that A and B are similar matrices.

Let $A = \begin{bmatrix} 1 & 1 \\ 0 & 1 \end{bmatrix}$ and $B = \begin{bmatrix} 11 & 25 \\ -4 & -9 \end{bmatrix}$, then $A \underset{s}{\sim} B$

$$B = \begin{bmatrix} 5 & -2 \\ -2 & 1 \end{bmatrix} \begin{bmatrix} 1 & 1 \\ 0 & 1 \end{bmatrix} \begin{bmatrix} 1 & 2 \\ 2 & 5 \end{bmatrix}, \text{ where } P = \begin{bmatrix} 1 & 2 \\ 2 & 5 \end{bmatrix}.$$

Definition 1.2.13. If A is a square matrix of order $n \times n$, then trace of A is defined as $\sum_{i=1}^{n} a_{ii}$, where $A = (a_{ij})$ and is denoted by $tr(A)$.

For example, if $A = \begin{bmatrix} 1 & 0 & 3 \\ 4 & 2 & -1 \\ -2 & 5 & -4 \end{bmatrix}$ then $tr(A) = 1 + 2 - 4 = -1$.

Theorem 1.2.14. If A and B are square matrices of order $n \times n$, then

(*i*) $tr(\alpha A) = atr(A)$

(*ii*) $tr(A + B) = tr(A) + tr(B)$

(*iii*) $tr(AB) = tr(BA)$

(*iv*) If C and D are similar matrices then, $tr(C) = tr(D)$.

Proof. (*i*) and (*ii*) are trivial. Let $A = (a_{ij})$ and $B = (b_{jk})$ be square matrices of order $n \times n$. Then

$(i, k)^{\text{th}}$ element of $AB = c_{ik}$ (say)

$$= \sum_{j=1}^{n} a_{ij} b_{jk}$$

$$\Rightarrow \quad c_{ii} = \sum_{j=1}^{n} a_{ij} b_{ji}$$

$$\Rightarrow \quad \sum_{i=1}^{n} c_{ii} = \sum_{i=1}^{n} \sum_{j=1}^{n} a_{ij} b_{ji}$$

$$\Rightarrow \quad tr(AB) = \sum_{i=1}^{n} \sum_{j=1}^{n} a_{ij} b_{ji} \qquad \ldots(1.2.2)$$

Now $(i, k)^{\text{th}}$ element of $BA = d_{ik}$ (say) $= \sum_{j=1}^{n} b_{ij}a_{ji}$

$$\Rightarrow \qquad \sum_{i=1}^{n} d_{ii} = \sum_{i=1}^{n}\sum_{j=1}^{n} b_{ij}a_{ji}$$

$$\Rightarrow \qquad tr(BA) = \sum_{i=1}^{n}\sum_{j=1}^{n} b_{ij}a_{ji} \qquad (1.2.3)$$

From (1.2.2) and (1.2.3) we see that $tr(AB) = tr(BA)$.

Now suppose that $C \underset{s}{\sim} D$

$$\Rightarrow \qquad D = P^{-1}CP$$

$$\Rightarrow \qquad tr(D) = tr(P^{-1}(CP))$$

$$= tr(CPP^{-1})$$

$$= tr(C).$$

Hence, the theorem is proved.

1.3. Determinant

In this section we shall only concentrate on working rule instead of going into the detailed theory of determinants. Throughout the entries of the matrices will be taken either real numbers or complex numbers.

Now suppose $A = \begin{bmatrix} a & b \\ c & d \end{bmatrix}$, then $ad - bc$ is symbolically written as $\begin{vmatrix} a & b \\ c & d \end{vmatrix} = ad - bc$ and say that the value of the determinant of A is $ad - bc$.

However, before giving a formal definition, we define permutation here and it is discussed in detail in Chapter 2.

A one-one onto function of $\{1, 2, \ldots, n\}$ onto itself is called *permutation.*

Suppose $f : \{1, 2, \ldots, n\} \to \{1, 2, \ldots, n\}$ is one-one onto function such that $f(i) = p_i$ then we write $f = \begin{pmatrix} 1 & 2 & \ldots & n \\ p_1 & p_2 & \ldots & p_n \end{pmatrix}$ or simply $p_1p_2 \ldots p_n$.

We also denote by S_n as collection of all permutations on n symbols $1, 2, \ldots, n$.

Now p_1 can take values $1, 2, \ldots, n$. Now, after fixing the position of p_1 by one $1, 2, \ldots, n$, we are left $n - 1$ number for the position of p_2, and so on.

Thus, total number of permutations on n-symbols 1, 2, ..., n are $n(n-1)(n-2)\ldots 3\cdot 2\cdot 1 = n!$

whenever we write $p_1p_2 \ldots p_n \in S_n$ we mean $1 \to p_1$, $2 \to p_2$, and so on. For example, if we write 5 3 1 2 4, then this means there exists f: {1, 2, 3, 4, 5} $\to$ {1, 2, 3, 4, 5} such that $f(1) = 5, f(2) = 3, f(3) = 1, f(4) = 2$ and $f(5) = 4$.

Now in S_2 there are 1 2 and 2 1. S_3 contains 1 2 3, 1 3 2, 2 1 3, 2 3 1, 3 1 2, 3 2 1 and the same we write $S_3 = \left\{\begin{pmatrix}1 & 2 & 3\\1 & 2 & 3\end{pmatrix}, \begin{pmatrix}1 & 2 & 3\\1 & 3 & 2\end{pmatrix}, \begin{pmatrix}1 & 2 & 3\\2 & 1 & 3\end{pmatrix},\right.$

$\left.\begin{pmatrix}1 & 2 & 3\\2 & 3 & 1\end{pmatrix}, \begin{pmatrix}1 & 2 & 3\\3 & 1 & 2\end{pmatrix}, \begin{pmatrix}1 & 2 & 3\\3 & 2 & 1\end{pmatrix}\right\}$. Thus whenever we write $f \in S_n$ we mean $f = t_1t_2 \ldots t_n$ such that $1 \to t_2$, $2 \to t_2$, $3 \to t_3$, ..., $n \to t_n$, where t_i's are from 1, 2, ..., n. If the number of pairs (t_i, t_j) such that $t_i > t_j$ and t_i precedes t_j is even (odd) then f is called even (odd). For example if 5 2 3 4 1 is a permutation on {1, 2, 3, 4, 5}, then (5, 2), (5, 3), (5, 4), (5, 1), (2, 1), (3, 1), (4, 1) are seven pairs, hence 5 2 3 4 1 is odd. Identity permutation 1 2 3 4 5 is always even. In S_3, for 3 1 2 we have only (3, 1), (3, 2) such pairs, so 3 1 2 is even.

We define

$$\text{sign}\, f = \begin{cases} 1 & \text{if } f \text{ is even} \\ -1 & \text{if } f \text{ is odd} \end{cases}$$

Now we formally give the definition of determinant.

Definition 1.3.1. Let $A = (a_{ij})$ be square matrix of order $n \times n$ then determinant of A is denoted by det (A) or $|A|$ and is defined as

$$\sum_{f \in S_n} \text{sign}\,(f)\, a_{1f(1)} a_{2f(2)} \cdots a_{nf(n)}$$

Thus we see that if entries of matrices are real numbers the determinant is a function from the collection of all square matrices to $\mathbb{R}$ (the set of real numbers).

Now let us illustrate this for

$$A = \begin{bmatrix} a_{11} & a_{12} & a_{13} \\ a_{21} & a_{22} & a_{23} \\ a_{31} & a_{32} & a_{33} \end{bmatrix}.$$

$$S_3 = \left\{ \begin{pmatrix} 1 & 2 & 3 \\ 1 & 2 & 3 \end{pmatrix}, \begin{pmatrix} 1 & 2 & 3 \\ 1 & 3 & 2 \end{pmatrix}, \begin{pmatrix} 1 & 2 & 3 \\ 2 & 1 & 3 \end{pmatrix}, \right.$$

$$\left. \begin{pmatrix} 1 & 2 & 3 \\ 2 & 3 & 1 \end{pmatrix}, \begin{pmatrix} 1 & 2 & 3 \\ 3 & 1 & 2 \end{pmatrix}, \begin{pmatrix} 1 & 2 & 3 \\ 3 & 2 & 1 \end{pmatrix} \right\}$$

$$= \{f_1, f_2, f_3, f_4, f_5, f_6\}.$$

Now by the definition

$$\begin{aligned}\det(A) = {} & \text{sign}(f_1)\, a_{1f_1(1)}a_{2f_1(2)}a_{3f_1(3)} + \text{sign}(f_2)\, a_{1f_2(1)}a_{2f_2(2)}a_{3f_2(3)} \\ & + \text{sign}(f_3)\, a_{1f_3(1)}a_{2f_3(2)}a_{3f_3(3)} + \text{sign}(f_4)\, a_{1f_4(1)}a_{2f_4(2)}a_{3f_4(3)} \\ & + \text{sign}(f_5)\, a_{1f_5(1)}a_{2f_5(2)}a_{3f_5(3)} + \text{sign}(f_6)\, a_{1f_6(1)}a_{2f_6(2)}a_{3f_6(3)}\end{aligned}$$

$$= a_{11}a_{22}a_{33} - a_{12}a_{21}a_{33} + a_{12}a_{23}a_{31} + a_{13}a_{21}a_{32} - a_{13}a_{22}a_{31} - a_{11}a_{23}a_{32}$$

$$= a_{11}(a_{22}a_{33} - a_{23}a_{32}) - a_{12}(a_{21}a_{33} - a_{31}a_{23}) + a_{13}(a_{21}a_{32} - a_{31}a_{22})$$

(1.3.1)

Similarly if $A = \begin{bmatrix} a_{11} & a_{12} \\ a_{21} & a_{22} \end{bmatrix}$ then $S_2 = \left\{ \begin{pmatrix} 1 & 2 \\ 1 & 2 \end{pmatrix}, \begin{pmatrix} 1 & 2 \\ 2 & 1 \end{pmatrix} \right\} = \{f_1, f_2\}$

and

$$\det A = \sum_{f \in S_2} \text{sign}(f)\, a_{1f(1)}a_{2f(2)}a_{3f(3)}$$

$$= \text{sign}(f_1)\, a_{1f_1(1)}a_{2f_1(2)} + \text{sign}(f_2)\, a_{1f_2(2)}a_{2f_2(2)}$$

$$= a_{11}a_{22} - a_{12}a_{21}$$

which is same as discussed in the very beginning.

Now if we again see the value of det (A), where

$$A = \begin{bmatrix} a_{11} & a_{12} & a_{13} \\ a_{21} & a_{22} & a_{23} \\ a_{31} & a_{32} & a_{33} \end{bmatrix}$$

then we find that

$$| A | = \det(A)$$

$$= a_{11} \begin{vmatrix} a_{22} & a_{23} \\ a_{32} & a_{33} \end{vmatrix} - a_{12} \begin{vmatrix} a_{21} & a_{23} \\ a_{31} & a_{33} \end{vmatrix} + a_{13} \begin{vmatrix} a_{21} & a_{22} \\ a_{31} & a_{32} \end{vmatrix}$$

$$= a_{11} \begin{vmatrix} a_{22} & a_{23} \\ a_{32} & a_{33} \end{vmatrix} - a_{21} \begin{vmatrix} a_{12} & a_{13} \\ a_{32} & a_{33} \end{vmatrix} + a_{31} \begin{vmatrix} a_{12} & a_{13} \\ a_{22} & a_{23} \end{vmatrix}$$

which we get after rearranging the terms in (1.3.1).

Thus, we get a rule to find the determinant of a matrix of order 3 × 3 by expanding either from first row or first column.

Now we illustrate with numerical example. Let

$$A = \begin{bmatrix} 1 & 2 & -1 \\ 3 & 0 & 5 \\ -2 & 1 & 4 \end{bmatrix}.$$

Let us calculate det (A) with respect to first row. Then

$$\begin{aligned} \det(A) &= 1(0-5) - 2(12 - 5(-2)) + (-1)(3-0) \\ &= -5 - 44 - 3 = -52 \end{aligned}$$

Now expanding with respect to first column, we have

$$\begin{aligned} \det(A) &= 1(0-5) - 3(8 - 1(-1)) + (-2)(10-0) \\ &= -5 - 27 - 30 = -52. \end{aligned}$$

Now let us consider a matrix of order 4 × 4

$$A = \begin{bmatrix} 1 & 2 & 0 & 5 \\ 0 & 4 & -1 & 1 \\ 0 & 1 & 2 & -1 \\ 2 & 0 & 1 & 3 \end{bmatrix}.$$

Expanding with respect to first column, we get

$$|A| = \det(A)$$

$$= 1\begin{vmatrix} 4 & -1 & 1 \\ 1 & 2 & -1 \\ 0 & 1 & 3 \end{vmatrix} - 0\begin{vmatrix} 2 & 0 & 5 \\ 1 & 2 & -1 \\ 0 & 1 & 3 \end{vmatrix} + 0\begin{vmatrix} 2 & 0 & 5 \\ 4 & -1 & 1 \\ 0 & 1 & 3 \end{vmatrix} - 2\begin{vmatrix} 2 & 0 & 5 \\ 4 & -1 & 1 \\ 1 & 2 & -1 \end{vmatrix}$$

$$= [4(6+1) - 1(-3-1) + 0] - 2[2(1-2) - 0 + 5(8+1)]$$

$$= 28 + 4 - 2[-2 + 45]$$

$$= 32 - 86 = -54.$$

Now if

$$A = \begin{bmatrix} a & b & c \\ d & e & f \\ g & h & a \end{bmatrix}$$

then to find the det (A), we can formulate as

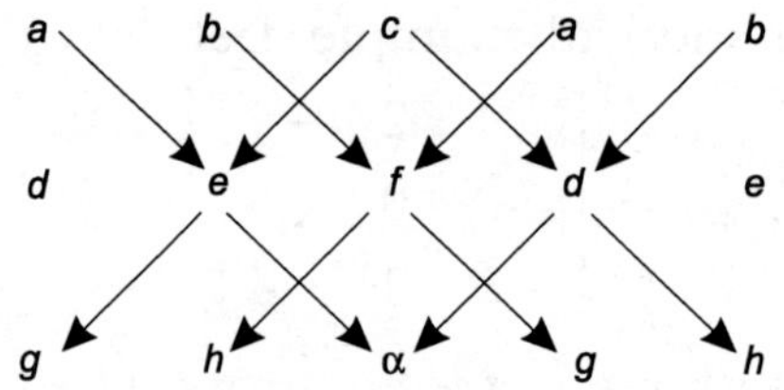

(1.3.2)

Repeating the first and second column of A as shown in (1.3.2) and multiplying the entries as shown by arrows from left to right we add them. Now again multiplying the entries as shown by the arrows from right to left we add them and then subtract this sum from the first sum to get the det (A). This is shown by the following example.

Let

$$A = \begin{bmatrix} 1 & 2 & -1 \\ 3 & 0 & 5 \\ -2 & 1 & 4 \end{bmatrix}.$$

Then to find the det (A), we can formulate as

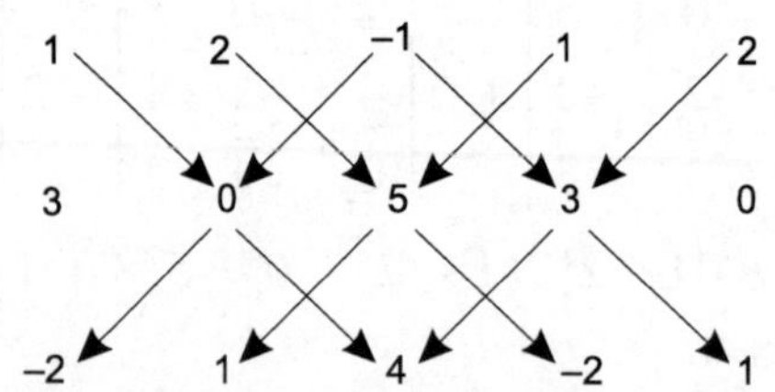

$$|A| = 0 - 20 - 3 - [0 + 5 + 24]$$
$$= -23 - 29 = -52.$$

Now we shall state some properties of determinant, the proofs (verification) may be seen from any book on matrices. Therefore, we shall omit the proofs.

Properties:

1. If A is a square matrix with a zero row (column), then $|A| = 0$.
2. If any two rows (columns) of a square matrix A are identical, then $|A| = 0$.
3. If B is a matrix obtained from a square matrix A by interchanging any two different rows (columns), then $|B| = -|A|$.
4. If B is a matrix obtained from a square matrix A after multiplying any row (column) by a scalar $\alpha \neq 0$, then $|B| = \alpha |A|$.
5. If B is a matrix obtained from a square matrix A after adding any multiple of any row (column) to any other row (column), then $|B| = |A|$.
6. If in a square matrix A each element (entry) in any row (column) consists of sum of p terms, then $|A|$ = sum of determinants of p matrices of same order.

   For example,

$$A = \begin{bmatrix} 2+3 & 1 & 0 \\ 1+4 & 0 & 1 \\ -1+5 & 2 & -1 \end{bmatrix}.$$

Then

$$|A| = \begin{vmatrix} 2 & 1 & 0 \\ 1 & 0 & 1 \\ -1 & 2 & -1 \end{vmatrix} + \begin{vmatrix} 3 & 1 & 0 \\ 4 & 0 & 1 \\ 5 & 2 & -1 \end{vmatrix}$$

For calculating determinant of square matrices with entries as larger numbers, we shall use the properties to reduce the entries smaller. For example, if

$$A = \begin{bmatrix} 87 & 88 & 89 \\ 88 & 89 & 90 \\ 89 & 90 & 91 \end{bmatrix},$$

then using property 5, we get

$$|A| = \begin{vmatrix} 87 & 88 & 89 \\ 1 & 1 & 1 \\ 2 & 2 & 2 \end{vmatrix} = 2\begin{vmatrix} 87 & 88 & 89 \\ 1 & 1 & 1 \\ 1 & 1 & 1 \end{vmatrix} = 0.$$

If A is an upper or lower triangular matrix, then by expanding either first row or first column, we get $|A|$ = product of diagonal entries of A.

Now we state the following, the proofs of which can be seen in any book on matrices.

Theorem 1.3.2. For any square matrices A and B of order $n \times n$, we have

(*i*) $|A^t| = |A|$

(*ii*) $|AB| = |A||B|$

(*iii*) $|\alpha A| = a^n |A|$.

Definition 1.3.3. If $A = (a_{ij})$ is a $n \times n$ matrix, then for any i and j delete i^{th} row and j^{th} column of A and denote by M_{ij}, the matrix thus obtained, then det (M_{ij}) is called the minor of a_{ij} and $(-1)^{-i+j}|M_{ij}|$ is called cofactor of a_{ij} and is denoted by A_{ij}.

Thus, cofactor of a_{ij} = cof $(a_{ij}) = (-1)^{i+j}|M_{ij}|$.

For example, if

$$A = \begin{bmatrix} 1 & 2 & 3 \\ 0 & -1 & -4 \\ -2 & 5 & 6 \end{bmatrix},$$

then

$$\text{cof}(-4) = (-1)^{2+3}\begin{vmatrix} 1 & 2 \\ -2 & 5 \end{vmatrix} = -(5+4) = -9$$

$$\text{cof}(-2) = (-1)^{3+1}\begin{vmatrix} 2 & 3 \\ -1 & -4 \end{vmatrix} = -(8+3) = -5,$$

and so on.

Applying the definition of cofactor and expanding the determinant along i^{th} row of $A = (a_{ij})$ for $1 \le i \le n$, we see that

$$|A| = a_{i1}A_{i1} + a_{i2}A_{i2} + \ldots + a_{in}A_{in} \qquad (1.3.3)$$

and similarly along j^{th} column $1 \le j \le n$, we have

$$|A| = a_{1j}A_{1j} + a_{2j}A_{2j} + \ldots + a_{nj}A_{nj} \qquad (1.3.4)$$

Now, replacing i^{th} row of A by the jth row of A, we get a new matrix B with two identical rows which implies that $|B| = 0$. But from (1.3.3)

$$|B| = a_{j1}A_{j1} + a_{j2}A_{j2} + \ldots + a_{jn}A_{in} \qquad \text{for } j \ne i$$

$$\Rightarrow \quad a_{j1}A_{j1} + a_{j2}A_{j2} + \ldots + a_{jn}A_{in} = 0.$$

Similarly, replacing j^{th} column of A by the ith column of A and using (1.3.4), we get

$$a_{1i}A_{1j} + a_{2i}A_{2j} + \ldots + a_{ni}A_{nj} = 0 \text{ for } i \neq j.$$

Definition 1.3.4. If $A = (a_{ij})$ is a square matrix of order $n \times n$ then we define adjoint of A as $[\text{cof}(a_{ij})]^t$ and is denoted by adj (A). Thus

$$\text{adj } A = \begin{bmatrix} \text{cof}(a_{11}) & \text{cof}(a_{21}) & \cdots & \text{cof}(a_{n1}) \\ \text{cof}(a_{12}) & \text{cof}(a_{22}) & \cdots & \text{cof}(a_{n2}) \\ \vdots & \vdots & \vdots & \vdots \\ \text{cof}(a_{1n}) & \text{cof}(a_{2n}) & \cdots & \text{cof}(a_{nn}) \end{bmatrix}$$

Now we consider an example. Let

$$A = \begin{bmatrix} 1 & 0 & 2 \\ -1 & 3 & -3 \\ 4 & -2 & -4 \end{bmatrix}$$

$$\text{cof}(1) = (-1)^{1+1} \begin{vmatrix} 3 & -3 \\ -2 & -4 \end{vmatrix}$$

$$= (-12 - 6) = -18$$

$$\text{cof}(0) = (-1)^{1+2} \begin{vmatrix} -1 & -3 \\ 4 & -4 \end{vmatrix}$$

$$= -(4 + 12) = -16$$

$$\text{cof}(2) = (-1)^{1+3} \begin{vmatrix} -1 & 3 \\ 4 & 2 \end{vmatrix}$$

$$= 2 - 12 = -10$$

$$\text{cof}(-1) = (-1)^{2+1} \begin{vmatrix} 0 & 2 \\ -2 & -4 \end{vmatrix}$$

$$= -(4) = -4$$

$$\text{cof}(3) = (-1)^{2+2} \begin{vmatrix} 1 & 2 \\ 4 & -4 \end{vmatrix}$$

$$= (-4 - 8) = -12$$

$$\text{cof}(-3) = (-1)^{2+3} \begin{vmatrix} 1 & 0 \\ 4 & -2 \end{vmatrix}$$

$$= -(-2) = 2$$

$$\text{cof}(4) = (-1)^{3+1}\begin{vmatrix} 0 & 2 \\ 3 & -3 \end{vmatrix} = -6$$

$$\text{cof}(-2) = (-1)^{3+2}\begin{vmatrix} 1 & 2 \\ -1 & -3 \end{vmatrix}$$

$$= -(-3+2) = 1$$

$$\text{cof}(-4) = (-1)^{3+3}\begin{vmatrix} 1 & 0 \\ -1 & 3 \end{vmatrix} = 0$$

Theorem 1.3.5. If A is a square matrix of order $n \times n$, then

$$A(\text{adj}(A)) = (\text{adj}(A))A = \det(A)I$$

Proof. Let $A = (a_{ij})$. Then

$$A(\text{adj}(A)) = \begin{bmatrix} a_{11} & a_{12} & \cdots & a_{1n} \\ a_{21} & a_{22} & \cdots & a_{2n} \\ \vdots & \vdots & \vdots & \vdots \\ a_{i1} & a_{i2} & \cdots & a_{in} \\ \vdots & \vdots & \vdots & \vdots \\ a_{n1} & a_{n2} & \cdots & a_{nn} \end{bmatrix} \begin{bmatrix} A_{11} & A_{21} & \cdots & A_{j1} & \cdots & A_{n1} \\ A_{12} & A_{22} & \cdots & A_{j2} & \cdots & A_{n2} \\ \vdots & \vdots & \vdots & \vdots & \vdots & \vdots \\ A_{1n} & A_{2n} & \cdots & A_{jn} & \cdots & A_{nn} \end{bmatrix}$$

and $(i, j)^{\text{th}}$ element of $A(\text{adj}(A))$ will be

$a_{i1}A_{j1} + a_{i2}A_{j2} + \ldots + a_{in}A_{in} = |A| \text{ if } i = j = 0 \text{ if } i \neq j.$

$$A(\text{adj}(A)) = \begin{bmatrix} |A| & 0 & \cdots & 0 \\ 0 & |A| & \cdots & 0 \\ \vdots & \vdots & \vdots & \vdots \\ 0 & 0 & \cdots & |A| \end{bmatrix}$$

$$= |A|\begin{bmatrix} 1 & 0 & \cdots & 0 \\ 0 & 1 & \cdots & 0 \\ \vdots & \vdots & \vdots & \vdots \\ 0 & 0 & \cdots & 1 \end{bmatrix} = |A|\,I$$

Similarly, we can show that $(\text{adj}(A))A = |A|\,I$. Hence, the result follows.

Corollary 1.3.6. $|\operatorname{adj}(A)| = |A|^{n-1}$, where A is a square matrix of the order $n \times n$ and $|A| \neq 0$.

$$|A||\operatorname{adj}(A)| = |A|^n$$

$$\Rightarrow \quad |\operatorname{adj}(A)| = |A|^{n-1}$$

Corollary 1.3.7. If A is a square matrix of order $n \times n$ and $|A| \neq 0$, then

$$|\operatorname{adj}(\operatorname{adj}(A))| = |A|^{(n-1)^2}.$$

Theorem 1.3.8. A square matrix A is non-singular if and only if $|A| \neq 0$.

Proof. Suppose A is non-singular. Then by Theorem 1.2.7, $A \underset{r}{\sim} I \Rightarrow$ there exist elementary matrices $E_1, E_2, \ldots E_p$ such that $E_p E_{p-1} \ldots E_2 E_1 A = I$.

Now since $|I| = I \Rightarrow |A| \neq 0$. Conversely, suppose $|A| \neq 0$, then by Theorem 1.3.5.

$$A \frac{\operatorname{adj}(A)}{|A|} = \frac{\operatorname{adj}(A)}{|A|} A = I$$

$$\Rightarrow \quad A^{-1} = \frac{\operatorname{adj}(A)}{|A|}$$

$\Rightarrow$ A is non-singular.

Corollary 1.3.9. If A is non-singular, then $|A^{-1}| = \dfrac{1}{|A|}$.

Corollary 1.3.10. Similar matrices have the same determinant.

If $A \underset{s}{\sim} B \Rightarrow B = P^{-1}AP \Rightarrow |B| = \dfrac{1}{|P|} \cdot |A||P| = |A|$.

If

$$A = \begin{bmatrix} 1 & 1 & 1 \\ a & b & c \\ a^2 & b^2 & c^2 \end{bmatrix}, \text{ then}$$

$$|A| = \begin{bmatrix} 1 & 0 & 0 \\ a & b-a & c-a \\ a^2 & b^2-c^2 & c^2-a^2 \end{bmatrix}$$

$$= (b-a)(c-a)\begin{bmatrix} 1 & 0 & 0 \\ a & 1 & 1 \\ a^2 & b+a & c+a \end{bmatrix}$$

$$= (b-a)(c-a)(c-b).$$

This determinant is called *Vandermonde determinant* of order 3.

Now, using Theorem 1.3.5, we calculate A^{-1}. Let

$$A = \begin{bmatrix} 1 & 0 & 2 \\ -1 & 3 & -3 \\ 4 & -2 & -2 \end{bmatrix},$$

then we have seen that

$$\text{adj}\,(A) = \begin{bmatrix} -18 & -4 & -6 \\ -16 & -12 & 1 \\ -10 & 2 & -4 \end{bmatrix}$$

Now, $$|A| = 1(-12-6) + 2(2-12)$$

$$= -18 - 20 = -38. \text{ Thus}$$

$$A^{-1} = \begin{bmatrix} \frac{9}{19} & \frac{2}{19} & \frac{3}{19} \\ \frac{8}{19} & \frac{6}{19} & -\frac{1}{38} \\ \frac{5}{19} & -\frac{1}{19} & -\frac{3}{38} \end{bmatrix}.$$

We can also verify that

$$AA^{-1} = \begin{bmatrix} 1 & 0 & 0 \\ 0 & 1 & 0 \\ 0 & 0 & 1 \end{bmatrix}.$$

1.4 System of Linear Equations

In this section we shall deal with some of the familiar methods to find the solutions for linear equations. We are familiar with the methods of solving two or three linear equations by eliminating the variable.

An equation of the type

$$ax + by = c, \tag{1.4.1}$$

where x and y are unknown variables and a, b, c are constants, is called a *linear equation.* We know that $ax + by = c$ represents a straight line and has infinite solutions. Also, $x = r$, $y = s$ is a solution of (1.4.1) if $ar + bs = c$. Now suppose we have

$$x + 2y = 9 \qquad (1.4.2)$$

$$2x + y = 3 \qquad (1.4.3)$$

then to find the solution we multiply (1.4.2) by 2 and subtract from (1.4.3) to get $y = 5$ by eliminating x. Substituting $y = 5$ in any one of (1.4.2) and (1.4.3), we get $x = -1$. Thus, we write equations (1.4.2) and (1.4.3) in matrix form

$$\begin{bmatrix} 1 & 2 \\ 2 & 1 \end{bmatrix} \begin{bmatrix} x \\ y \end{bmatrix} = \begin{bmatrix} 9 \\ 3 \end{bmatrix}$$

$$AX = b$$

Now, we write $\begin{bmatrix} 1 & 2 & : & 9 \\ 2 & 1 & : & 3 \end{bmatrix}$.

Applying $R_2 \to R_2 + R_1(-2)$, we get

$$\begin{bmatrix} 1 & 2 & : & 9 \\ 0 & -3 & : & -15 \end{bmatrix}$$

$\Rightarrow$ $$-3y = -15$$

$\Rightarrow$ $$y = 5 \text{ and } x + 2y = 9$$

$\Rightarrow$ $$x = -1$$

$\Rightarrow$ we eliminate variables by elementary row operations to get the solutions. An equation $\sum_{i=1}^{n} a_i x_i = b$ is called linear equation in n-variables.

Now let us consider

$$\begin{matrix} a_{11}x_1 & + & a_{12}x_2 & + & \cdots & + & a_{1n}x_n & = & b_1 \\ a_{21}x_1 & + & a_{22}x_2 & + & \cdots & + & a_{2n}x_n & = & b_2 \\ \vdots & & \vdots & & \cdots & \vdots & \vdots & & \\ a_{m1}x_1 & + & a_{m2}x_2 & + & \cdots & + & a_{mn}x_n & = & b_m \end{matrix} \qquad (1.4.4)$$

Equation (1.4.4) is known as system of linear equations in n-variables $x_1, x_2, \ldots, x_n$. We can write system of equations (1.4.4) in matrix form as

$$\begin{bmatrix} a_{11} & a_{12} & \cdots & a_{1n} \\ a_{21} & a_{22} & \cdots & a_{2n} \\ \vdots & \vdots & \cdots & \vdots \\ a_{m1} & a_{m2} & \cdots & a_{mn} \end{bmatrix} \begin{bmatrix} x_1 \\ x_2 \\ \vdots \\ x_n \end{bmatrix} = \begin{bmatrix} b_1 \\ b_2 \\ \vdots \\ b_n \end{bmatrix}$$

$\Rightarrow$ $$AX = b,$$

where A is called coefficient matrix of (1.4.4) and

$$\begin{bmatrix} a_{11} & a_{12} & \cdots & a_{1n} & : & b_1 \\ a_{21} & a_{22} & \cdots & a_{2n} & : & b_2 \\ \vdots & \vdots & \cdots & \vdots & : & \vdots \\ a_{m1} & a_{m2} & \cdots & a_{mn} & : & b_m \end{bmatrix}$$

is known as augmented matrix of (1.4.4).

Thus, for every system of equations we can get coefficient matrix as well as augmented matrix. Conversely, if augmented matrix is known, then we can find the system of linear equations. For example, if

$$\begin{bmatrix} 1 & 0 & 3 & 4 & 0 & : & 3 \\ 2 & 1 & 0 & -1 & 3 & : & 4 \\ 0 & 0 & 5 & 2 & 1 & : & 5 \end{bmatrix}$$

is augmented matrix then the system of equations is given by

$$x_1 + 3x_3 + 4x_4 = 3$$
$$2x_1 + x_2 - 3x_5 = 4$$
$$5x_3 + 2x_4 + x_5 = 5$$

Now, if we see the system of equations (1.4.4) then either $m < n$ or $m > n$ or $m = n$. To discuss these cases we shall restrict to the following well known methods:

1. Gauss Elimination Method.
2. Gauss-Jordan Elimination Method.
3. Cramer's Method.
4. Matrix Inverse Method.
5. LU-Factorizationd Method.

If the augmented matrices $[A : b]$ and $[C : d]$ of two m-systems of equations in n unknowns are row equivalent, then both the systems of equations have exactly the same solutions.

Definition 1.4.1. A linear system of m-equations in n variables is called consistent if it has at least one solution otherwise it is called inconsistent.

In other words, if $r(A) = r(|\ A : b\ |)$ then it is consistent, otherwise inconsistent.

For example, $x + 2y = 4$ and $2x + y = 3$ are consistent while $x + 2y = 4$ and $2x + 4y8 = 10$ are inconsistent.

Example 1. Solve

$$\begin{aligned} x + 2y - z + 3w &= 3 \\ 2x + 4y + 4z + 3w &= 9 \\ 3x + 6y - z + 8w &= 10 \end{aligned}$$

Augmented matrix

$$[A : b] = \begin{bmatrix} 1 & 2 & -1 & 3 & : & 3 \\ 2 & 4 & 4 & 3 & : & 9 \\ 3 & 6 & -1 & 8 & : & 10 \end{bmatrix}$$

$$\sim \begin{bmatrix} 1 & 2 & -1 & 3 & : & 3 \\ 0 & 0 & 6 & -3 & : & 3 \\ 0 & 0 & 2 & -1 & : & 1 \end{bmatrix} \text{ by } R_2 \to R_2 + R_1(-2)$$

$$R_3 \to R_3 + R_1(-3)$$

$$\sim \begin{bmatrix} 1 & 2 & -1 & 3 & : & 3 \\ 0 & 0 & 2 & -3 & : & 1 \\ 0 & 0 & 2 & -1 & : & 1 \end{bmatrix} \text{ by } R_2 \to R_2\left(\frac{1}{3}\right)$$

$$\sim \begin{bmatrix} 1 & 2 & -1 & 3 & : & 3 \\ 0 & 0 & 2 & -1 & : & 1 \\ 0 & 0 & 0 & 0 & : & 0 \end{bmatrix} \text{ by } R_3 \to R_3 + R_1(-1)$$

$$\sim \begin{bmatrix} 1 & 2 & -1 & 3 & : & 3 \\ 0 & 0 & 1 & -\frac{1}{2} & : & \frac{1}{2} \\ 0 & 0 & 0 & 0 & : & 0 \end{bmatrix} \text{ by } R_2 \to R_2\left(\frac{1}{2}\right)$$

This implies that $z - \frac{1}{2}w = \frac{1}{2}$

$$\Rightarrow \qquad z = \frac{1 + w}{2}. \text{ Also}$$

$$x + 2y - z + 3w = 3$$

$$\Rightarrow \qquad x + 2y = -3w + 3 + z = -3w + 3 + \frac{1+w}{2}$$

$$= \frac{-6w + 6 + 1 + w}{2}$$

$$= \frac{-5w + 7}{2}$$

Now choosing $w = r$ and $y = s$ for some r and s, we get

$$x = \frac{-5w+7}{2} - 2s = \frac{-5w+7-4s}{2}$$

Thus $w = r$, $z = \dfrac{1+r}{2}$, $y = s$, $x = \dfrac{-5w+7-4s}{2}$, if $r = 2$, $s = 1$ then $w = 2$, $z = \dfrac{3}{2}$, $y = 1$, $x = -\dfrac{7}{2}$. Thus, the system has no unique solution and has infinite solutions.

Example 2. For what value of a the equations $x + y + z = 1$, $x + 2y + 4z = a$ and $x + 4y + 10z = a^2$ is consistent.

Augmented matrix

$$[A : b] = \begin{bmatrix} 1 & 1 & 1 & : & 1 \\ 1 & 2 & 4 & : & a \\ 1 & 4 & 10 & : & a^2 \end{bmatrix}$$

$$\sim \begin{bmatrix} 1 & 1 & 1 & : & 1 \\ 0 & 1 & 3 & : & a-1 \\ 0 & 0 & 0 & : & a^2-1 \end{bmatrix} \text{ by } R_2 \to R_2 + R_1(-1)$$

$$R_3 \to R_3 + R_1(-1)$$

$$\sim \begin{bmatrix} 1 & 1 & 1 & : & 1 \\ 0 & 1 & 3 & : & a-1 \\ 0 & 0 & 0 & : & a^2-1-3a+3 \end{bmatrix}$$

$$\text{by } R_3 \to R_3 + R_2(-3)$$

This implies that

$$0x + 0y + 0z = a^2 - 3a + 2$$

$$\Rightarrow \qquad a^2 - 3a + 2 = 0$$

$$\Rightarrow \qquad (a-2)(a-1) = 0$$

$$\Rightarrow \qquad a = 2, a = 1$$

Now for $a = 1$, we have $y + 3z = 0 \Rightarrow y = -3z$ and $x + y + z = 1 \Rightarrow x = 1 - y - z = 1 + 3z - z = 1 + 2z$. Take $z = r$, then $y = -3r$ and $x = 1 + 2r$ for any value of $r \Rightarrow$ the given system of equations have infinite number of solutions.

Example 3. Find the solution of equations $x + 2y - 5z = -9$, $3x - y + 2z = 5$, $2x + 3y - z = 3$ and $4x - 5y + z = -3$.

Augmented matrix

$$[A : b] = \begin{bmatrix} 1 & 2 & -5 & : & -9 \\ 3 & -1 & 2 & : & 5 \\ 2 & 3 & -1 & : & 3 \\ 4 & -5 & 1 & : & -3 \end{bmatrix}$$

$$\sim \begin{bmatrix} 1 & 2 & -5 & : & -9 \\ 0 & -7 & 17 & : & 32 \\ 0 & -1 & 9 & : & 21 \\ 0 & -13 & 21 & : & 33 \end{bmatrix} \text{ by } R_2 \to R_2 + R_1(-3)$$

$$R_3 \to R_3 + R_1(-2)$$
$$R_4 \to R_4 + R_1(-4)$$

$$\sim \begin{bmatrix} 1 & 2 & -5 & : & -9 \\ 0 & 0 & 46 & : & -115 \\ 0 & -1 & -9 & : & -21 \\ 0 & 0 & -96 & : & -240 \end{bmatrix}$$

$$\text{by } R_2 \to R_2 + R_3(-7)$$
$$R_3 \to R_3(-1)$$
$$R_4 \to R_4 + R_3(-13)$$

$$\sim \begin{bmatrix} 1 & 2 & -5 & : & -9 \\ 0 & 1 & -9 & : & -21 \\ 0 & 0 & -46 & : & -115 \\ 0 & 0 & -96 & : & -240 \end{bmatrix} \text{ by } R_2 \leftrightarrow R_3$$

$$\sim \begin{bmatrix} 1 & 2 & -5 & : & -9 \\ 0 & 1 & -9 & : & -21 \\ 0 & 0 & 2 & : & 5 \\ 0 & 0 & 2 & : & 5 \end{bmatrix} \text{ by } R_3 \to R_3\left(-\frac{1}{23}\right)$$

$$\sim \begin{bmatrix} 1 & 2 & -5 & : & -9 \\ 0 & 1 & -9 & : & -21 \\ 0 & 0 & 1 & : & \frac{5}{2} \\ 0 & 0 & 0 & : & 0 \end{bmatrix} \text{ by } \begin{array}{l} R_4 \to R_4\left(-\frac{1}{48}\right) \\ R_3 \to R_3\left(\frac{1}{2}\right) \\ R_4 \to R_4 + R_3(-1) \end{array}$$

$\Rightarrow$ $$r([A : b]) = r(A)$$

$\Rightarrow$ Solution exists and

$$z = \frac{5}{2}$$

$$y - 9z = -21$$

$\Rightarrow$ $$y = -21 + 9z = -21 + \frac{45}{2} = \frac{3}{2}$$

and $$x = -9 - 2y + 5z = -9 - 3 + \frac{25}{2} = \frac{1}{2}$$

2 GAUSS-JORDAN ELIMINATION METHOD

To solve the linear system of equations we reduce the augmented matrix to row reduced echelon form and the rest of technique is the same. However, it may be noted again that if there are m-equations in n-unknowns and if augmented matrix has r non-zero rows in echelon form, then its rank will be r.

If $r = r(A)$, then the system is consistent otherwise it is consistent as noted earlier.

If $r = n$, then the system will have unique solution. If $r < n$ or $r \le m$ then the system will have infinite solutions.

Now we illustrate with examples.

In Example (3) the augmented matrix

$$[A : b] \sim \begin{bmatrix} 1 & 2 & -5 & : & -9 \\ 0 & 1 & -9 & : & -21 \\ 0 & 0 & 1 & : & \frac{5}{2} \\ 0 & 0 & 0 & : & 0 \end{bmatrix}$$

$$\sim \begin{bmatrix} 1 & 0 & 13 & : & 33 \\ 0 & 1 & -9 & : & -21 \\ 0 & 0 & 1 & : & \frac{5}{2} \\ 0 & 0 & 0 & : & 0 \end{bmatrix} \text{ by } R_1 \to R_1 + R_2(-2)$$

$$\sim \begin{bmatrix} 1 & 0 & 0 & : & \frac{1}{2} \\ 0 & 1 & 0 & : & -\frac{3}{2} \\ 0 & 0 & 1 & : & \frac{5}{2} \\ 0 & 0 & 0 & : & 0 \end{bmatrix} \begin{array}{l} \text{by } R_1 \to R_1 + R_3(-13) \\ \quad R_2 \to R_2 + R_3(-9) \end{array}$$

$$\Rightarrow \qquad x = \frac{1}{2}, y = \frac{3}{2}, z = \frac{5}{2}.$$

Now if we take in Example (3) the last equation as $4x - 5y + z = 11$, then it will be easy to see that $r(A) = 3$ and $r([A : b]) = 4$ and therefore the system will not have solutions.

Example 4. Let

$$\begin{aligned} x + y + 2z - w &= 3 \\ 2x + 5y - z + w &= 4 \\ 2x + y + z + 3w &= 5 \\ x - y + 2z + 5w &= 2 \end{aligned}$$

Augmented matrix

$$[A : b] = \begin{bmatrix} 1 & 1 & 2 & -1 & : & 3 \\ 2 & 5 & -1 & 1 & : & 4 \\ 2 & 1 & 1 & 3 & : & 5 \\ 1 & -1 & 2 & 5 & : & 2 \end{bmatrix}$$

$$\sim \begin{bmatrix} 1 & 1 & 2 & -1 & : & 3 \\ 2 & 5 & -1 & 1 & : & 4 \\ 0 & -4 & 2 & 2 & : & 1 \\ 0 & -2 & 0 & 6 & : & -1 \end{bmatrix} \begin{array}{l} \text{by } R_3 \to R_3 + R_2(-1) \\ \quad R_4 \to R_4 + R_1(-1) \end{array}$$

$$\sim \begin{bmatrix} 1 & 1 & 2 & -1 & : & 3 \\ 0 & 3 & -5 & 3 & : & -2 \\ 0 & -2 & 1 & 1 & : & \frac{1}{2} \\ 0 & 1 & 0 & -3 & : & \frac{1}{2} \end{bmatrix} \text{ by } R_2 \to R_2 + R_1(-2), \; R_3 \to R_3\left(\frac{1}{2}\right), \; R_4 \to R_4\left(-\frac{1}{2}\right)$$

$$\sim \begin{bmatrix} 1 & 1 & 2 & -1 & : & 3 \\ 0 & 1 & -4 & 4 & : & -\frac{3}{2} \\ 0 & -2 & 1 & 1 & : & \frac{1}{2} \\ 0 & 1 & 0 & -3 & : & \frac{1}{2} \end{bmatrix} \text{ by } R_2 \to R_2 + R_3(1)$$

$$\sim \begin{bmatrix} 1 & 1 & 2 & -1 & : & 3 \\ 0 & 1 & -4 & 4 & : & -\frac{3}{2} \\ 0 & 0 & -7 & 9 & : & -\frac{5}{2} \\ 0 & 0 & 4 & -7 & : & 2 \end{bmatrix} \text{ by } R_3 \to R_3 + R_2(2), \; R_4 \to R_4 + R_2(-1)$$

$$\sim \begin{bmatrix} 1 & 0 & 6 & -5 & : & \frac{9}{2} \\ 0 & 1 & -4 & 4 & : & -\frac{3}{2} \\ 0 & 0 & 1 & -5 & : & \frac{3}{2} \\ 0 & 0 & 4 & -7 & : & 2 \end{bmatrix} \text{ by } R_1 \to R_1 + R_2(-1), \; R_3 \to R_3 + R_4(2)$$

$$\sim \begin{bmatrix} 1 & 0 & 6 & -5 & : & \frac{9}{2} \\ 0 & 1 & -4 & 4 & : & -\frac{3}{2} \\ 0 & 0 & 1 & -5 & : & \frac{3}{2} \\ 0 & 0 & 0 & 13 & : & -4 \end{bmatrix} \text{ by } R_4 \to R_4 + R_3(-4)$$

$$\sim \begin{bmatrix} 1 & 0 & 0 & 25 & : & -\frac{9}{2} \\ 0 & 1 & 0 & -16 & : & \frac{9}{2} \\ 0 & 0 & 1 & -5 & : & \frac{3}{2} \\ 0 & 0 & 0 & 1 & : & -\frac{4}{13} \end{bmatrix} \begin{matrix} \text{by } R_1 \to R_1 + R_3(-6) \\ R_2 \to R_2 + R_3(4) \\ R_4 \to R_4\left(\frac{1}{13}\right) \end{matrix}$$

$$\sim \begin{bmatrix} 1 & 0 & 0 & 0 & : & -\frac{9}{2}+\frac{100}{13} \\ 0 & 1 & 0 & 0 & : & \frac{9}{2}-\frac{64}{13} \\ 0 & 0 & 1 & 0 & : & \frac{3}{2}-\frac{20}{13} \\ 0 & 0 & 0 & 1 & : & -\frac{4}{13} \end{bmatrix} \begin{matrix} \text{by } R_1 \to R_1 + R_4(-25) \\ R_2 \to R_2 + R_4(16) \\ R_3 \to R_3 + R_4(5) \end{matrix}$$

$$\Rightarrow x = -\frac{9}{2}+\frac{100}{13},\ y = \frac{9}{2}-\frac{64}{13},\ z = \frac{3}{2}-\frac{20}{13} \text{ and } w = -\frac{4}{13}.$$

We may do the verification.

Example 5. Let $2x - y + z = 4$, $3x - y + z = 6$, $4x - y + 2z = 7$ and $-x + y - z = 9$.

Augmented matrix

$$[A : b] = \begin{bmatrix} 2 & -1 & 1 & : & 4 \\ 3 & -1 & 1 & : & 6 \\ 4 & -1 & 2 & : & 7 \\ -1 & 1 & -1 & : & 9 \end{bmatrix}$$

$$\sim \begin{bmatrix} 1 & -1 & 1 & : & -9 \\ 3 & -1 & 1 & : & 6 \\ 4 & -1 & 2 & : & 7 \\ 2 & -1 & 1 & : & 4 \end{bmatrix} \text{ by } R_4 \to R_4(-1) \text{ and } R_4 \leftrightarrow R_1$$

$$\sim \begin{bmatrix} 1 & -1 & 1 & : & -9 \\ 0 & 2 & -2 & : & 33 \\ 0 & -3 & -2 & : & 43 \\ 0 & 1 & -1 & : & 22 \end{bmatrix} \begin{matrix} \text{by } R_2 \to R_2 + R_1(-3) \\ R_3 \to R_3 + R_1(-4) \\ R_4 \to R_4 + R_1(-2) \end{matrix}$$

$$\sim \begin{bmatrix} 1 & -1 & 1 & : & -9 \\ 0 & 0 & 0 & : & -11 \\ 0 & -3 & -2 & : & 43 \\ 0 & 1 & -1 & : & 22 \end{bmatrix} \text{ by } R_2 \to R_2 + R_4(-2)$$

$\Rightarrow 0 = -11$ not possible. Further $r(A) = 3$ and $r([A : b]) = 4 \Rightarrow$ the system of equations is inconsistent.

Example 6. Solve the equations $x - 2y + z = a$, $2x + 7y - 3z = b$ and $3x + 5y - 2z = c$ if it exists.

Augmented matrix

$$[A : b] = \begin{bmatrix} 1 & -2 & 1 & : & a \\ 2 & 7 & -3 & : & b \\ 3 & 5 & -2 & : & c \end{bmatrix}$$

$$\sim \begin{bmatrix} 1 & -2 & 1 & : & a \\ 0 & 11 & -5 & : & b-2a \\ 0 & 11 & -5 & : & c-3a \end{bmatrix} \text{ by } \begin{array}{l} R_2 \to R_2 + R_1(-2) \\ R_3 \to R_3 + R_1(-3) \end{array}$$

$$\sim \begin{bmatrix} 1 & -2 & 1 & : & a \\ 0 & 11 & -5 & : & b-2a \\ 0 & 0 & 0 & : & c-3a-b+2a \end{bmatrix}$$

by $R_3 \to R_3 + R_2(-1)$

$\Rightarrow$ for consistency $c - 3a - b + 2a = 0 \Rightarrow c - a - b = 0 \Rightarrow c = a + b$. For inconsistency $c \neq a + b$.

For example, if $c = 8$ and $a = 2$, $b = 6$ or $a = 1$, $b = 7$, and so on, then equations have solutions. However, if $c = 8$, $a = 5$ and $b = 5$, then the above equations have no solutions.

Now, we shall give some examples of homogeneous linear equations *i.e.*, equations of the type $a_1x_1 + a_2x_2 + \ldots + a_nx_n = 0$, and so on, *i.e.*, the matrix equation is of the type $AX = O$.

The method of solving homogeneous equations by Gauss elimination and Guass-Jordan elimination method is same. However, homogeneous equations will have always at least one solution *i.e.*, the trivial solution. It may have nontrivial solutions. For example, $x + 2y = 0$ has trivial solution $x = 0$, $y = 0$ and has finite number or non-trivial solutions. However, a homogeneous system of m equations in n unknowns will always have a nontrivial solution if $m < n$.

Example 7. Let

$$\begin{aligned} x - 2y + 3z - 2w &= 0 \\ 3x - 7y - 2z + 4w &= 0 \\ 4x + 3y + 5z + 2w &= 0 \Rightarrow AX = O \\ x + y + 4z - w &= 0 \end{aligned}$$

Augmented matrix

$$[A : b] = \begin{bmatrix} 1 & -2 & 3 & -2 & : & 0 \\ 3 & -7 & -2 & 4 & : & 0 \\ 4 & 3 & 5 & 4 & : & 0 \\ 1 & 1 & 4 & -1 & : & 0 \end{bmatrix}$$

$$\sim \begin{bmatrix} 1 & -2 & 3 & -2 & : & 0 \\ 0 & -1 & -11 & 10 & : & 0 \\ 0 & 11 & -7 & 12 & : & 0 \\ 0 & 3 & 1 & 1 & : & 0 \end{bmatrix} \begin{matrix} \text{by } R_2 \to R_2 + R_1(-3) \\ R_3 \to R_3 + R_1(-4) \\ R_4 \to R_4 + R_1(-1) \end{matrix}$$

$$\sim \begin{bmatrix} 1 & -2 & 3 & -2 & : & 0 \\ 0 & -1 & -11 & 10 & : & 0 \\ 0 & 0 & -128 & 122 & : & 0 \\ 0 & 0 & -32 & 31 & : & 0 \end{bmatrix} \begin{matrix} \text{by } R_3 \to R_3 + R_2(11) \\ R_4 \to R_4 + R_2(3) \end{matrix}$$

$$\sim \begin{bmatrix} 1 & -2 & 3 & -2 & : & 0 \\ 0 & -1 & -11 & 10 & : & 0 \\ 0 & 0 & -128 & 122 & : & 0 \\ 0 & 0 & 0 & \frac{1}{2} & : & 0 \end{bmatrix} \text{by } R_4 \to R_4 + R_3\left(-\frac{1}{4}\right)$$

$\Rightarrow \qquad r(A) = r([A : b]) = 4$

$\Rightarrow$ has only the trivial solution $x = y = z = w = 0$.

Example 8. Let

$$\begin{aligned} x + 2y - 5z + 4w &= 0 \\ 2x - 3y + 2z + 3w &= 0 \\ 4x - 7y + z - 6w &= 0 \end{aligned}$$

Then, as remarked earlier, nontrivial solution will exist. The augmented matrix is

$$[A : b] = \begin{bmatrix} 1 & 2 & -5 & 4 & : & 0 \\ 2 & -3 & 2 & 3 & : & 0 \\ 4 & -7 & 1 & -6 & : & 0 \end{bmatrix}$$

$$\sim \begin{bmatrix} 1 & 2 & -5 & 4 & : & 0 \\ 0 & -7 & 12 & -5 & : & 0 \\ 0 & -15 & 21 & -22 & : & 0 \end{bmatrix} \text{ by } R_2 \to R_2 + R_1(-2), \quad R_3 \to R_3 + R_1(-4)$$

$$\sim \begin{bmatrix} 1 & 2 & -5 & 4 & : & 0 \\ 0 & -7 & 12 & -5 & : & 0 \\ 0 & -1 & -3 & -12 & : & 0 \end{bmatrix} \text{ by } R_3 \to R_3 + R_2(-2)$$

$$\sim \begin{bmatrix} 1 & 2 & -5 & 4 & : & 0 \\ 0 & 0 & 33 & 79 & : & 0 \\ 0 & -1 & -3 & -12 & : & 0 \end{bmatrix} \text{ by } R_2 \to R_2 + R_1(-7)$$

$$\sim \begin{bmatrix} 1 & 2 & -5 & 4 & : & 0 \\ 0 & 1 & 3 & 12 & : & 0 \\ 0 & 0 & 33 & 79 & : & 0 \end{bmatrix} \text{ by } R_3 \to R_3(-1) \text{ and } R_3 \to R_2$$

$$\sim \begin{bmatrix} 1 & 2 & -5 & 4 & : & 0 \\ 0 & 1 & 0 & \frac{53}{11} & : & 0 \\ 0 & 0 & 1 & \frac{79}{33} & : & 0 \end{bmatrix} \text{ by } R_2 \to R_2 + R_3\left(-\frac{1}{11}\right)$$

$$R_3 \to R_3\left(\frac{1}{33}\right)$$

This gives that $z + \frac{79}{33}w = 0 \Rightarrow z = -\frac{79}{33}w$, $y + \frac{53}{11}w = 0 \Rightarrow y = -\frac{53}{11}w$

and $x = -2y + 5z - 4w$. Let $w = r$

$$= \frac{106}{11}w - \frac{395}{33}w - 4r$$

$$= \frac{(318 - 395 - 132)r}{33} = -\frac{209}{33}r$$

$$\Rightarrow x = -\frac{209}{33}r \; , y = -\frac{53}{11}r \; , z = -\frac{79}{33}r \; , w = r \text{ for any } r.$$

Now we shall discuss the methods of solving linear homogeneous and non-homongeneous equations where number of equations and number of unknowns are the same.

3 CRAMER'S RULE

If

$$\begin{array}{ccccccccc} a_{11}x_1 & + & a_{12}x_2 & + & \cdots & + & a_{1n}x_n & = & b_1 \\ a_{21}x_1 & + & a_{22}x_2 & + & \cdots & + & a_{2n}x_n & = & b_2 \\ \vdots & & \vdots & & \cdots & \vdots & \vdots & & \\ a_{n1}x_1 & + & a_{n2}x_2 & + & \cdots & + & a_{nn}x_n & = & b_n \end{array} \qquad (1.4.5)$$

then $AX = b$ where A is the coefficient matrix of (1.4.5). Let A be non-singular then $| A | = D$ (say) $\neq 0$, then using the properties of determinant it is easy to see that $x_i = \dfrac{D_i}{D}$ for every i, where

$$D_i = \left\| \begin{bmatrix} a_{11} & a_{12} & \cdots & a_{1i-1} & b_1 & a_{1i+1} & \cdots & a_{1n} \\ a_{21} & a_{22} & \cdots & a_{2i-1} & b_2 & a_{2i+1} & \cdots & a_{2n} \\ \vdots & \vdots & \cdots & \vdots & \vdots & \vdots & \cdots & \vdots \\ a_{n1} & a_{n2} & \cdots & a_{ni-2} & b_n & a_{ni+1} & \cdots & a_{nn} \end{bmatrix} \right\|$$

Example 9. Consider $x + 2y - z = 2$, $x - y + 3z = 3$, $2x + y + z = 4$. Then

$$D = \begin{vmatrix} 1 & 2 & -1 \\ 1 & -1 & 3 \\ 2 & 1 & 1 \end{vmatrix}$$

$$= 1(-1-3) - 2(1-6) - 1(1+2)$$
$$= -4 + 10 - 3 = 3$$

$$D_1 = \begin{vmatrix} 2 & 2 & -1 \\ 3 & -1 & 3 \\ 4 & 1 & 1 \end{vmatrix}$$

$$= 2(-1-3) - 2(3-12) - 1(3+4)$$
$$= -8 + 18 - 7 = 3$$

$$D_2 = \begin{vmatrix} 1 & 2 & -1 \\ 1 & 3 & 3 \\ 2 & 4 & 1 \end{vmatrix}$$

$$= 1(3 - 12) - 2(1 - 6) - 1(4 - 6)$$

$$= -9 + 10 + 2 = 3$$

$$D_3 = \begin{vmatrix} 1 & 2 & 2 \\ 1 & -1 & 3 \\ 2 & 1 & 4 \end{vmatrix}$$

$$= 1(-4 - 3) - 2(4 - 6) + 2(1 + 2)$$

$$= -7 + 4 + 6 = 3$$

$$\Rightarrow x = \frac{D_1}{D} = 1,\ y = \frac{D_2}{D} = 1,\ z = \frac{D_3}{D} = 1.$$

4 MATRIX INVERSE METHOD

Consider the matrix equation $AX = b$, where A is $n \times n$, X is $n \times 1$ and b is $n \times 1$. If A is non-singular, then $A(A^{-1}b) = (AA^{-1})b = Ib \Rightarrow A^{-1}b$ is a solution of $AX = b$.

Consider the same system of equations as in Example (9), then

$$A = \begin{bmatrix} 1 & 2 & -1 \\ 1 & -1 & 3 \\ 2 & 1 & 1 \end{bmatrix} \text{ and } |A| = 3.$$

cof (1) = $(-1 - 3) = -4$, cof (2) = $-(1 - 6) = 5$, cof $(-1) = (1 + 2) = 3$

cof (1) = $-(2 + 1) = -3$, cof $(-1) = (1 + 2) = 3$, cof (3) = $-(1 - 4) = 3$

cof (2) = $(6 - 1) = 5$, cof (1) = $-(3 + 1) = -4$, cof (1) = $(-1 - 2) = -3$

$$\Rightarrow \qquad A^{-1} = \frac{1}{3}\begin{bmatrix} -4 & -3 & 5 \\ 5 & 3 & -4 \\ 3 & 3 & -3 \end{bmatrix}$$

$$\Rightarrow \qquad \begin{bmatrix} x \\ y \\ z \end{bmatrix} = \frac{1}{3}\begin{bmatrix} -4 & -3 & 5 \\ 5 & 3 & -4 \\ 3 & 3 & -3 \end{bmatrix}\begin{bmatrix} 2 \\ 3 \\ 4 \end{bmatrix} = \frac{1}{3}\begin{bmatrix} 3 \\ 3 \\ 3 \end{bmatrix}$$

$$\Rightarrow \qquad x = 1, y = 1, z = 1.$$

5 LU-FACTORIZATION METHOD

Consider the system of n linear equations in n unknowns and the matrix equation is $AX = b$. If A is written as product of lower and upper triangular matrices L and U then the equation $AX = b$ is same as $L(UX) = b$. Suppose $UX = C$, then to find the solution for $LC = b$ is easy and after finding the solution C we can find the solution X. However, in decomposing the matrix $A = LU$, we should be cautious to find the main diagonal entries to be nonzero because if some diagonal entries become zero, then the method will not work. In such cases we have to rearrange the system of equations.

It is also important to see that the decomposition may not be unique. To find the decomposition we always apply elementary row operation of the type $R_i \to R_i + R_j(\alpha)$ or similarly the elementary column operation.

For example, if we take $A = \begin{bmatrix} 2 & 1 \\ 3 & 5 \end{bmatrix}$, then

$$\begin{bmatrix} 2 & 0 \\ 3 & \frac{7}{2} \end{bmatrix}\begin{bmatrix} 1 & \frac{1}{2} \\ 0 & 1 \end{bmatrix} = \begin{bmatrix} 2 & 1 \\ 3 & 5 \end{bmatrix}$$

where prefactor is lower triangular and post factor is upper triangular. Also

$$\begin{bmatrix} 2 & 1 \\ 3 & 5 \end{bmatrix} = \begin{bmatrix} 1 & 0 \\ \frac{3}{2} & 1 \end{bmatrix}\begin{bmatrix} 2 & 1 \\ 0 & \frac{7}{2} \end{bmatrix} = LU.$$

Therefore, we see that both are not the same. Let

$$A = \begin{bmatrix} 1 & 0 & -2 \\ 2 & 1 & 3 \\ 4 & 2 & 5 \end{bmatrix}, \text{ then } A = \begin{bmatrix} 1 & 0 & 0 \\ 2 & 1 & 0 \\ 4 & 2 & -1 \end{bmatrix}\begin{bmatrix} 1 & 0 & -2 \\ 0 & 1 & 7 \\ 0 & 0 & 1 \end{bmatrix} = LU.$$

Let

$$A = \begin{bmatrix} 2 & -2 & 6 \\ 3 & 1 & 4 \\ 1 & 5 & 7 \end{bmatrix},$$

then $$A = \begin{bmatrix} 2 & 0 & 0 \\ 3 & 1 & 0 \\ 1 & \frac{3}{2} & 1 \end{bmatrix}\begin{bmatrix} 1 & -1 & 3 \\ 0 & 4 & -5 \\ 0 & 0 & \frac{23}{2} \end{bmatrix} = L_1U_1.$$

Also

$$A = \begin{bmatrix} 1 & 0 & 0 \\ \frac{3}{2} & 1 & 0 \\ \frac{1}{2} & \frac{3}{2} & 1 \end{bmatrix} \begin{bmatrix} 2 & -2 & 6 \\ 0 & 4 & -5 \\ 0 & 0 & \frac{23}{2} \end{bmatrix} = L_2 U_2.$$

$\Rightarrow$ LU-decomposition is not unique.

Now we shall solve linear equation using LU-decomposition. Let

$$2x - 2y + 6z = 4$$
$$3x + y + 4z = 7$$
$$x + 5y + 7z = 9$$

then $AX = b$, where

$$A = \begin{bmatrix} 2 & -2 & 6 \\ 3 & 1 & 4 \\ 1 & 5 & 7 \end{bmatrix}, X = \begin{bmatrix} x \\ y \\ z \end{bmatrix}, b = \begin{bmatrix} 4 \\ 7 \\ 9 \end{bmatrix}.$$

Then

$$A = \begin{bmatrix} 2 & 0 & 0 \\ 3 & 1 & 0 \\ 1 & \frac{3}{2} & 1 \end{bmatrix} \begin{bmatrix} 1 & -1 & 3 \\ 0 & 4 & -5 \\ 0 & 0 & \frac{23}{2} \end{bmatrix} = L_1 U_1$$

$$\Rightarrow \quad AX = (L_1 U_1)X = b$$

$$\Rightarrow \quad L_1(U_1 X) = b.$$

Let

$$U_1 X = C = \begin{bmatrix} \\ \\ \end{bmatrix}$$

$$\Rightarrow \quad \begin{bmatrix} 1 & -1 & 3 \\ 0 & 4 & -5 \\ 0 & 0 & \frac{23}{2} \end{bmatrix} \begin{bmatrix} x \\ y \\ z \end{bmatrix} = \begin{bmatrix} c_1 \\ c_2 \\ c_3 \end{bmatrix}$$

$$\Rightarrow \quad \frac{23}{2} z = c_3 \text{ and } 4y - 5y = c_2$$

$$\Rightarrow \quad y = \frac{c_2 + 5z}{4}$$

and

$$x - y + 3z = c_1$$

$$\Rightarrow \qquad x = y - 3z + c_1.$$

Now

$$L_1(C) = \begin{bmatrix} 4 \\ 7 \\ 9 \end{bmatrix}$$

$$\Rightarrow \qquad \begin{bmatrix} 2 & 0 & 0 \\ 3 & 1 & 0 \\ 1 & \frac{3}{2} & 1 \end{bmatrix} \begin{bmatrix} c_1 \\ c_2 \\ c_3 \end{bmatrix} = \begin{bmatrix} 4 \\ 7 \\ 9 \end{bmatrix}$$

$$\Rightarrow \qquad 2c_1 = 4 \Rightarrow c_1 = 2,\ 3c_1 + c_2 = 7 \Rightarrow c_2 = 1.$$

Also $\quad c_1 + \frac{3}{2}c_2 + c_3 = 9$

$$\Rightarrow \qquad c_3 = 9 - 2 - \frac{3}{2} = \frac{11}{2}.$$

Therefore $z = \dfrac{11}{23}, y = \dfrac{1}{4}\left(1 + \dfrac{55}{23}\right) = \dfrac{39}{46}, x = \dfrac{39}{46} - \dfrac{33}{23} + 2 = \dfrac{65}{46}.$

Now, let us consider the other LU-decomposition of A.

$$AX = (L_2U_2)X = b$$

$$\Rightarrow \qquad AX = L_2(U_2X) = b.$$

Let

$$U_2X = D = \begin{bmatrix} d_1 \\ d_2 \\ d_3 \end{bmatrix}$$

$$\Rightarrow \qquad L_2D = b$$

$$\Rightarrow \qquad \begin{bmatrix} 1 & 0 & 0 \\ \frac{3}{2} & 1 & 0 \\ \frac{1}{2} & \frac{3}{2} & 0 \end{bmatrix} \begin{bmatrix} d_1 \\ d_2 \\ d_3 \end{bmatrix} = \begin{bmatrix} 4 \\ 7 \\ 9 \end{bmatrix}$$

$$\Rightarrow \qquad d_1 = 4,\ \frac{3}{2}d_1 + d_2 = 7$$

$$\Rightarrow \qquad d_2 = 1.$$

Also
$$\frac{1}{2}d_1 + \frac{3}{2}d_2 + d_3 = 9$$

$$\Rightarrow \qquad 2d_3 = 18 - 4 - 3$$

$$\Rightarrow \qquad d_3 = \frac{11}{2}.$$ Now

$$\begin{bmatrix} 2 & -2 & 6 \\ 0 & 4 & -5 \\ 0 & 0 & \frac{23}{2} \end{bmatrix} \begin{bmatrix} x \\ y \\ z \end{bmatrix} = \begin{bmatrix} 4 \\ 1 \\ \frac{11}{2} \end{bmatrix}$$

$$\Rightarrow \qquad \frac{23}{2}z = \frac{11}{2}$$

$$\Rightarrow \qquad z = \frac{11}{23},$$

$$4y - 5z = 1$$

$$\Rightarrow \qquad 4y = 1 + 5z = \frac{78}{23}$$

$$\Rightarrow \qquad y = \frac{39}{46}$$

and

$$2x - 2y + 6z = 4$$

$$\Rightarrow \qquad x = \frac{1}{2}\left(4 + \frac{39}{23} - \frac{66}{23}\right) = \frac{65}{46}.$$

Now, as the solution is same, we are free to choose any LU-decomposition. In other words, the solutions of system of linear equations do not depend upon the choice of LU-decomposition.

Now we state a rule to find LU-decomposition.

1. First we rearrange (if needed) the system of linear equations to get $(1, 1)^{\text{th}}$ entry of the coefficient matrix, a nonzero scalar.
2. Apply elementary row operation $R_i \to R_i + R_j\,(\alpha)$ to make all the rest entries of the first column to be zero. Similarly, make all the entries of 2nd column below $(2, 2)^{\text{th}}$ entry to be zero, and so on. Ultimately we shall get upper triangular matrix.
3. To get the lower triangular, start with identity matrix and apply the inverse operations on I in the same sequel below the $(i, i)^{\text{th}}$ entry only.

For example, if

$$A = \begin{bmatrix} 1 & 3 & 1 \\ 2 & -1 & 3 \\ -4 & 2 & 5 \end{bmatrix},$$

then apply $R_2 \to R_2 + R_1(-2)$ and $R_3 \to R_3 + R_1(4)$, to get

$$A \sim \begin{bmatrix} 1 & 3 & 1 \\ 0 & -7 & 1 \\ 0 & 14 & 9 \end{bmatrix} = C.$$

Applying inverse operation on

$$I = \begin{bmatrix} 1 & 0 & 0 \\ 0 & 1 & 0 \\ 0 & 0 & 1 \end{bmatrix}$$

we get

$$\begin{bmatrix} 1 & 0 & 0 \\ 2 & 1 & 0 \\ -4 & 0 & 1 \end{bmatrix} = D.$$

Applying $R_3 \to R_3 + R_2(2)$ on C, we get

$$\begin{bmatrix} 1 & 3 & 1 \\ 0 & -7 & 1 \\ 0 & 0 & 11 \end{bmatrix} = U.$$

Now we apply the inverse of this operation below $(2, 2)^{\text{th}}$ entry of D to get

$$\begin{bmatrix} 1 & 0 & 0 \\ 2 & 1 & 0 \\ -4 & -2 & 1 \end{bmatrix} = L.$$

$$LU = \begin{bmatrix} 1 & 0 & 0 \\ 2 & 1 & 0 \\ -4 & -2 & 1 \end{bmatrix} \begin{bmatrix} 1 & 3 & 1 \\ 0 & -7 & 1 \\ 0 & 0 & 11 \end{bmatrix}$$

$$= \begin{bmatrix} 1 & 3 & 1 \\ 2 & -1 & 3 \\ -4 & 2 & 5 \end{bmatrix}.$$

The LU-decomposition method is commonly used by computer scientists by developing the algorithm from triangular form.

EXERCISE

1. If A is a square matrix, then show that $A+\overline{A^t}$, $A\overline{A^t}$, $\overline{A^t}A$ are Hermitian and $A-(\overline{A})^t$ is skew-Hermitian.
2. Show that every square matrix is uniquely expressible as the sum of Hermitian and skew-Hermitian matrix.
3. Show that a square matrix A is Hermitian iff iA is skew-Hermitian.
4. Show that a square matrix A is skew-Hermitian iff iA is Hermitian.
5. Show that every square matrix A can be uniquely written as $P + iQ$, where P and Q are Hermitian.
6. Prove or disprove that $\frac{1}{2}\begin{bmatrix} 1+i & -1+i \\ 1+i & 1-i \end{bmatrix}$ and $\begin{bmatrix} \frac{1}{\sqrt{2}} & \frac{i}{\sqrt{2}} \\ -\frac{i}{\sqrt{2}} & -\frac{1}{\sqrt{2}} \end{bmatrix}$ are unitary matrices.
7. Show that $\frac{1}{6}\begin{bmatrix} -2 & 4 & 4 \\ 4 & -2 & 4 \\ 4 & 4 & -2 \end{bmatrix}$ is orthogonal.
8. If X_1 and X_2 are solutions of $AX = b$, then show that $rX_1 + sX_2$ such that $r + s = 1$ is also a solution.
9. Using Gauss-Jordan method find the solutions of:
   (*a*) $x + y + z - 2w = 3$, $2x + y + 3z + 2w = 5$, $-y + z + 6w = 3$.
   (*b*) $$\begin{aligned} 6x - 2y - 4z + 4w &= 2 \\ 3x - 3y - 6z + w &= -4 \\ -12x + 8y + 21z - 8w &= 8 \\ -6x - 10z + 7w &= -43 \end{aligned}$$
10. Using inverse matrix method, find the solutions of:
    (*a*) $x + 2y + z = 4$, $2x + y - z = 2$, $3x - y + 4z = 6$.
    (*b*) $x + y - z = 0$, $2x + y + 2z = 11$, $4x + y + 3z = 15$.
11. Solve the equations in Exercise 10 by LU-decomposition method.
12. Solve $2x + 2y - z = 3$, $-8x - 11y + 5z = -14$, $4x + 13y - 7z = -5$ using LU-decomposition method.

2

Sets and Algebraic Structures

1 SETS AND FUNCTIONS

The theory of sets and functions is the backbone and foundation stone of mathematics. In this preliminary chapter we shall discuss sets and functions in brief. In our daily life we come across many things which after long way of the process of definition new things come to our mind and we ultimately say that those are God given things. In other words, in any logical investigation, either certain terms remain undefined or any attempt to define every term leads to circular definitions. Although the term 'set' is undefined, but for the sake of simplicity we define set as a collection of well defined and distinct objects. If an object a is present in a set A then we say that $a \in A$ or a is an element of A or a is a member of A. If a does not belong to A, we write $a \in A$ (read, "a is not in A"). If a_1, a_2, …, a_n are objects, then the set consisting of these objects will be written as $A = \{a_1, a_2, \ldots, a_n\}$. The set with no element is called *empty set* or *void set* or *vacuous set* or *null set* and is denoted by $\emptyset$. If A and B are any two sets then A is said to be subset of B if each $x \in A$ also belongs to B and we denote $A \subseteq B$ (read also "A is contained in B). If $A \subseteq B$ and $B \subseteq A$, then $A = B$. We denote $P(A)$ as the collection of all subsets of A and is said to be power set of A. The intersection of two sets A and B, denoted as $A \cap B$, is the set of all elements which belong to both A and B. The union of A and B, denoted as $A \cup B$, is the set of elements belonging either to A or B or both A and B. If $A \cap B = \emptyset$, then A and B are disjoint sets. A meets B if $A \cap B \neq \emptyset$.

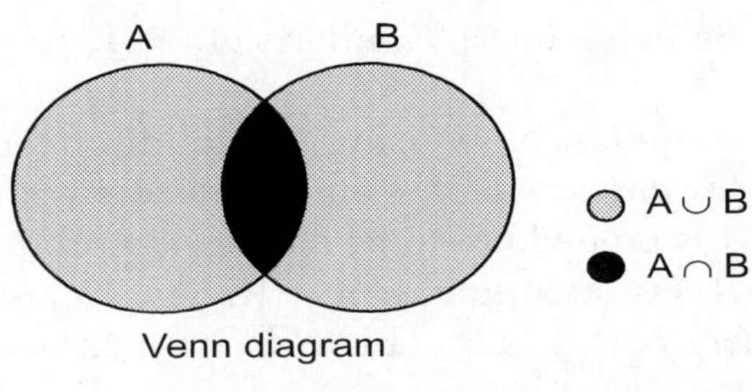

Venn diagram

If X is a set and A and B subsets of X then complement of A is $\{x \in X \mid x \notin A\}$ and is denoted by A^c. $A - B$ is defined as $\{x \in A \mid x \notin B\}$ and is called relative complement of B in A. Now it is trivial to see that $A \cup B = B$ iff $A \subseteq B$. $A \cup B = A \cup (B - A)$. $A \cap (B \cup C) = (A \cap B) \cup C$ iff $C \subseteq A$. Further, if A is the empty collection of subset of A then $\cup A = \emptyset$ and $\cap A = A$. Also $(A^c)^c = A$.

Demorgan's Laws $(A \cup B)^c = A^c \cap B^c$ and $(A \cap B)^c = A^c \cup B^c$. Also if A is any set and $B_i \subseteq A$, then for each i

$A - (\cup B_i) = \cap(A - B_i)$ and $A - (\cap B_i) = \cup(A - B_i)$.

The proofs are straightforward, so they are omitted. Further, we also mention the following which can be easily verified:

1. $A - B = A \cap B^c$.
2. $A \cap (\cup B_i) = \cup(A \cap B_i)$.
3. $A \cup (\cap B_i) = \cap(A \cup B_i)$.
4. $A \subset B^c$ iff $A \cap B = \emptyset$.
5. $A^c \subseteq B$ iff $A \cup B = X$, where X is the universal set taken into consideration.
6. $A \subset B^c$ iff $B \subseteq A^c$.
7. $A \subseteq B$ iff $B^c \subseteq A^c$.
8. $A \cap (B \cup C) = B \cup (A \cap C)$ if $B \subseteq A$.
9. If $A \subseteq B \subseteq C$, then $C - (B - A) = A \cup (C - B)$.
10. $A \cup B = A \cup (B - A)$.

1.1 Cartesian Product

If $A = \{x_1, x_2\}$ and $B = \{x_2, x_1\}$ then $A = B$, so it is useful to have a device for getting the priority of the members of the set. It is given by the notion of ordered pairs (x_1, x_2). Two ordered pairs (x_1, x_2) and (y_1, y_2) are equal if $x_1 = y_1, x_2 = y_2$. If A and B are two sets then the cartesian product $A \times B = \{(a, b) \mid a \in A, b \in B\}$ *i.e.,* $A \times B$ is the set of all ordered pairs. It is trivial to see that the number of elements in $A \times B =$ (number of elements in A) $\cdot$ (number of elements in B). Similarly, $A_1 \times A_2 \times \ldots \times A_n = \prod_{i=1}^{n} A_i$ are called ordered n-tupples. Now using the definiton we can assert that $A \times \varphi = \varphi \times A = \varphi$ for any set A. If A and B are sets with more than two elements then $X \subset A \times B$ need not to be a cartesian product of subsets of A with a subset of B. For example, if $A = \{a_1, a_2, a_3\}$ and $B = \{b_1, b_2\}$ then $X = \{(a_1, b_1), (a_2, b_2)\} \subseteq A \times B$ but X is not a cartesian product of a subset of A with any subset of B. We can see that $A \times B \neq B \times A$ in general.

1.2 Function

If A and B are any two sets, then a function f from A to B is a subset of $A \times B$ such that for every $a \in A$ there exists $b \in B$ with $(a, b) \in f$ and if $(a, b_1), (a, b_2) \in f$, then $b_1 = b_2$ and we denote it by $f : A \rightarrow B$. A function is also called *mapping*. The definition shows that a function is a rule which associates each element of A to a unique element of B. If A is the set of students who appeared in the examination of Linear Algebra carrying 50 marks and $B = \{0, 1, 2, \ldots, 50\}$, then $f : A \rightarrow B$ is given by the rule of getting marks. We can see that f is a function. If A and B are sets of polynomials of degree $\leq n$, then $\frac{d}{dx} : A \rightarrow B$ is a function, but the integration is not a function as the constant of integration varies. However, if A is the set of continuous function defined on $[\alpha, \beta]$, then $\int_\alpha^\beta f(x)\, dx$ is a unique real number, so $\int_\alpha^\beta : A \rightarrow$ is a function, where $\mathbb{R}$ is the set of real numbers.

Now, if $f : A \rightarrow B$ is a function such that $(a, b) \in f$, then we write $f(a) = b$ and say that b is image of a under f or value of f at a. The set $\{f(a) \mid a \in A\} = f(A)$ is called range of f and A is domain of f. If $f(A) = B$ then f is onto and if $f(a) = f(a')$ for $a, a' \in A$ implies $a = a'$ then f is called one-one. Further in order to get well defined function we have to check that equal elements must have equal images. If $A \neq \emptyset$ then a function $f : A^n \rightarrow A$ is called n-ary operation. For $n = 1, 2, 3$ we get unary, binary and ternary operations, respectively. We now discuss some well known operations with which we all are well acquainted, for example, addition and multiplication. Throughout we shall adapt the following notations:

$\mathbb{N}$ $\mathbb{N}$: set of natural numbers.

$\mathbb{Z}$ $\mathbb{Z}$: set of integers.

$\mathbb{Q}$: set of rational numbers.

$\mathbb{Q}$ $\mathbb{Q}'$: set of irrational numbers.

$\mathbb{R}$ $\mathbb{R}$: set of real numbers.

$\mathbb{C}$ $\mathbb{C}$: set of complex numbers.

$\mathbb{Q}$ $\mathbb{Q}''$: nonzero rationals.

$\mathbb{R}$ $\mathbb{R}''$: nonzero reals.

It is trivial to see that +, · denote by f and g, respectively, are binary operations on $\mathbb{N}$, $\mathbb{Z}$, $\mathbb{Q}$, $\mathbb{R}$ and $\mathbb{C}$ but f and g are not binary operations on $\mathbb{Q}'$. For example, $f(\sqrt{2}, -\sqrt{2}) \notin \mathbb{Q}'$. Further, $f: \mathbb{Z} \times \mathbb{Z} \to \mathbb{Z}$ given as $f(m, n) = m + n$ is not one-one but in onto. Similarly, $g: \mathbb{Z} \times \mathbb{Z} \to \mathbb{Z}$ given as $g(m, n) = mn$ is not one-one but is onto. We can check that f is never one-one unless it is a map $\{0\} \times \{0\} \to \{0\}$. Similarly, we can see for multiplication *i.e.* for g. However, $g : N_1 \times N_1 \to N_1 \to N_1$ is neither one-one nor onto where $N_1 = \{2, 3, 4, \ldots\}$. If $f: A \to B$ is a function such that $f(a) = \alpha \in B$ for all $a \in A$, then f is called *constant function*. It is neither one-one nor onto provided A an B are not singleton sets. For any set A if we define $f: P(A) \to P(A)$, $g: P(A) \to P(A)$, $h: P(A) \times P(A) \to P(A)$, given as $f(B) = B^c$, $g(B, C) = B \cup C$, $h(B, C) = B \cap C$, $B, C \subseteq A$ then f is a unary operation, g and h are binary operations. It is trivial to see that f is one-one and onto. A function $f: A \to A$ such that $f(a) = a$ for all $a \in A$ is called *identity function* and will be denoted by I. For any real number $x \in R$, we define

$$|x| = \begin{cases} x & \text{if } x \geq 0 \\ -x & \text{if } x < 0 \end{cases}$$

then $|\cdot| : \mathbb{R} \to \mathbb{R}$ is a function known as *absolute value function.* Also we can see that if A is the set of square matrices over $\mathbb{R}$ then $\det : A \to \mathbb{R}$ is a function which is not one-one but is onto. If $f: A \to B$ is a function then for any two subsets A', A'' of A we can easily show that $f(A' \cup A'') = f(A') \cup f(A'')$ and $f(A' \cap A'') \subseteq f(A') \cap f(A'')$. Trivially taking the absolute value function on $\mathbb{R}$, we can check that equality may not hold in general, but if the function is one-one then the equality is always valid.

Now consider $X = \{\pm 1, \pm 2, \pm 3, \pm 4\}$, $Y = \{1, 2, 3, 4\}$ and $A = \{1, 2\}$, then $A \subseteq X$. Define $f: X\ Y$ such that $f(x) = |x|$, then $f(A^c) = Y$ and $(f(A))^c = \{3, 4\}$ so $(f(A))^c \subseteq f(A^c)$. But if we take $Y = \{1, 2, 3, 4, 5\}$ then $g: X \to Y$ such that $g(x) = |x|$. We get $g(A^c) = \{1, 2, 3, 4\}$ and $(g(A))^c = \{3, 4, 5\}$ which shows that none of them is included in the other. Therefore, we can say that if $f: A \to B$ is function then $f(D^c) \neq (f(D))^c$ for $D \subseteq A$ but, however, if f is one-one and onto then the equality follows. Till now we have seen that given a function $f: X \to Y$, for any subset A of X, $f(A)$ is

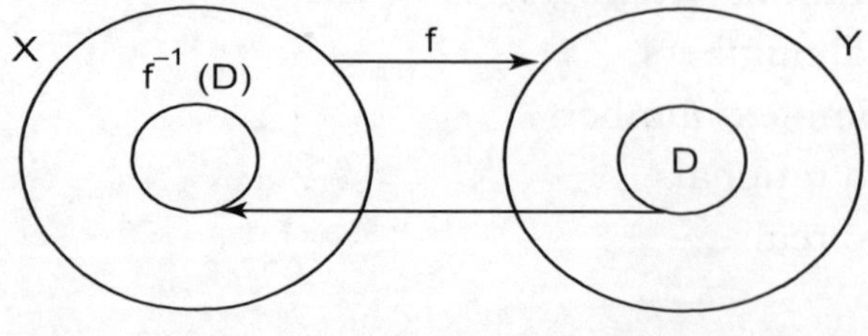

a subset of Y. For the inverse process we define for any subset D of Y, $f^{-1}(D) = \{x \in X \mid f(x) \in D\}$ and it is called the *inverse image* of D. In the above first example if we take $D = \{1, 3\}$ then $f^{-1}(D) = \{\pm 1, \pm 3\}$. If $f: \mathbb{R} \to \mathbb{R}$ such that $f(x) = x^2$, then $f^{-1}\{-1, -2, -3, -4\} = \emptyset$. The following results can be easily proved.

Theorem 1.2.1. If $f: X \to Y$ and $A, B \subseteq Y$, then $f^{-1}(A \cup B) = f^{-1}(A) \cup f^{-1}(B)$, $f^{-1}(A \cap B) = f^{-1}(A) \cap f^{-1}(B)$ and $f^{-1}(A^c) = (f^{-1}(A))^c$.

If A and B are sets then for any $(a, b) \in A \times B$ we get elements $a \in A$, $b \in B$. Now consider $\pi_1(a, b) = a$, $\pi_2(a, b) - b$ then π_1 and π_2 are called projections on A and B, respectively. Trivially $\pi_1 : A \times B \to A$ is onto if $A \neq 0$ and $\pi_2 : A \times B \to B$ is onto if $B \neq 0$. If $a \in A$ and $b \in B$ then $\pi_1^{-1}\{a\} = \{a\} \times B$ and $\pi_1^{-1}\{b\} = A \times \{b\}$.

Now it is very natural to think of the injection.

For any $a \in A$, $i_a : B \to A \times B$ defined as $i_a(b) = (a, b) \in A \times B$.

Similarly, for any $b \in B$, $i_b : A \to A \times B$ defined as $i_b(a) = (a, b)$ are injection maps from B into $A \times B$ and A into $A \times B$, respectively. Now, if $f: X \to Y$ and $g: Y \to Z$ then the composition $gof: X \to Z$ is given as $(gof)(x) = g(f(x))$, $x \in X$. It can be noted that $gof \neq fog$ in general.

If A and B are sets then any subset of $A \times B$ is called a relation from A to B and any subset of $A \times A$ is called a relation on A. If $R \subseteq A \times A$ and $(a, b) \in R$ then we denote it by aRb and read it 'a is related to b' where $a, b \in A$. One should not confuse R with the set of real numbers. Here it is only a symbol and sometimes it is denoted by $\leq$ *i.e.*, if $(a, b) \in \leq$, then we write $a \leq b$. A relation R on A is reflexive if aRa for all $a \in R$; symmetric $bRc \Rightarrow aRc$ for all $a, b, c \in R$. A relation which is reflexive, symmetric and transitive is called equivalence relation on A. Each equivalence relation partitions the set into mutually disjoint classes. For example, if we define R on $\mathbb{Z}$ such that aRb iff 2 divides $a - b$, then it is trivial to see that R is an equivalence relation and equivalence classes are $[0] = \{0, \pm 2, \pm 4, \ldots\}$ and $[1] = \{\pm 1, \pm 3, \pm 5, \ldots\}$. A relation R on A is called partially ordered (p.o.) if R is reflexive, antisymmetric and transitive. For example, for any set X define R on $P(X)$ such that ARB iff $A \subseteq B$, $A, B \in P(X)$, then R is a p.o. relation on X. The set of real numbers $\mathbb{R}$ with usual $\leq$ 'less than or equal', is a p.o. set.

For any $a, b \in \mathbb{N}$ if we define aRb iff a divides b then we can check that R is p.o. relation on $\mathbb{N}$.

A set A with a p.o. relation R is a chain (totally ordered) if for any $a, b \in A$ either aRb or bRa or $a = b$. In the above example if we take $A = \{2, 4, 6\}$ and $B = \{2, 4, 8\}$, then A is not totally ordered while B

is totally ordered under the relation defined on ℕ. The sets ℝ, ℚ, ℤ are totally ordered under usual ≤, while for any set S, ($P(S)$ ⊆) is not a chain unless S is singleton set.

Zorn's Lemma: If each chain in a nonempty partially ordered set A has an upper bound then A has a maximal element.

2 ALGEBRAIC STRUCTURES

In this section we shall deal with only those algebraic structures and the related elementary concepts which will be used in the subsequent chapters. Although the title 'algebraic structures' is a very vast topic and requires volumes and volumes, but here only very restricted concepts, results and examples will be given for subsequent use.

Definition 2.2.1. A nonempty set A together with one or more binary operations and/or one or more relatins is called an *algebraic structure.*

Now taking only one binary operation we define.

Definition 2.2.2. If $G \neq \emptyset$ (G is a nonempty set) and $*$ is a binary operation on G such that following hold:

(*i*) For all $a, b, c \in G$, $(a * b) * c = a * (b * c)$ (Associative law)

(*ii*) There exists $e \in G$ such that
$a * e = e * a = a$ for all $a \in G$ (Identity law)

(*iii*) For every $a \in G$ there exists $b \in G$ such that
$a * b = b * a = e$ (Inverse law)

then $(G, *)$ is called a *group.*

Now onwards, whenever we say that G is a group then we mean that some binary operation on G is defined and instead of denoting by $*$, we shall replace $a * b$ by ab meaning that a is operating on b under the binary operation defined on G.

Example 2.2.3

1. (ℤ, +) is a group. Here for any $a, b \in$ ℤ, ab means $a + b$.
2. (ℝ, +), (ℚ, +), (ℂ, +) are groups.
3. (N, +) is not a group.
4. (ℕ, +), (ℚ*, ·), (ℂ*, ·) are also groups, where '·' is usual multiplication.
5. $(M_{r \times s}(\mathbb{R}), +)$ is a group, where $(a_{ij}) + (b_{ij}) = (a_{ij} + b_{ij})$.
6. $(M^*_n(\mathbb{R}), +)$ is a group under matrix multiplication, where $M^*_n(\mathbb{R})$ is the set of non-singular matrices of order n.

The set $\mathbb{R}^+ = \{x \in \mathbb{R} \mid x > 0\}$ and $\mathbb{Q}^+ = \{x \in \mathbb{Q} \mid x > 0\}$ are groups under multiplication.

The set of all one-one and onto functions from {1, 2, 3} to {1, 2, 3} is also a group under composition of maps and this group is called symmetric group on three symbols. The functions are $f_1 : 1 \to 1, 2 \to 2$ and $3 \to 3; f_2 : 1 \to 2, 2 \to 1$ and $3 \to 3; f_3 : 1 \to 1, 2 \to 3$ and $3 \to 2; f_4 : 1 \to 3, 3 \to 1$ and $2 \to 2; f_5 : 1 \to 3, 3 \to 2$ and $2 \to 1; f_6 : 1 \to 2, 2 \to 3$ and $3 \to 1$. The set of all these is denoted by

$$S_3 = \left\{\begin{pmatrix}1 & 2 & 3\\1 & 2 & 3\end{pmatrix}, \begin{pmatrix}1 & 2 & 3\\2 & 1 & 3\end{pmatrix}, \begin{pmatrix}1 & 2 & 3\\1 & 3 & 2\end{pmatrix}, \begin{pmatrix}1 & 2 & 3\\3 & 2 & 1\end{pmatrix}, \begin{pmatrix}1 & 2 & 3\\3 & 1 & 2\end{pmatrix}, \begin{pmatrix}1 & 2 & 3\\2 & 3 & 1\end{pmatrix}\right\}$$

which we can again write as

$$S_3 = \{(1), (12), (23), (13), (132), (123)\}$$

If $f : A \to A$, $A \neq 0$, is one-one and onto function, then f is called permutation. So, if $A = \{a, b, c\}$ then the total number of permutations will be $3! = 6$. Similarly, if A has n elements then the total number of permutations will be $n!$. However, we shall not go into details to know how to multiply as it can be easily checked through the composition of the functions defined. For the sale of simplicity, we write the elements of S_4 = {(1), (12), (13), (23), (24), (34), (123), (124), (132), (134), (142), (143), (234), (243), (12)(34), (13)(24), (14)(23), (1234), (1243), (1324), (1342), (1423), (1432)}

The group S_n is also called *permutation group* or *symmetric group* on n-symbols.

If G is a group and $e \in G$ is such that $ea = ae = a$ for all $a \in G$ then e is called *identity*. If for $a \in G$ there exists $b \in G$ such that $ab = ba = e$ then b is called inverse of a and is denoted by a^{-1} (read as inverse of a).

If G is a group then we can define may binary operations on G to get every element of G as identity element. Let $a \in G$, then define $\varepsilon_a : G \times G \to G$ such that $\varepsilon_a(x, y) = \dfrac{xy}{a}$, $x, y \in G$. We can easily see that G is a group under the operation ε_a and a is identity of G and $x \in G$ has $\dfrac{a^2}{x}$ as its inverse.

Consider $(\mathbb{R}^*, \cdot)$ and choose $10 \in \mathbb{R}^*$ then under ε_{10} operation 10 is identity of $\mathbb{R}^*$ and inverse of $-2 \in \mathbb{R}^*$ is -50, since $(-2)\ \varepsilon_{10}\ (-50) =$

$$\frac{(-2)(-50)}{10} = \frac{100}{10} = 10 \text{ and } (-2)\,\varepsilon_{10}\,(10) = \frac{-20}{10} = -2.$$

Similarly, consider ($\mathbb{Z}$, +) and 15 $\in \mathbb{Z}$, define $m\,\varepsilon_{15} n = m + n - 15$ as the operation in $\mathbb{Z}$ is addition. Now we see that $(3)\,\varepsilon_{15}\,(15) = 3 + 15 - 15 = 3$ and $(-7)\,\varepsilon_{15}\,(37) = -7 + 37 - 15 = 15$.

Thus in a group G, every element of G can be made to be identity under a new binary operation. In the previous example, (Z, +), (Z, .15) have different group structures though the set is same.

The set $\mathbb{R}^n = (x_1, x_2, ..., x_n)\ (x_1 \in \mathbb{R})$ can be made to be a group under coordinatewise addition given as $(x_1, x_2, \ldots, x_n) + (y_1, y_2, \ldots, y_n) = (x_1 + y_1, x_2 + y_2, \ldots, x_n + y_n)$. Here + signs are not the same as in one case it is addition of n-tupples while in the other it is usual addition of real numbers.

Now if we consider the solution of $x^2 = 1$ then we have $x = 1, -1$. The set $G = \{1, -1\}$ is a group under multiplication. Also $G \times G = \{(1, 1), (1, -1), (-1, 1), (-1, -1)\}$ is a group under $(a, b)\,(c, d) = (ac, bd)$. The equation $x^3 = 1$ bas $1, \omega, \omega^2$ as solution, where $\omega = \dfrac{-1+\sqrt{-3}}{2}$. Then $G = \{1, \omega, \omega^2\}$ is a group under multiplication. It is called the group of 3rd root of unity. Similarly $x^4 = 1$ will have $\pm 1, \pm i$ as its roots and $G = \{1, -1, i, -i\}$ is a group under multiplication and is called group of 4th root of unity.

If G is a group then

$$A = \left\{\begin{bmatrix} a & 0 \\ 0 & 0 \end{bmatrix} \middle|\, a \in G\right\},\ B = \left\{\begin{bmatrix} a & a \\ 0 & 0 \end{bmatrix} \middle|\, a \in G\right\},$$

$$C = \left\{\begin{bmatrix} a & 0 \\ a & 0 \end{bmatrix} \middle|\, a \in G\right\},\ D = \left\{\begin{bmatrix} 0 & 0 \\ a & a \end{bmatrix} \middle|\, a \in G\right\}$$

and $$E = \left\{\begin{bmatrix} a & 0 \\ 0 & b \end{bmatrix} \middle|\, a, b \in G\right\}$$

are groups under matrix multiplication. Also $F = \left\{\begin{bmatrix} a & a \\ a & a \end{bmatrix} \middle|\, a \in \mathbb{R}\right\}$ is a group under matrix multiplication and has $\begin{bmatrix} \frac{1}{2} & \frac{1}{2} \\ \frac{1}{2} & \frac{1}{2} \end{bmatrix}$ as its identity.

Definition 2.2.4. A group G is called *abelian* (commutative) if $ab = ba$ for all $a, b \in G$.

If for some $x, y \in G, xy \neq yx$, then G is called non-abelian (non-commutative).

We can see that $\mathbb{Z}$, $\mathbb{Q}$, $\mathbb{R}$, $\mathbb{C}^*$, $\mathbb{R}^*$, $\mathbb{Q}^*$, $\mathbb{R}^n$ are all abelian groups, but S_3, S_4 are non-abelian. The group $M^*_n(\mathbb{R})$ is also non-abelian.

The set

$$G = \left\{ \begin{bmatrix} 1 & 0 \\ 0 & 1 \end{bmatrix}, \begin{bmatrix} 0 & 1 \\ -1 & 0 \end{bmatrix}, \begin{bmatrix} 0 & -1 \\ 1 & 0 \end{bmatrix}, \begin{bmatrix} -1 & 0 \\ 0 & -1 \end{bmatrix}, \right.$$

$$\left. \begin{bmatrix} i & 0 \\ 0 & -i \end{bmatrix}, \begin{bmatrix} 0 & i \\ i & 0 \end{bmatrix}, \begin{bmatrix} -i & 0 \\ 0 & -i \end{bmatrix}, \begin{bmatrix} 0 & -i \\ -i & 0 \end{bmatrix} \right\}$$

is also a group under matrix multiplication and is non-abelian. Let $G = \{\pm 1, \pm i, \pm j, \pm k\}$ such that $i^2 = j^2 = k^2 = -1$ and $ij = k, jk = k, ki = j$,

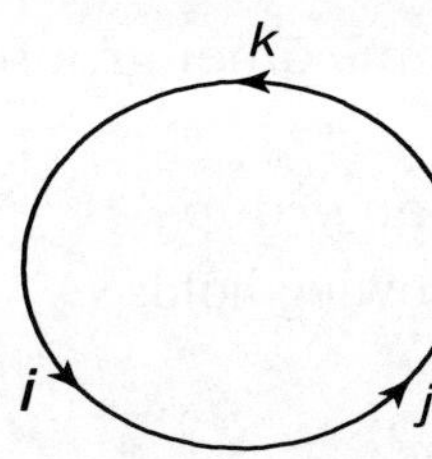

.	1	i	j	k
1	1	i	j	k
i	i	-1	k	$-j$
j	j	$-k$	-1	i
k	k	j	$-i$	-1

then G is non-abelian group which is the same group discussed above. This is also called *quaternian group.* If we take

$$G = \left\{ \begin{bmatrix} 1 & 0 \\ 0 & 1 \end{bmatrix}, \begin{bmatrix} 1 & 0 \\ 0 & -1 \end{bmatrix}, \begin{bmatrix} -1 & 0 \\ 0 & 1 \end{bmatrix}, \begin{bmatrix} -1 & 0 \\ 0 & -1 \end{bmatrix} \right\},$$

then G under matrix multiplication is an abelian group known as Klein four-group.

In additive groups, generally we talk of abelian and in multiplicative groups we talk of cummutative.

3 ADDITION AND MULTIPLICATION MODULO *N* OPERATIONS

For any natural number $n \geq 2$, we define addition modulo n and multiplication modulo n denoted by $\oplus_n$ and $\oplus_n$, respectively as $a \oplus_n b = r$, where r is the remainder obtained after dividing the ordinary sum of a and b by n, here $a, b \in \{0, 1, 2, 3, \ldots\}$. Similarly, $a \oplus_n b = r$, where r is the remainder obtained after dividing the ordinary product of a and b by n. Although this concept involves definition of congruence relation to get the equivalence classes but for the time being we do not go into detail. If $G = \{0, 1, 2, ..., m - 1\}$ is equipped with $\oplus_m$ then G is a group. Also if $G = \{1, 2, \ldots, p - 1\}$ is equipped with $\oplus_p$, then G is a group. The sets $G = \{0, 1, 2, \ldots, 7\}$, $A = \{0, 2, 4, 6\}$, $B = \{0, 4\}$ are groups under addition modulo $8(\oplus_8)$ with 0 as identity and 2 as inverse of 6, 4 as inverse of 4, 3 as inverse of 5, and so on.

Note that 0 has no inverse under $\otimes_m$. However, if m is a composite number then $A = \{1, 2, \ldots, m - 1\}$ will not form a group as $m = tr$ for some $r, t \in A$ and $t \otimes_m r = 0 \notin A$. Therefore, more often on the set of integers modulo p which is equal to $\{0, 1, \ldots, p - 1\}$ we discard 0 and take p to be prime to get taking p to be prime we can define $\otimes_p$ to get a group. For example, $A = \{2, 4\}$ is a group under $\otimes_6$, where 4 is the identity and 2 is inverse of itself. Also $G = \{0, 2, 4, 6, 8\}$ is a group under $\oplus_{10}$ and $B = \{2, 4, 6, 8\}$ is a group under $\otimes_{10}$ with 6 as identity.

Definition 2.3.1. A group G is finite if number of elements denoted as $O(G)$ is finite otherwise it is an infinite group.

$\mathbb{Z}, \mathbb{Q}, \mathbb{R}, \mathbb{C}$ are infinite groups. S_3, S_4 and some other groups discussed above are finite groups.

Now we shall prove some elementary results on groups.

Theorem 2.3.2. If G is any group then the following hold:

(*i*) Identity element is unique.

(*ii*) For any $a \in G$, the inverse of a is unique.

(*iii*) For any $a, b \in G$, $(ab)^{-1} = b^{-1}a^{-1}$.

(*iv*) For any $a \in G$, $(a^{-1})^{-1} = a$.

Proof.

(*i*) Let e and e' be two identities of G, then

$$ee' = e' \text{ (taking } e \text{ to be identity)}$$

and $$ee' = e \text{ (considering } e' \text{ as identity)}$$

$\Rightarrow$ $$e = e'.$$

(*ii*) Let x and y be two inverse of a, then $ax = xa = e$, $ya = ay = e$.

Hence $x = xe = x(ay) = (xa)y = ey = y$

$\Rightarrow$ Inverse of a is unique.

(*iii*) $(ab)(b^{-1}a^{-1}) = a(b(b^{-1}\,a^{-1})) = a((bb^{-1})a^{-1}) = a(ea^{-1}) = aa^{-1} = e$ and since the inverse is unique

$\Rightarrow$ $$(ab)^{-1} = b^{-1}\,a^{-1}.$$

(*iv*) By definition $a^{-1}a = e$ and the inverse is unique $\Rightarrow a = (a^{-1})^{-1}$.

Theorem 2.3.3. For a group G, the following hold:

(*i*) G is abelian iff $(ab)^2 = a^2b^2$.

(*ii*) If $a^2 = e$ for all $a \in G$, then G is abelian.

Proof.

(*i*) Suppose G is abelian, then

$$(ab)^2 = (ab)(ab) = a(b(ab)) = a((ba)b) = a(ab)b = a^2b^2.$$

Conversely

$$(ab)^2 = a^2b^2$$

$$\Rightarrow \quad a^{-1}((ab)(ab)) = a^{-1}(a^2b^2)$$

$$\Rightarrow \quad (a^{-1}(ab))ab = (a^{-1}a^2)b^2$$

$$\Rightarrow \quad (eb)ab = (ea)b^2$$

$$\Rightarrow \quad b(ab) = ab^2.$$

Again, operating by b^{-1}, we get

$$(b(ab)b^{-1}) = ((ab)b)b^{-1}$$

$$\Rightarrow \quad (ba)bb^{-1} = (ab)bb^{-1}$$

$$\Rightarrow \quad (ba)e = (ab)e$$

$$\Rightarrow \quad (eb)ab = (ea)b^2$$

$$\Rightarrow \quad ba = ab$$

$\Rightarrow G$ is abelian.

(*ii*) Since $a^2 = e$

$$\Rightarrow \quad a = a^{-1} \text{ for all } a \in G, \text{ hence, for any}$$

$$a, b \in G, ab = (ab)^{-1} = b^{-1}a^{-1} = ba$$

$\Rightarrow G$ is abelian.

Note: Converse of (*ii*) is not true. For example, $(\mathbb{Z}, +)$ is abelian $2m \neq 0$ for all $m \in \mathbb{Z}$.

Theorem 2.3.4. For any $a \in G$, the following hold:

(*i*) $(a^n)^{-1} = (a^{-1})^n$

(*ii*) $a^n a^m = a^{n+m}$

(*iii*) $(a^n)^m = a^{nm}$

(*iv*) $a^0 = e$.

Proof. It trivially follows from the definition $a^n = \underbrace{a \dots a}_{n\text{-times}}$.

Definition 2.3.5. For an element $a \in G$, define $a^{-n} = (a^{-1})^n$, where n is a positive integer.

Theorem 2.3.6. A group G is abelian iff $(ab)^n = a^n b^n$ for all $a, b \in G$ and $n \in \mathbb{Z}$.

Proof. If $(ab)^n = a^n b^n$ for all n, then $(ab)^2 = a^2 b^2$. Hence, by Theorem 2.3.3, G is abelian.

Conversely, taking n to be positive integer as we know that G is abelian $\Rightarrow (ab)^2 = a^2 b^2$. Suppose $(ab)^k = a^k b^k$, where k is positive integer. Then $(ab)^{k+1} = (a^k b^k)(ab) = a^{k+1} b^{k+1}$

as G is abelian. Hence, by induction $(ab)^n = a^n b^n$ for all positive integers n.

Now suppose $n = -m$, $m > 0$, then

$$(ab)^n = (ab)^{-m} = ((ab)^{-1})^m$$
$$=(a^{-1} b^{-1})^m = (a^{-1})^m (b^{-1})^m = a^{-m} b^{-m} = a^n b^n$$

$\Rightarrow$ If G is abelian then $(ab)^n = a^n b^n$ for all $n \in \mathbb{Z}$.

Subgroup: A nonempty subset H of a group G is called subgroup if H is again a group under the same binary operation of G.

It may be noted that the subset may be grouped without being subgroup. For example, consider $(\mathbb{Z}, +)$ and $H = \{1, -1\}$ then $H \subseteq \mathbb{Z}$ but H is a group under multiplication without being a subgroup of $\mathbb{Z}$. We can easily check that $(\{0, 1, 2, \ldots, 7\}, \oplus_8)$ is a group and $H = \{0, 2, 4\}$ is a group under $\oplus_6$ but is not a group.

Theorem 2.3.7. The identity of the subgroup coincides with the identity of the group.

Proof. Let H be a subgroup of a group G and e' and e be the identities of H and G, respectively. Then $e\,e = e'$ as $e' \in H \subseteq G$. Also $e\,h = he' = h$ for all $h \in H$ and $he = eh = h$ as $h \in H \subseteq G \Rightarrow e$ is identity of H but identity is unique $\Rightarrow e = e'$.

Now onwards we shall denote $\mathbb{Z}_m = \{0, 1, 2, \ldots, m-1\}$. If $G = (\mathbb{Z}_8, \oplus_8)$, then $H = \{0, 2, 4, 6\}$, $K = \{0, 4\}$ are subgroups of G.

Theorem 2.3.8. A nonempty subset H of a group G is subgroup iff for all $a, b \in H$, $ab^{-1} \in H$.

Proof. If H is a subgroup of G, then trivially $ab^{-1} \in H$. Conversely, for $a \in H$, we get $aa^{-1} \in H \Rightarrow e \in H$, also $ea^{-1} = a^{-1} \in H$. Also for any $a, b \in H$, $b^{-1} \in H$, hence by hypothesis $a(b^{-1})^{-1} = ab \in H$. Associativity is trivial.

If H is a subgroup of K and K is a subgroup of a group G, then trivially H is a subgroup of G. In $(\mathbb{Z}, +)$,

$$2\mathbb{Z} = \{0, \pm 2, \pm 4, \pm 6, \ldots\}$$

$$3\mathbb{Z} = \{0, \pm 3, \pm 6, \pm 9, \ldots\}$$
$$\mathbb{M} \qquad \mathbb{M} \qquad \mathbb{M}$$
$$m\mathbb{Z} = \{0, \pm m, \pm 2m, \pm 3m, \ldots\}$$

are subgroups of $\mathbb{Z}$. In S_3, $H = \{(1), (12)\}$, $K = \{(1), (13)\}$, $L = \{(1), (23)\}$, $P = \mathrm{P}\{(1), (123), (132)\}$, are subgroups of S_3.

For a subgroup $H \subseteq G$, define $H^{-1} = \{x^{-1} \mid x \in H\}$ then H^{-1} is also a subgroup of G. For $a, b \in H^{-1}$ is a subgroup of G. It is trivial to see that $H^{-1} = H$. Consider $(\mathbb{Z}, +)$ and $H = \{0, 1, 2\}$, then $H^{-1} = \{0, -1, -2\} = -H$ is not a subgroup of $\mathbb{Z}$. Define $aH = \{ah \mid h \in H\}$, where H is a subgroup of a group G and $a \in G$, then aH is called a left coset of H determined by $a \in G$.

Similarly, the right coset $H\alpha$ can be defined. Let
$G = (\mathbb{Z}, +)$ and $H = 3\mathbb{Z} = \{0, \pm 3, \pm 6, \ldots\}$

then

$$1 + H = \{\ldots, -5, -2, 1, 4, 7, 10, \ldots\}$$
$$2 + H = \{\ldots, -4, -1, 2, 5, 8, 11, \ldots\}$$

We can check that $G = H \cup (1 + H) \cup (2 + H)$ and $3 + H = H$, $4 + H = 1 + H$, $5 + H = 2 + H$, …, and H, $1 + H$, $2 + H$ are mutually disjoint.

Theorem 2.3.9. If G is a group and H is a subgroup then the following hold:

(*i*) $hH = H, \forall h \in H$ and $HH = \{h_1h_2 \mid h_1, h_2 \in H\} = H$.

(*ii*) For any $a, b \in G$ either $aH = bH$ or $aH \cap bH = \emptyset$.

(*iii*) There exists one-one and onto map from H to aH.

Proof. (*i*) Obviously $hH \subseteq H$, let $x \in H$, then $x = hh^{-1}x \in hH \Rightarrow hH = H$. Similarly, $HH = H$ follows as for any $h \in H$, $h = eh \in H^2 \Rightarrow H \subseteq H^2$.

(*ii*) Let $aH \neq bH$ and suppose $aH \cap bH \neq \emptyset$ then there exists $x \in aH \cap bH \Rightarrow x = ah_1$, $x = bh_2$ for some h_1 and $h_2 \in H \Rightarrow a^{-1}b = h_1h_2^{-1} \in H \Rightarrow (a^{-1}b)H = H \Rightarrow aH = bH$, a contradiction $\Rightarrow$ either $aH = bH$ or $aH \cap bH = \emptyset$.

(*iii*) Define $f : H \to aH$ given as $f(h) = ah$. Then trivially f is onto. Let $f(h_1) = f(h_2) \Rightarrow ah_1 = ah_2 \Rightarrow h_1 = h_2 \Rightarrow f$ is one-one.

Hence, we can say that $O(aH) = OH = O(Ha)$. Further, we can say that $aH = H$ iff $a \in H$.

Now we prove the well known Lagrange's theorem.

Theorem 2.3.10. If G is a finite group and H is a subgroup of G, then under of H divides the order of G.

Proof. Since G is finite, appealing to Theorem 2.3.9, we can get $G = H \cup (a_1H) \cup (a_2H) \cup \ldots \cup (a_tH)$ and get the result.

If G is abelian then left coset and right coset are same but in non-abelian case it is not true. For example, in S_3, take $H = \{(1), (12)\}$ then $(13)H = \{(13), (132)\}$ while $H(13) = \{(13), (123)\}$ which are not the same. It is also easy to check that $(123)H \neq Ha$ for any $a \in S_3$.

Hence, we can think of the necessity of considering those subgroups for which $Ha = aH$ for all $a \in G$ and this originates the concept of normal subgroups which plays an important role in non-abelian structures because in abelian structures $Ha = aH$ is always alid.

Definition 2.3.11. A subgroup H of a group G is normal if for all $a \in G$, $Ha = aH$ and we write for this as $H \Delta G$.

Thus, in abelian groups all subgroups are normal. In S_3 we can see that $\{(1), (123), (132)\}$ is normal while the subgroups $\{(1), (12)\}$, $\{(1), (13)\}$ and $\{(1), (23)\}$ are not normal.

Theorem 2.3.12. A subgroup H of G is normal iff $x^{-1} hx \in H$ for all $x \in G$ and $h \in H$.

Proof. Let $H \Delta G$ then $Hx = xH$ for all $x \in G \Rightarrow hx = xh_1$ for all $h \in H$ and $h_1 \in H \Rightarrow x^{-1} hx \in H$ for all $x \in G$ and $h \in H$.

Conversely, suppose $x^{-1} hx \in H$ for all $x \in G$ and $h \in H$. Now let

$$
\begin{aligned}
& z \in xH \\
\Rightarrow\quad & z = xh_1 \text{ for some } h_1 \in H \\
\Rightarrow\quad & z = xhx^{-1} x \\
& (\text{since } x \in G \Rightarrow x^{-1} \in G \Rightarrow xhx^{-1} \in H, \forall\, h \in H) \\
\Rightarrow\quad & z \in Hx \\
\Rightarrow\quad & xH \subseteq Hx.
\end{aligned}
$$

Similarly, we can show that $Hx \subseteq xH \Rightarrow xH = Hx \Rightarrow H \Delta G$.

Theorem 2.3.13. A subgroup H of G normal iff $(xH)(yH) = (xy)H$ for all $x, y \in G$.

Proof. Suppose $H \Delta G$ and $z \in (xH)(yH)$, then $z = xh_1yh_2$ for some $h_1, h_2 \in H$, then $z = xyy^{-1} h_1yh_2 \in (xy)H \Rightarrow (xH)(yH) \subseteq (xy)H$.

Now let

$$
\begin{aligned}
& w \in (xy)H \\
\Rightarrow\quad & w = (xy)h \text{ for some } h \in H \\
& \quad = xe \cdot yh \in (xH)(yH) \\
\Rightarrow\quad & (xH)(yH) = (xy)H \text{ for all } x, y \in G.
\end{aligned}
$$

Conversely, suppose $(xH)(yH) = (xy)H$ for all $x, y \in G$.

Hence, $(xH)(x^{-1}H) = (xx^{-1})H = eH = H$

$\Rightarrow \qquad xHx^{-1} \subseteq H$

$\Rightarrow \qquad xH \subseteq Hx.$

Similarly, $(x^{-1}H)(xH) = H$

$\Rightarrow \qquad x^{-1}Hx \subseteq H$, implying $Hx \subseteq xH$

$\Rightarrow \qquad xH = Hx \ \forall\ x \in G \Rightarrow H \Delta G.$

Now due to Theorem 2.3.13, binary operation known as coset product can be defined on $\{aH \mid a \in G \text{ and } H \Delta G\}$. It is trivial to see that if we consider S_3 and $H = \{(1), (12)\}$, then $(13)H(23)H \neq ((13)(23))H$. The set $\{aH \mid a \in G, H \Delta G\}$ is denoted by G/H. It is trivial to see that H is identity of G/H and aH has inverse $a^{-1}H$, $a \in G$, also $((aH)(bH))(cH) = (aH)((bH)(cH))$ for all $a, b, c \in G$. Therefore, G/H is a group, known as quotient group or difference group. For example, if $G = (\mathbb{Z}, +)$ and $H = 4\mathbb{Z}$ then $G/H = \{H, 1 + H, 2 + H, 3 + H\}$ and in S_3 we have since $H = \{(1), (123), (132)\}$ as only proper normal subgroup $\Rightarrow S_3/H = \{H, (12)H\} = \{H, (23)H\} = \{H, (13)H\}$. For any group G, $E = \{e\}$ and G are improper normal subgroups of G.

If H and K are subgroups of G then $HK = \{hk \mid h \in H, k \in K\}$ need not be subgroup of G. For example, in S_3, $H = \{(1), (12)\}$ and $K = \{(1), (13)\}$ are subgroups of S_3 but HK is not a subgroup of S_3, but, however, it is easy to see that if any one of H or K is normal in G, then HK is a subgroup.

4 HOMOMORPHISM

Definition 2.4.1. If G and G' are any two groups (not necessarily with the same binary operations) then a map $f: G \to G'$ is called *homomorphism* if $f(ab) = f(a)(b)$ for all $a, b \in G$.

Examples 2.4.2.

1. $f: M_n^*(\mathbb{R}) \to \mathbb{R}^*$ given as $f(A) = \det(A)$ is a homogeneous.
2. $f: \mathbb{R}^* \to \mathbb{R}^*$ such that $f(x) = |x|$ is a homogeneous.
3. $f: \mathbb{R} \to \mathbb{R}$ such that $f(x) = |x|$ is not a homomorphism.
4. $f: \mathbb{Z} \to \mathbb{Z}$ given as $f(x) = mx$ is a homomorphism.
5. $f: \mathbb{R}^2 \to \mathbb{R}^2$ such that $f(x, y) = x$ is a homomorphism.
6. $f: \mathbb{Z} \to G = (\mathbb{Z}_2, \oplus_2)$ such that

$$f(x) = \begin{cases} 0 & \text{if } x \text{ is even} \\ 1 & \text{if } x \text{ is odd} \end{cases}$$

is homomorphism.

7. $f : \mathbb{Z} \to G = (\mathbb{Z}_4, \oplus_4)$ given as

$$f(x) = \begin{cases} 2 & \text{if } x \text{ is odd} \\ 0 & \text{if } x \text{ is even} \end{cases}$$

is a homomorphism.

We can construct many examples. Now we prove some very elementary properties of homomorphism.

Theorem 2.4.3. If $f : G \to G'$ is a homomorphism then the following hold:

(*i*) $f(e) = e'$, where e and e' are identities of G and G'.

(*ii*) $f(a^{-1}) = (f(a))^{-1}$ for all $a \in G$.

(*iii*) $f(a^n) = (f(a))^n$.

(*iv*) $f(G)$ is a subgroup of G'.

Proof. (*i*) Since $e = e$

$\Rightarrow \quad f(ee) = f(e)$

$\Rightarrow \quad f(e)f(e) = f(e)$ but identity of G' is unique

$\Rightarrow \quad f(e) = e'$.

(*ii*) Since

$$aa^{-1} = e$$

$\Rightarrow \quad f(aa^{-1}) = f(e) = e'$

$\Rightarrow \quad f(a)f(a^{-1}) = e'$ but inverse is unique.

$\Rightarrow \quad f(a^{-1}) = (f(a))^{-1}$.

(*iii*) $f(a^n) = f(a \ldots a) = f(a)f(a) \ldots f(a) = (f(a))^n$.

(*iv*) Let $x, y \in f(G)$

$\Rightarrow \quad x = f(a)$ and $y = f(b)$ for some $a, b \in G$

$\Rightarrow \quad xy^{-1} = f(a)\,(f(b))^{-1} = f(ab^{-1}) \in f(G)$

$\Rightarrow f(G)$ is a subgroup of G'.

Definition 2.4.4. If $f : G \to G'$ is a homomorphism then $\{x \in G \mid f(x) = e\}$ is called kernel of f and is denoted as $\ker f$.

In Example 2.4.2

1. $\ker f = \{A \in M_n^* (\mathbb{R}) \mid \det (A) = 1\}$.
2. $\ker f = \{1, -1\}$.
3. $\ker f = \{0\}$.
4. $\ker f = \{(0, y) \mid y \in \ \}$.
6. $2\mathbb{Z}$.

More examples can be worked out.

Theorem 2.4.5. If $f: G \to G'$ is a homomorphism then $\ker f \Delta G$.

Proof. Let $x, y \in \ker f \Rightarrow f(x) = e', f(y) = e'$

$\Rightarrow \quad f(xy^{-1}) = f(x)\,(f(y))^{-1} = e'$

$\Rightarrow \quad xy^{-1} \in \ker f$

$\Rightarrow \ker f$ is a subgroup of G.

Now, let $a \in G$ and $x \in \ker f$ then $f(a^{-1}xa) = (f(a))^{-1} f(x) f(a)$

$= (f(a))^{-1}\, e f(a) = e'$

$\Rightarrow \quad a^{-1}xa \in \ker f$

$\Rightarrow \ker f \Delta G$.

Cyclic group: For any group G if $a \in G$ then for any $n \in \mathbb{Z}$, $a^n \in G$ but, however, if $G = \{a^n \mid n \in \mathbb{Z}\}$ then G is called *cyclic group* generated by a and we write $G = (a)$ and a is called generator of G. If for some $s \in \mathbb{Z}$, $a^s = e \in G$ then G is finite cyclic group otherwise it is infinite cyclic group.

If G is a group and $a \in G$ and s is the least positive integer such that $a^s = e \in G$, then s is called order of a and we write $O(a) = s$. Now using Lagrange's theorem we can say that if G is a finite group and $a \in G$ then $O(a)$ divides $O(G)$. Now we give some examples.

$(\mathbb{Z}, +)$ is cyclic generated by 1 and – 1 but $\mathbb{R}$, $\mathbb{Q}$, $\mathbb{R}^*$, $\mathbb{Q}^*$ are not cyclic. If $G = (\mathbb{Z}_m, \oplus_m)$ then G is cyclic. For instance, in $G = (\mathbb{Z}_8, \oplus_8)$, 1, 3, 5, 7 are generaters of G, but 2 generates $\{0, 2, 4, 6\}$ are 4 generates $\{0, 4\}$. However, S_3 is not cyclic as every cyclic group is abelian. $G = \{1, -1, i, -i\}$ is cyclic generated by i, but if we take $G = (\mathbb{Z}_2, \oplus_2)$ then $G \times G = \{(0, 1), (0, 0), (1, 0), (1, 1)\}$ is an abelian group but not cyclic. It can be easily seen that subgroup of cyclic is cyclic.

Definition 2.4.6. A homomorphism $f: G \to G'$ is called *isomorphism* if f is one-one and onto and we write $G \cong G'$.

If G and G' are isomorphic *i.e.,* if there exists an isomorphism from G onto G' then exactly $G \neq G'$ in general but the basic properties of G and G' are preserved *i.e.,* if G is finite then G' is also finite, if G is cyclic then G' is also cyclic, and so on.

Examples 2.4.7

1. Consider $G = (\mathbb{Z}, +)$ and $G' = (m\mathbb{Z}, +)$, where $m \in \mathbb{Z}$, then define $f: G \to G'$ such that $f(x) = mx$ then f is an isomorphism.
2. If $f: \mathbb{Z}^2 \to \mathbb{C}$ such that $f(x, y) = x + iy$, then f is an isomorphism.
3. S_3 is never isomorphic to $(\mathbb{Z}_6, +)$ as $(\mathbb{Z}_6, +)$ is abelian and S_3 is non-abelian.

4. $G = \{1, -1, i, -i\} \cong \{(0, 1), (1, 0), (0, 0), (1, 1)\} = \mathbb{Z}_2 \times \mathbb{Z}_2 = G'$ as G is cyclic but G' is not cyclic, through their orders are same.
5. If $G = (\mathbb{R}, +)$ and $G' = \{x \in \mathbb{R} \mid x > 0\}$, then G' is a group under multiplication and $f: \mathbb{R} \to G'$ such that $f(x) = e^x$, is an isomorphism.

For any group G and $H \Delta G$, a map $f: G \to G/H$ is always homomorphism given as $f(x) = xH$. Since $f(xy) = (xy)H = (xH)(yH) = f(x) f(y)$ and f is onto. This map is called natural homomorphism. We can easily check that isomorphism is an equivalence relation *i.e.*, $G \cong G$, if $G \cong G'$, then $G' \cong G$ and if $G \cong G'$ and $G' \cong G''$, then $G \cong G''$.

Theorem 2.4.8. A homomorphism $f: G \to G'$ is one-one iff $\ker f = \{e\}$.

Proof. Let f be one-one and $x \in \ker f$

$\Rightarrow$ $f(x) = e' = f(e)$

$\Rightarrow$ $x = e$

$\Rightarrow$ $\ker f = \{e\}$.

Conversely, suppose $f(x) = f(y)$

$\Rightarrow$ $f(x)\,(f(y))^{-1} = e'$

$\Rightarrow$ $f(x) f(y^{-1}) = e'$

$\Rightarrow$ $f(xy^{-1}) = e'$

$\Rightarrow$ $xy^{-1} \in \ker f = \{e\}$

$\Rightarrow$ $xy^{-1} = e$

$\Rightarrow$ $x = y$

$\Rightarrow f$ is one-one.

5 RINGS, INTEGRAL DOMAIN AND FIELDS

Now we shall introduce the structures, where we shall not restrict to only one binary operation and do very elementary properties of the same for future purposes.

Definition 2.5.1. A nonempty set R with binary operations $R \times R \xrightarrow{+} R$, $(a, b) \to a + b$ and $R \times R \to R$, $(a, b) \to a \cdot b$ such that

(*i*) $(R, +)$ is abelian group.

(*ii*) $a(bc) = (ab)c$ for all $a, b, c \in R$.

(*iii*) $a(b + c) = ab + ac$, $(a + b)c = ac + bc$

is called a *ring*.

The binary operations '+' and '·' are not necessarily addition and multiplication but we have only adopted this notation.

$\mathbb{Z}, \mathbb{R}, \mathbb{Q}, \mathbb{C}$ are rings under usual addition and multiplication.

Sometimes the third property known as distributive law, comes into mind that it is normally true but this is not the case. For example, consider $\mathbb{Z}$ under multiplication and define $\oplus_{15}$ as $m \oplus_{15} n = m + n - 15$, then $(\mathbb{Z}, \oplus_{15})$ is abelian group and $a(bc) = a(bc)$ but $a(b \oplus_{15} c) \neq ab \;_{15}\; ac$. Similarly considering $\mathbb{Q}$, we know that $(\mathbb{Q}, +)$ is abelian group but if we define for any $a, b \in \mathbb{Q}$, $a \,\mathbb{Q}_5\, b = \dfrac{ab}{5}$ then $a \,\mathbb{Q}_5\, (b \,\mathbb{Q}_5\, c) = (a \,\mathbb{Q}_5\, b) \,\mathbb{Q}_5\, c$ but $a \,\mathbb{Q}_5\, (b + c) \neq a \,\mathbb{Q}_5\, b + a \,\mathbb{Q}_5\, c$. Thus, we see that $a(b + c) = ab + ac$ is not redundant.

Note.

1. Every abelian group $(G, +)$ is a ring if we define for any $x, y \in G$, $xy = 0$, 0 identity of G under + and it is called zero ring.
2. For ring R, R must have at least zero element (the identity under +). Therefore, $R = \{0\}$ is always a ring.

Example 2.5.2

1. $M_{s \times s}(R)$ is also a ring under usual addition of matrices and matrix multiplication and we denote it by $M_s(R)$.
2. $(\mathbb{Z}_m, \oplus_m, \otimes_m)$ is a ring.
3. $(\mathbb{Z}_m, +, \cdot)$ is a ring where + is coordinatewise and $(x_1, x_2, \ldots, x_n) \cdot (y_1, y_2, \ldots, y_n) = (x_1y_1, x_2y_2, \ldots, x_ny_n)$
4. $S = \{a + bi + cj + dk \mid a, b, c, d \in R\}$ under the binary operation + (denoted) given as $(a + bi + cj + dk) + (x + yi + zj + wk) = (a + x) + (b + y)i + (c + z)j + (d + w)k$ is an abelian group. If we consider the multiplication operation as defined in Section 3, we see that S is a ring, known as ring of quanternions.
5. $[i] = \{a + bi \mid a, b \in \mathbb{Z}\}$ is a ring under usual addition and multiplication and this ring is called ring of Gaussian integers.
6. $S = \{a + b\sqrt{p} \mid a, b \in \mathbb{Q}$ and p is a prime number$\}$, then under usual addition and multiplication S is a ring.
7. For any $m \in \mathbb{Z}$, $(m\mathbb{Z}, +, \cdot)$ is a ring. If $m = 2$ then it is called ring of even integers.
8. If X is any set, then $P(X)$, under the binary operations are given as: For $A, B \in P(X)$, define $A + B = (A \cup B) - (A \cap B)$ and $A \cdot B = A \cap B$, is a ring.
9. $\mathbb{Z} \times \mathbb{Z} = \{(a, b) \mid a, b \in \mathbb{Z}\}$ denoted as $\mathbb{Z} \oplus \mathbb{Z}$ is also a ring under the same operation as in 3.
10. The set $P(X)$ of polynomials in x is a ring under usual addition and multiplication of polynomials.

More examples of rings can be given.

5.1 Some Elementary Properties

Here we shall state very few elementary properties of rings:

Theorem 2.5.3. For any ring R the following hold:

(*i*) $a0 = 0a = 0$ for all $a \in R$.

(*ii*) $a(-b) = (-a)b = -(ab)$, $a, b \in R$.

(*iii*) $(-a)(-b) = ab$, $a, b \in R$.

(*iv*) $m(ab) = (ma)b = a(mb)$, $a, b \in R$, $m \in \mathbb{Z}$.

(*v*) $a + b = a + c \Rightarrow b = c$ for all $a, b, c \in \mathbb{Z}$.

Proof. (*i*) $0 + 0 = 0$

$$
\begin{aligned}
\Rightarrow \quad & a(0+0) = a0 \\
\Rightarrow \quad & a0 + a0 = a0 \\
\Rightarrow \quad & (a0 + a0) + (-a0) = a0 + (-a0) = 0 \\
\Rightarrow \quad & a0 + (a0 + (-a0)) = 0 \\
\Rightarrow \quad & a0 + 0 = 0 \\
\Rightarrow \quad & a0 = 0.
\end{aligned}
$$

Similarly, the other proofs can be trivially given.

Definition 2.5.4. A ring R is called commutative if $ab = ba$ for all $a, b \in R$.

Definition 2.5.5. A ring R is said to have identity 1 (symbolically) if $a1 = 1a = a$ for all $a \in R$.

Definition 2.5.6. An element $a \in R$ is called *zero divisor* if there exists $b \neq 0$ in R such that $ab = 0$.

Since $0a = 0$

$\Rightarrow$ 0 is always a zero divisor.

Example 2.5.2 (1 and 4) are non-commutative but have the identity.

1. The rest rings in the above examples are commutative and $\mathbb{R}$, $\mathbb{Q}$, $\mathbb{C}$ are commutative rings with identity and $(4\mathbb{Z}, +, \cdot)$ is a ring without identity. In example 2.5.2 (8) $P(X)$ is a commutative ring with X as identity. $S = \begin{bmatrix} R & R \\ O & Q \end{bmatrix}$ is a ring with identity $\begin{bmatrix} 1 & 0 \\ 0 & 1 \end{bmatrix}$ while $P = \begin{bmatrix} R & R \\ O & 2Z \end{bmatrix}$ is a ring without identity.

Throughout the notation $\begin{bmatrix} R & R \\ O & Q \end{bmatrix}$ means as follows:

$$\begin{bmatrix} R & R \\ O & Q \end{bmatrix} = \left\{ \begin{bmatrix} a & b \\ 0 & x \end{bmatrix} \middle| \, a, b \in R \text{ and } x \in Q \right\}.$$

The set $\begin{bmatrix} R & R \\ Q & O \end{bmatrix}$ is abelian group but not a ring under matrix addition and matrix multiplication.

$\mathbb{Z}$, $\mathbb{R}$, $\mathbb{Q}$ and $\mathbb{C}$ have no common divisors. In $M_2(\mathbb{R})$,

$$\begin{bmatrix} 1 & 0 \\ 0 & 0 \end{bmatrix}\begin{bmatrix} 0 & 0 \\ 1 & 0 \end{bmatrix} = \begin{bmatrix} 0 & 0 \\ 0 & 0 \end{bmatrix},$$

so $M_2(\mathbb{R})$ has nonzero divisors., In $(\mathbb{Z}_6, \oplus_6, \otimes_6)$, $3 \otimes_6 2 = 0$, so $\mathbb{Z}_6$ has nonzero divisors.

Definition 2.5.7. A nonempty subset S of a ring R is subring if S is again a ring under the same operation.

As done in section 2, it is easy to see that a subset S of R is a subring iff $x - y \in S$ and $xy \in S$ for all $x, y \in S$. For example, $\mathbb{Z} \subseteq \mathbb{R} \subseteq \mathbb{Q} \subseteq \mathbb{C}$ are subrings in the next ones.

The subring need not contain the identity of the ring. Also the subring may contain different identity. For example, $\mathbb{Z}$ has identity 1 but $2\mathbb{Z}$, $4\mathbb{Z}$, ... are subrings of $\mathbb{Z}$ without identity. $(\mathbb{Z}_6, \oplus_6, \otimes_6)$ is a ring with identity 1 but $\{0, 3\}$ is a subring of $\mathbb{Z}_6$ with 3 as identity. It can be seen that the ring may not contain identity but the subring may contain identity. For example, $S = \mathbb{Z} \times 2\mathbb{Z}$ is a ring without identity but $\mathbb{Z} \times \{0\}$ is a subring of S with $(1, 0)$ as identity.

If the ring is commutative then it may have commutative subrings. For example, $S = \begin{bmatrix} \mathbb{R} & \mathbb{R} \\ \mathbb{R} & \mathbb{R} \end{bmatrix}$, where $\mathbb{R}$ is the set of real numbers, is non-commutative but $S' = \begin{bmatrix} \mathbb{R} & O \\ O & \mathbb{R} \end{bmatrix}$ is commutative subring of S.

If R is a ring and S and T are subrings of R then $S + T = \{s + t \mid s \in S, t \in T\}$ is a subgroup of R but need not to be subring. For example, in matrix ring $A = \begin{bmatrix} R & R \\ R & R \end{bmatrix}$, $S = \begin{bmatrix} R & R \\ O & O \end{bmatrix}$ and $T = \begin{bmatrix} \mathbb{R} & O \\ \mathbb{R} & O \end{bmatrix}$ are subringys of A but $S + T$ is not a subring of A.

As done in section 2, define for any subring S of a ring R, $a + S = \{a + s \mid s \in S\}$, where $a \in R$, then we can think of $R/S = \{a + S \mid a \in R\}$. The question aries whether R/S can be made a nonzero ring? However the answer is that R/S is abelian group under $(a + S) + (b + S) = a + b + S$, but $(a + S)(b + S) = ab + S$ is not defined in general. For example $\mathbb{Z}$ is a subring of $\mathbb{Q}$ and $\left(\frac{3}{2}+\mathbb{Z}\right)\left(\frac{2}{3}+\mathbb{Z}\right) = 1 + \mathbb{Z} = \mathbb{Z}$ is not true. Therefore, to overcome this problem we define a subring S of R to be an ideal if for every $x \in S$ and $r \in R$ $rx \in S$ and $xr \in S$. As done above $\mathbb{Z}$ is not an ideal of $\mathbb{Z}$ and $\mathbb{Z}$ is not an ideal of $\mathbb{Z}$ and $\mathbb{Z}$ is not an ideal of $\mathbb{Z}$, but $m\mathbb{Z}$ is an ideal of $\mathbb{Z}$ for all $m \in \mathbb{Z}$. Now as considered earlier $\mathbb{Z} \times \mathbb{Z} = \{(m, n) \mid m, n \in \mathbb{Z}\}$ is a ring and $A = \{(m, 0) \mid m \in \mathbb{Z}\}$ and $B = \{(0, n) \mid n \in \mathbb{Z}\}$ are ideals of $\mathbb{Z} \times \mathbb{Z}$, but $C = \{(m, m) \mid m \in \mathbb{Z}\}$ is not ideal of $\mathbb{Z} \times \mathbb{Z}$.

Now if S is an ideal of R then $R/S = \{a + S \mid a \in R\}$ can be made a ring under operations $(a + S) + (b + S) = a + b + S$ and $(a + S)(b + S) = ab + S$ and the ring R/S is called *quotient ring* or *difference ring*. A map $f: R \to R'$ is called ring homomorphism or simply homomorphism if $f(a + b) = f(a) + f(b)$ and $f(ab) = f(a)f(b)$ for all $a, b \in R$ and rings R and R'. The rings R and R' may have different operations. For example, consider $\mathbb{Z}$ and $R = \left\{\begin{bmatrix} m & 0 \\ 0 & 0 \end{bmatrix} \middle| m \in \mathbb{Z}\right\}$ then the map $f: \mathbb{Z} \to \mathbb{Z}$ given as $f(m) = \begin{bmatrix} m & 0 \\ 0 & 0 \end{bmatrix}$ is a homomorphism. For any ideal S of R, the ring R/S is called homomorphic image of R.

Definition 2.5.8. A commutative ring R with identity 1 is called integral domain if R is free from nonzero divisors.

$\mathbb{Z}$, $\mathbb{Q}$, $\mathbb{R}$, $\mathbb{C}$ are integral domains. $(\mathbb{Z}_6, \oplus_6, \otimes_6)$ is not integral domain but $(\mathbb{Z}_p, \oplus_p, \otimes_p)$ is integral domain for each prime p. Other examples can also be given.

Theorem 2.5.9. If $\mathbb{R}$ is an integral domain then $ab = ac$ for $a, b, c \in \mathbb{R}$ imply either $a = 0$ or $b = c$.

Proof. If $a = 0$, we are done. Let $a \neq 0$ then $ab = ac \Rightarrow a(b - c) = 0$ but $\mathbb{R}$ is integral domain $\Rightarrow b - c = 0 \Rightarrow b = c$.

Definition 2.5.10. A nonempty set F together with two binary operations denoted by $+$, $\cdot$; is called a field if (*i*) $(F, +)$ is abelian group.

(*ii*) $(F^*, \cdot)$ is commutative group, where $F^* = F - \{0\}$.

(*iii*) $a(b + c) = ab + ac$ and $(a + b)c = ab + bc$ for all $a, b, c \in F$

$\mathbb{Q}$, $\mathbb{R}$, $\mathbb{C}$ and $(\mathbb{Z}_p, \oplus_p, \otimes_p)$ are fields.

Trivially every field is integral domain as for any $a \in F$, $a \neq 0$ if $ab = 0$ for some $b \in F \Rightarrow a^{-1}(ab) = (a^{-1}a)b = 1b = b = a^{-1}0 = 0 \Rightarrow b = 0 \Rightarrow$ F has no nonzero zero divisors.

We already know that $\mathbb{Z}$ is integral domain but is not a field, it is easy to see from the definition that the field must contain at least two elements 0 and 1.

Examples 2.5.11

1. $(\mathbb{Z}_4, \oplus_4, \otimes_4)$ is a ring neither integral domain nor field.
2. $(\mathbb{Z}_6, \oplus_6, \otimes_6)$ is not integral domain but $F = \{0, 3\}$ and $K = \{0, 2, 4\}$ are fields with 3 and 4 as identities.
3. $(\mathbb{Z}_8, \oplus_8, \otimes_8)$ is a ring neither integral domain nor has any field structure.
4. $(\mathbb{Z}_{10}, \oplus_{10}, \theta_{10})$ is not integral domain but has $F_1 = \{0, 5\}$, $F_2 = \{0, 2, 4, 6, 8\}$ as fields with 5 and 6 as respective identities.
5. $(\mathbb{Z}_{12}, \oplus_{12}, \theta_{12})$ is not integral domain but has $F = \{0, 4, 8\}$ as field with 4 as identity.
6. $(\mathbb{Z}_{14}, \oplus_{14}, \theta_{14})$ is not integral domain but has $F_1 = \{0, 7\}$, and $F_2 = \{0, 2, 4, 6, 8, 10, 12\}$ as fields with 7 and 8 as respective identities, and so on, we can construct fields in integers modulo m.
7. $\mathbb{R}^2 = \{(x, y) \mid x, y \in \mathbb{R}\}$ define $(x, y) + (w, t) = (x + w, y + t)$ and $(x, y)(w, t) = (xy - wt, xt + yw)$ then $\mathbb{R}^2$ is a field. For any $(x, y) \in \mathbb{R}^2$ such that $(x, y) \neq (0, 0)$ we can check that $(x, y)^{-1} = \left(\dfrac{x}{x^2 + y^2}, \dfrac{-y}{x^2 + y^2}\right)$.

Using the example we can construct fields of order 4, 9 etc.

Earlier we defined another operation under which $\mathbb{R}^2$ cannot be a field but now defining the other binary operation $\mathbb{R}^2$ becomes a field. Therefore, it is the binary operation which is playing important role.

Theorem 2.5.12. Every finite integral domain is a field.

Proof. Let $R = \{0, 1, a_1, a_2, \ldots, a_t\}$ be an integral domain, then for any $a \neq 0, 1$ and $a \in R$, we have $aa_i \neq aa_j$ for $i \neq j \Rightarrow aa_i = 1$ for some $i \Rightarrow a$ is invertible $\Rightarrow R$ is a field.

6 LINEAR SPACE

Till now we have discussed either one or two binary operations on a nonempty set and discussed the structures. Now we mention few structures for which we require more operations.

Definition 2.6.1. Let R be a ring and $M \neq \emptyset$ together with a binary operation denoted as $+ : M \times M \rightarrow M$ such that $(x, y) \xrightarrow{+} x + y$ and a scalar operation denoted by $\cdot : R \times M \rightarrow M$ such that $(r, x) \xrightarrow{\cdot} rx$. Then M is called a left R-module if

(*i*) $(M, +)$ is abelian group.

(*ii*) $r(x + y) = rx + ry$.

(*iii*) $(r_1 + r_2)x = r_1x + r_2x$.

(*iv*) $(r_1r_2)x = r_1(r_2x)$.

We may note that the operation $+, \cdot$ in module M and ring R may be different.

Every ring over itself is module. Every abelian group G is a module over ring of integers and these are known as $\mathbb{Z}$ modules. If the ring R contains $1 \in R$ and $1x = x$ for all $x \in M$, then M is called unital module over R. In $\mathbb{R}^2$ if we define for any $(x_1, x_2) \in \mathbb{R}^2$ and $\lambda \in \mathbb{R}$, $\lambda(x_1, x_2) = (\lambda x_1, 0)$ then $\mathbb{R}^2$ is a left $\mathbb{R}$-module but it is not a unital module. Submodules, module homomorphism can be similarly defined as done in the case of rings. A map $f : M_1 \rightarrow M_2$ where M_1 and M_2 are left R-module is a homomorphism if $f(x + y) = f(x) + f(y)$ and $f(\lambda x) = \lambda f(x)$, $x, y \in M_1$ and $\lambda \in \mathbb{R}$. The sum of submodules S and T of a module M is defined as $\{s + t \mid s \in S, t \in T\} = S + T$ and is again a submodule. If S is a submodule of M then $M/S = \{x + S \mid x \in M\}$ is again a module over $\mathbb{R}$ under $(x + S) + (y + S) = x + y + S$ and $r(x + S) = rx + S$ and is called *quotient* or *difference module.*

Definition 2.6.2. Let F be a field and $V \neq \emptyset$ together with a binary operation denoted as $+ : V \times V \rightarrow VG$ such that $(x, y) \xrightarrow{+} x + y$ and a scalar operation (external operation) denoted as $\cdot : F \times V \rightarrow V$ such that $(\lambda, x) \xrightarrow{\cdot} \lambda x$, then V is called a vector space (linear space) over F if

(*i*) $(V, +)$ is abelian group.

(*ii*) $\lambda(x + y) = \lambda x + \lambda y$.

(*iii*) $(\lambda + \mu)x = \lambda x + \mu x$.

(*iv*) $(\lambda\mu)x = \lambda(\mu x)$.

(*v*) $1x = x$ for all $x, y \in V$ and $\lambda, \mu \in F$.

Definition 2.6.3. Let F be a field and $A \neq \emptyset$, then A is called an algebra over F if

(*i*) A is a ring.

(*ii*) A is a vector space over F.

(*iii*) $\alpha(xy) = (\alpha x)y = x(\alpha y)$ for all $x, y \in A, \alpha \in F$.

The examples and other details are given in the following chapters.

EXERCISE

1. Which of the following are equivalence relations on $\mathbb{R}$:
   (*a*) $a \leq b$ iff $a - b$ is rational.
   (*b*) $a \leq b$ iff $a - b$ is irrational.
   (*c*) $a \leq b$ iff $a - b$ is an integer.
   (*d*) $a \leq b$ iff $| a - b | \leq 1$.
2. If $f : A \to B$ and $D \subseteq A, C \subseteq B$, then show that $ff^{-1}(C) \neq C$ and $f^{-1}(f(D)) \neq D$ in general. Under what condition the equalities follow.
3. Prove that intersection of any number of subgroups is again a subgroup, but the union of two subgroups need to be a subgroup.
4. Show that $Ha = Hb$ iff $ab^{-1} \in H$ and $O(Ha) = O(Ha^{-1})$.
5. Prove or disprove that if $H \subseteq G$ and $hH = H$ for all $h \in H$, then H is a subgroup.
6. Prove that if H or K is normal in G then HK is a subgroup of G and HK is a subgroup iff $HK = KH$.
7. Prove or disprove that if G/H and H is cyclic, then G is cyclic.
   Hint. Consider S_3.
8. Prove that $\mathbb{Q} = \bigcup_n \left(\frac{1}{n!}\right)$.
9. Prove that a subgroup of cyclic is always cyclic.
10. If $f : G \to G'$ is a onto homomorphism, then prove that $G/\ker f \cong G'$.
11. If $a, x \in G$, then show that $O(a) = O(a^{-1}) = O(x^{-1}\, ax)$.
12. If $f : G \to G$ is given as $f(x) = a^{-1}\, xa$ for a fixed $a \in G$, then show that f is an isomorphism.
13. Prove or disprove that if homomorphism image of a group is finite, then the group itself is finite.
    Hint. Consider $\mathbb{Z}$.
14. Prove or disprove that homomorphism image of an integral domain is an integral domain.

15. If F is a field and $f : F \xrightarrow{\text{onto}} R$ is a ring homomorphism, where R is a ring, then show that either $R = (0)$ or $R \cong F$.
16. For any ring R, prove that $R/\{0\} \cong R$ and $R/R \cong (0)$.
17. Prove that F is a field iff F has no proper ideals.
18. If $f : \mathbb{Z} \to \mathbb{Z}$ is a ring homomorphism, then show that either $f = I$ or $f = O$.
19. If $f : M \to N$ is a module homomorphism, then for any submodule $A \subseteq M$ and $B \subseteq N$, prove that $f^{-1}(f(A)) = A + \ker f$ and $f(f^{-1}(B)) = B \cap f(M)$.
20. Show that $M_n(\mathbb{R})$ is an algebra over $\mathbb{R}$.
21. Prove that for any subgroup A of G, $A^{-1} = A$.
22. Prove or disprove that if G/H is finite for any normal subgroup H of G, then G is always finite.
23. Prove or disprove that the sets

$$\left\{\begin{bmatrix} x & x \\ x & x \end{bmatrix} \middle| x \in \mathbb{R}, x \neq 0\right\} \text{ and } \left\{\begin{bmatrix} x & x \\ x & -x \end{bmatrix} \middle| x \in \mathbb{R}, x \neq 0\right\}$$

form a group.

3

Vector Spaces

1 INTRODUCTION

The concept of vector spaces is the fundamental of linear algebra. It is a basic tool for those doing algebra but is more basic and essential for those carrying the study of functional analysis because without vector spaces, the functional analysis is zero. It is commonly used in all the branches of mathematics and statistics. The beginning of vector spaces starts from $\mathbb{R}^n$ which has been discussed in Chapter 2. First, we give again the definition of vector space.

Definition 3.1.1. Let F be a field and $V \neq \emptyset$ together with a binary operation '+' known as addition and external operation '·' known as scalar multiplication: $F \times V \rightarrow V$ given as $(\lambda, x) \rightarrow \lambda x$ such that

1. $(V, +)$ is abelian group
2. $\lambda (x + y) = \lambda x + \lambda y$
3. $(\lambda + \mu)x = \lambda(\mu x)$
4. $(\lambda\mu)x = \lambda(\mu x)$
5. $1x = x$

For all $x, y \in V$ and $\lambda, \mu \in F$, where 1 is identity of F.

Example 3.1.2

(*i*) Every field F over itself a vector space.

(*ii*) F^n is a vector space over F.

(*iii*) $M_{m \times n}(\mathbb{R})$ is a vector space over $\mathbb{R}$.

(*iv*) $\mathbb{C}$ is a vector space over $\mathbb{R}$.

(*v*) $\mathbb{R}$ is a vector space over $\mathbb{Q}$.

(*vi*) The set of polynomials in a variable x with coefficients in $\mathbb{R}$ is a vector space.

(*vii*) The set $P_n(x)$ of polynomial of degree $\leq n$ with coefficients in $\mathbb{R}$ is vector space.

(*viii*) The singleton set $\{0\}$ is a vector space over R.

(*ix*) For any set $A \neq \emptyset$ and a field F, S, the set of functions $f : A \rightarrow F$ is a vector space under

$$(f_1 + f_2)(x) = f_1(x) + f_2(x)$$

and $$(\lambda f)(x) = \lambda\,(f(x)).$$

(*x*) $(V = \{0, 2, 4, 6, 8\}, \oplus_{10}, \otimes_{10})$ is a vector space over itself and $(V, \oplus_{10})$ is also a vector space over $(\mathbb{Z}_5, \oplus_{10}, \otimes_{10})$.

(*xi*) The set of all continuous functions $f: \mathbb{R} \to \mathbb{R}$ is a vector space over $\mathbb{R}$. Similarly, the differentiable functions also form a vector space.

(*xii*) $S = \{(x_1, x_2, \ldots) \mid x_i \in R\}$ is also a vector space known as vector space known as vector space of infinite sequences.

Once can state more examples of sequences which converge, and so on.

Theorem 3.1.3. If V is a vector space over F then the following hold:

(*i*) $0x = 0$ for all $x \in V$.

(*ii*) $\lambda 0 = 0$ for all $\lambda \in F$

(*iii*) $(\lambda)x = \lambda\,(-x) = -(\lambda\,x)$.

(*iv*) If $\lambda\,x = 0$ either $\lambda = 0$ or $x = 0$

Proof. Left to readers.

Note.

1. We should be cautious about 0 as 0 is zero element of vector space and zero element of field, but keeping the definition of the vector space in mind it is easy to differentiate between the zero element of the vector space and the zero element of the field.
2. $(1 + 1)(x + y) = 1(x + y) + 1(x + y) = x + y + x + y;\ x, y \in V$.

Also $$(1 + 1)(x + y) = (1 + 1)\,x + (1 + 1)\,y$$
$$= x + x + y + y;\ \text{using } 1x = x$$

$\Rightarrow$ $$x + y + x + y = x = x = y = y$$

$\Rightarrow$ $$y + x = x + y \text{ for all } x, y \in V.$$

$\Rightarrow$ Using conditions (2), (3) and (5) from definition 3.1.1 we get $(V, +)$ to be abelian.

$\Rightarrow$ We can say that if $(V, +)$ is a group along with the conditions (2), (3), (4) and (5) of definition 3.1.1 then we get a vector space. Therefore, abelian part is redundant.

2 SUBSPACES AND QUOTIENT SPACES

Definition 3.2.1. Let V be a vector space over F then a nonempty subset S of V is a subspace if S is again a vector space under the same operations.

Theorem 3.2.2. A nonempty subset S of V is a subspace iff $x - y \in S$ and $\lambda x \in S$ for all $x, y \in S$ and $\lambda \in F$.

Proof. If S is a subspace of V then trivially $x - y \in S$ and $\lambda x \in S$ for all $x, y \in S$ and $\lambda \in F$.

Conversely, if $x - y \in S$ for all $x, y \in S$ then as done in groups $(S, +)$ is a subgroup and being a subset of V, is also abelian. Since all the elements of S are also elements of V and V is a vector space, therefore, all the required conditions of the definition of vector space are satisfied.

$\Rightarrow S$ is a subspace of V.

In $\mathbb{R}^2$ if we take $W_1 = \{(x, 0) \mid x \in \mathbb{R}\}$, $W_2 = \{(0, y) \mid y \in \mathbb{R}\}$, $W_3 = \{(x, mx) \mid x \in \mathbb{R}, m \in \mathbb{Z}\}$ then it is easy to see that these are subspaces of $\mathbb{R}^2$. In $\mathbb{R}^3$ if we take $W = \{x, y, z \mid ax + by + cz = 0, a, b, c \in \mathbb{R}\}$ then it is easy to see that W is a subspace of $\mathbb{R}^3$. If V is the set of Cauchy sequences (x_n) and W is the set of convergent sequences (x_n) then we can check that W is a subspace of the vector space V.

It is easy to see that $\mathbb{R} \subseteq \mathbb{R}^2 \subseteq \mathbb{R}^3 \subseteq \ldots$. is not true but later we will show that $\mathbb{R}^n$ has an isomorphic copy of $\mathbb{R}^{n-1}$.

As discussed earlier, if $V_1 + V_2 + \ldots + V_t$ are the subspace of V then $V_1 + V_2 + \ldots + V_t = \{v_1 + v_2 + \ldots + v_t \mid v_i \in V_i\}$ can be seen to be a subspace of V and we state

Theorem 3.2.3. If V is a vector space over F and V_i's are subspaces of V then $\cap V_i$ is a subspace of V and $\Sigma_{i=1}^{n} V_i$ is also a subspace of V.

Proof. Let $x, y \in \cap V_i$ and $\lambda \in F$ then $x, y \in V_i$ for all i.

$\Rightarrow x - y \in V_i$ for all i and $\lambda x \in V_i$ for all i.

$\Rightarrow x - y \in \cap V_i$ and $\lambda x \in V_i$

$\Rightarrow \cap V_i$ is a subspace of V.

Let $x, y \in \cap \Sigma_{i=1}^{n} V_i\ V_i$

$\Rightarrow$ $$x = x_1 + x_2 + \ldots + x_n$$

and $$y = y_1 + y_2 + \ldots + y_n \text{ where } x_i, y_i\ V_i$$

$\Rightarrow$ $$x - y = x_1 - y_1 + x_2 - y_2 + \ldots + x_n - y_n \in \Sigma_{i=1}^{n} V_i\,.$$

Also $$\lambda x = \lambda x_1 + \lambda x_2 + \ldots + \lambda x_n \in \Sigma_{i=1}^{n} V_i$$

$\Rightarrow \Sigma_{i=1}^{n} V_i$ is a subspace of V.

Now if V is a vector space over F and $x \in V$ then $Fx = \{\alpha x \mid \alpha \in F\}$ is a subspace of V and $Fx \neq 0$. Hence, now onwards we shall consider only nonzero elements $x \in V$ for considering the subspace Fx and vector space V will mean over F, denoted as V_F.

Definition 3.2.4. If $V_1, V_2, \ldots, V_t$ are subspaces of V_F then $V_1, V_2, \ldots, V_t$ are called independent if $V_i \cap \sum_{j \neq 1}^{n} V_j = 0$ for each i.

Theorem 3.2.5. Let V_F be a vector space then the following hold.

(*i*) Subspaces $V_1, V_2, \ldots, V_n$ are independent iff $x_1 + x_2 + \ldots + x_n = 0$ $\Rightarrow x_i = 0$ for each i, where $x_i \in V_i$.

(*ii*) For any $x, y \in V, x \neq y \neq 0$ either $Fx = Fy$ or $Fx \cap Fy = 0$.

Proof.

(*i*) Let $V_1, V_2, \ldots, V_n$ be independent and $x_1 + x_2 + \ldots + x_n = 0, x_i \in V_i$ then $x_i = -x_1 - x_2 - \ldots - x_{i-1} + x_{i+1} - \ldots - x_n \in V_i \cap \sum_{j \neq 1}^{n} V_j = 0$

$\Rightarrow x_i = 0$ for each i.

Conversely, suppose $w \in V_i \cap \Sigma_{j \neq i}^{n} V_j$ then $w \in V_i$ and $w = v_1 + v_2 + \ldots + v_{i-1} + v_{i+1} + \ldots + v_n$, where $v_i \in V_i$ then $v_1 + v_2 + \ldots + v_{i-1} + \ldots v_n = 0$.

Hence, by given condition $w = 0$ and $v_i = 0$.

$$\Rightarrow \qquad V_i \cap \Sigma_{j \neq i}^{n} V_j = 0$$

$\Rightarrow V_1, V_2, \ldots, V_n$ are independent.

(*ii*) If $Fx = Fy$ we have done. Suppose $Fx \neq Fy$. Let $w \in Fx \cap Fy$ the $w = \alpha x = \beta y$ for some $\alpha, \beta \in F$.

$$\Rightarrow \qquad x \in Fy \text{ and } y \in Fx$$

$$\Rightarrow \qquad Fx \subseteq Fy \text{ and } Fy \subseteq Fx$$

$$\Rightarrow \qquad Fx = Fy, \text{ a contradiction}$$

$$\Rightarrow \qquad Fx \cap Fy = 0.$$

Definition 3.2.6

Let V_F be a vector space and U and W be a subspace of V then V is called direct sum of U and W if $V = U + W$ and $U \cap W = 0$ *i.e.*, U and W are independent and we denoted it by $V = U$?? W and U is called direct summand of V.

Theorem 3.2.7

$V = U \oplus$ W iff every $x \in V$ can be uniquely written as $u + w$, $u \in U$, $w \in W$.

Proof Let $V = U$?? W and $x \in V$ then $x = u + w$, $u \in U$, $w \in W$. Suppose $x = u_1 + w_1$, $u_1 \in U$, $w_1 \in W$ then $u - u_1 = w_1 - w \in U \cap W = 0 \Rightarrow u = u_1$ and $w = w_1$.

Conversely, suppose $x \in V$ has unique representation as $x = u + w$, $u \in U$, $w \in W$, then $V = U + W$. Let $z \in U \cap W$ then $z \in U$ and $z \in W$

$$\Rightarrow \qquad z = z + 0,\ 0 \in W$$

$$= 0 + z,\ 0 \in U.$$

$$\Rightarrow \qquad z = 0 \text{ (due to unique representation).}$$

$$\Rightarrow \qquad V = U \oplus W.$$

Definition 3.2.8

Let U be a subspace of V then a subspace N of V is called complement of U if N is maximal with the property that $N \cap U = 0$.

Note The existence is guaranteed due to Zorn's lemma.

Theorem 3.2.9

If U is a subspace of V then U is a direct summand of V.

Proof Let W be a complement of U then $U \cap W = 0$

Let $S = U \oplus W$, $S \subseteq V$. Now we show that $V = S$.

Let $x \in V$ such that $x \in S$

$\Rightarrow x \in W$.

Let $z \in U \cap (W + Fx)$ then $z \in U$,

$$Z = w + \alpha x,\ \alpha \in F,\ w \in W$$

$$\Rightarrow \qquad z - w = \alpha x$$

$$\Rightarrow \qquad \alpha^{-1} z - \alpha^{-1} w = x \in U + W = S, \text{ a contradiction.}$$

$\Rightarrow U \cap (W + Fx) = 0$ but this again contradicts the maximality of W.

$$\Rightarrow \qquad x \in S$$

$$\Rightarrow \qquad V = U \oplus W.$$

Corollary 3.2.10

The subspaces of $\mathbb{R}^2$ represented by straight lines passing through origin are direct summands of $\mathbb{R}^2$.

Note.

1. In (3.2.9) we can get infinite number of direct summands.
2. Union of two subspaces need not be a subspace.
3. In $\mathbb{R}^3$, $W = \{(x, y, z) \mid 2x + y - z = 0\}$ is a direct summand.

Now if U_α are subspaces of V_F, then

$$\sum_{\alpha \in I} U_\alpha = \{x_{\alpha 1} + \ldots + x_{\alpha n} \mid x_{\alpha i} \in U_{\alpha i},\ \alpha_i \in I\}$$

can be seen to be also a subspace of V. But, since we shall be dealing with only finite dimensional vector spaces, therefore, we do not develop the theory of infinite sum.

Definition 3.2.11

If U is a subspace of V_F then $S = \{x + U \mid x \in V\}$ under $(x + U) + (y + U) = x + y + U$ and $\lambda\,(x + U) = \lambda x + U$ is again a vector space over F (an exercise).

Now denote S by V/U which is called *quotient* or *difference space*. The elements $x + U$ are called cosets and sometimes also called translation of U by x.

In $\mathbb{R}^2$ if $U = \{(x, 0) \mid x \in \mathbb{R}\}$

then $\mathbb{R}^2/U = \{(0, y) + U \mid y \in \mathbb{R}\}$

Note.

1. $x + U = U$ iff $x \in U$
2. U is zero element of V/U.
3. $x + U = y + U$ iff $x - y \in U$.
4. $V/V = \{V\}$, $V/0 = V$.

3 LINEAR INDEPENDENCE AND BASIS

For nonzero elements x, y in a vector space V_F, we have already seen that either $Fx = Fy$ or $Fx \cap Fy = 0$, *i.e.*, either $Fx = Fy$ or Fx and Fy are independent. Trivially, for any $x \neq 0$ in V, Fx is independent (See Theorem 3.2.5)

Definition 3.3.1

A subset $\{x_1, x_{2\,\ldots,}\, x_n\}$ of V such that $x_i \neq 0$ is called linearly independent if the subspaces $Fx_1, Fx_2, \ldots, Fx_n$ are independent.

For example, let $x_1 = (1, 2)$ and $x_2 = (2, 1)$ in $\mathbb{R}^2$. Consider $\mathbb{R}x_1$ and $\mathbb{R}x_2$ and let $z \in \mathbb{R}x_1 \cap \mathbb{R}x_2$ then $z = \lambda\,(1, 2)$ and $z = \alpha\,(2, 1)$ for some $\lambda, \alpha \in \mathbb{R}$ $\Rightarrow \lambda = 2\alpha$ and $2\lambda = \alpha$ which give $\lambda = 0, \alpha = 0 \Rightarrow \mathbb{R}x_1 \cap \mathbb{R}x_2 = 0 \Rightarrow \mathbb{R}x_1$ and $\mathbb{R}x_2$ are independent $\Rightarrow \{x_1, x_2\}$ is linearly independent.

Theorem 3.3.2

A subset $\{x_1, x_2, \ldots x_n\}$ is linearly independent iff $\Sigma_{i=1}^{n}\alpha_i x_i = 0$, implies $\alpha_i = 0$ for all i, where $\alpha_i \in F$.

Proof

Let $\{x_1, x_2, \ldots x_n\}$ be linearly independent and $\Sigma_{i=1}^{n}\alpha_i x_i = 0$, $\alpha_i \in F$. Since $Fx_i \cap \left(\Sigma_{j\neq 1}^{n} Fx_j\right) = 0 \Rightarrow \alpha_i x_i \in Fx_i \cap \left(\Sigma_{j\neq 1}^{n} Fx_j\right) = 0$ for each $i \Rightarrow \alpha_i\, x_i = 0 \Rightarrow \alpha_i = 0$ for each i.

Conversely, suppose $z \in Fx_i \cap \left(\Sigma_{j\neq 1}^{n} Fx_j\right)$, then

$$z = \alpha x_i$$

and $\quad z = (\alpha_1 x_1 + \alpha_2 x_2 + \ldots + \alpha_{i-1} x_{i-1} + \alpha_{i+1} x_{i+1} + \ldots \alpha_n x_n$

$\Rightarrow \quad \alpha_1 x_1 + \alpha_2 x_2 + \ldots + \alpha_{i-1} x_{i-1} - \alpha_{xi} + \alpha_{i+1} x_{i+1} + \ldots + \alpha_n x_n = 0$

$\Rightarrow \quad \alpha_i = 0 \ \forall_i$ and $\alpha = 0$

$$\Rightarrow \quad Fx_i \cap \left(\sum_{j\neq 1}^{n} Fx_j\right) = 0$$

$\Rightarrow \{x_1, x_2, \ldots, x_n\}$ is linearly independent.

Corollary 3.3.3

1. Subsets of linearly independent set is linearly independent.
2. Set containing zero element is never linearly independent.

   Now before proceeding further, we do linearly dependence. As we have seen earlier that for $x, y \in V$ if $Fx \cap Fy \neq 0$, then $Fx = Fy$. This originates the concept of linearly dependence.

Definition 3.3.4

A subset $B = \{x_1, x_2, \ldots, x_n\}$ is called linearly dependent if for some i

$$Fx_i \cap \left(\sum_{j\neq 1}^{n} Fx_j\right) \neq 0$$

We may note $Fx_i \cap \left(\Sigma_{j\neq 1}^{n} Fx_j\right) \neq 0$ does not always imply $Fx_i = Fx_k$ for some i and k. For example, let (1, 3, 2), (1, 1, 0) and (0, 1, 1) $\in \mathbb{R}^3$ then $\mathbb{R}(1, 3, 2) \cap (\mathbb{R}(1, 1, 0) + \mathbb{R}(0, 1, 1)) \neq 0$ but neither $\mathbb{R}(1, 3, 2) = \mathbb{R}(1, 1, 0)$ nor $\mathbb{R}(1, 3, 2) = \mathbb{R}(0, 1, 1)$. Further, if $x = (1, 3, 2)$, $y = (1, 1, 0)$ and $z = (0, 1, 1)$ then for any $\mathbb{R}y \cap (\mathbb{R}x + \mathbb{R}z) \neq 0$. Therefore, $\{(1, 3, 2), (1, 1, 0), (0, 1, 1)\}$ is linearly dependent. Further, it is easy to see that if $Fx_i = Fx_j$ for $i \neq j$ then $\{x_1, x_2 \ldots, x_n\}$ is linearly dependent.

Theorem 3.3.5

A subset $\{x_1, x_2 \ldots, x_n\}$ of V is linearly dependent iff there exist scalars α_i, not all zero (at least two of them are nonzero) such that $\Sigma_{j\neq 1}^{n} \alpha_i x_i = 0$

Proof

Let $B = \{x_1, x_2 \ldots, x_n\}$ be linearly dependent $\Rightarrow$ for some i $Fx_i \cap \left(\Sigma_{j\neq 1}^{n} Fx_j\right) \neq 0$

$\Rightarrow \exists z \neq 0$ such that $z \in Fx_i \cap \left(\Sigma_{j\neq 1}^{n} Fx_j\right)$ and hence

$z = \alpha x_1$ and $z = \alpha_1 x_1 + \alpha_2 x_2 + \ldots + \alpha_{i-1} x_{i-1} + \alpha_{i+1} x_{i+1} + \ldots + \alpha_n x_n$ such that $\alpha \neq 0$ and at least one $\alpha_k \neq 0$, $k \in \{1, 2, \ldots, i-1, i+1, \ldots, n\}$.

Conversely, suppose αk and $\alpha t \neq 0$, then

$$Fx_k \cap \left(\sum_{j \neq 1}^{n} Fx_j \right) \neq 0$$

$\Rightarrow \{x_1, x_2, \ldots, x_n\}$ is linearly dependent.

Corollary 3.3.6.

(*i*) Every superset of linearly dependent is linearly dependent.

(*ii*) Every subset containing 0 is linearly dependent.

(*iii*) $\{x, y\}$ is linearly dependent iff $Fx = Fy$.

(*iv*) A subset $S \subseteq V$ is linearly dependent iff S contains a linearly dependent subset.

Theorem 3.3.7

A subset $B = \{x_1, x_2 \ldots, x_n\}$ is linearly dependent iff $x_i = \Sigma_{j \neq i}^{n} \alpha_j x_j$ for some i.

Proof

Let $\{x_1, x_2 \ldots, x_n\}$ be linearly dependent.

$\Rightarrow Fx_i \cap \left(\Sigma_{j \neq i}^{n} Fx_j \right) \neq 0$, for some i.

$\Rightarrow \alpha x_i = \alpha_1 x_1 + \alpha_2 x_2 + \ldots + \alpha_{i-1} x_{i-1} + \alpha_{i+1} x_{i+1} + \ldots + \alpha_n x_n$

where $\alpha \neq 0$ and at least one $\alpha_j \neq 0$

$\Rightarrow x_i = \alpha^{-1} \alpha_1 x_1 + \alpha^{-1} \alpha_2 x_2 + \ldots + \alpha^{-1} \alpha_{i-1} x_{i-1} + \alpha^{-1} \alpha_{i+1} x_{i+1} + \ldots + \alpha^{-1} \alpha_n x_n.$

Conversely, if $x_i = \Sigma_{j \neq i}^{n} \alpha_j x_j$ for some i, then

$$x_i \in Fx_i$$

$$\Rightarrow \qquad Fx_i \cap \left(\sum_{j \neq i}^{n} Fx_j \right) \neq 0$$

$\Rightarrow \{x_1, x_2, \ldots, x_n\}$ is linearly dependent.

Definition 3.3.8

If S is a subset of V_F then $\Sigma_{x \in S} Fx$ is called space generated by S or spanned by S is denoted as $[S]$ or FS.

Trivially $S \subseteq \Sigma_{x \in S} Fx$ and if $S \subseteq K \subseteq \Sigma_{x \in S} Fx$ for some subspace K of V then $Fx \subseteq K$ for every $x \in S \Rightarrow \Sigma_{x \in S} Fx \subseteq K \Rightarrow K = \Sigma_{x \in S} Fx$. Hence, $[S]$ is the smallest subspace of V containing S.

In $\mathbb{R}$, $\mathbb{R}\{2\} = \mathbb{R}$, $\mathbb{R}\{1, 2\} = \mathbb{R}$, and so on. If $S = \{2, 5), (1, 3), (2, 5)\}$ then $[S] = \mathbb{R}^2$.

Theorem 3.3.9

If S is a subset of V_F then $[S] = \cap K$, where K are subspaces of V containing S.

Proof

Since $\Sigma_{x \in S} Fx$ is a subspace containing $S \Rightarrow \cap K \subseteq \Sigma_{x \in S}\ Fx = [S]$.

Further, for every $x \in S$, $Fx \subseteq K$ as K is a subspace containing S

$$\Rightarrow \quad \sum_{x \in S} Fx \subseteq K \text{ for every subspace } K \text{ containing } S$$

$$\Rightarrow \quad \sum_{x \in S} Fx \subseteq \cap K$$

$$\Rightarrow \quad [S] = \cap K.$$

Corollary 3.3.10

If U is a subspace of V then $[U] = U$ and $[\emptyset] = \{0\}$.

Example 3.3.11.

(*i*) If $A = \{x, y \in \mathbb{R}^2 \mid x^2 + y^2 = a, a \in \mathbb{R}, a > 0\}$ then since there is no proper subspace of R^2 containing $A \Rightarrow [A] = \mathbb{R}^2$.

(*ii*) If $B = \{(x, y, z) \mid 2x + 3y + z = 4]$ then $[B] = \mathbb{R}^3$.

(*iii*) If $C = \{(x, y) \in \mathbb{R}^2 \mid 2x + 3y = 2\}$, then $[C] = \mathbb{R}^2$.

(*iv*) If $D = \{(x, y) \in \mathbb{R}^2 \mid x = 1\}$, then $[D] = \mathbb{R}^2$.

(*v*) If $E = \{(1, 1), (3, 3)\}$, then $[E] = [(1, 1)]$.

Note.

1. An element $x \in V$ such that $x = \alpha_1 x_1 + \alpha_2 x_2 + \ldots + \alpha_n x_n$, $x_i \in V$ is called a linear combination of $x_1, x_2, \ldots, x_n$.
2. Vector space $[S]$ is also denoted by $L(S)$, known as linear span of S.

Definition 3.3.12

A vector space V is called finite dimensional if V has a finite subset S such that $[S] = V$ and if for any finite subset S of V, $[S] \neq V$ then V is called infinite dimensional.

Example 3.3.13

(*i*) $\mathbb{R}^n$ is finite dimensional vector space.

(*ii*) $M_n(\mathbb{R})$, $P_n(x)$, F (any field), $\mathbb{C}$ over $\mathbb{R}$ are finite dimensional vector spaces.

(*iii*) R over Q is infinite dimensional vector space.

(*iv*) $P(x)$, the set of polynomials in x is infinite dimensional vector space.

(*v*) The space of infinite sequences $(x) = (x_1, x_2, \ldots, x_n, \ldots)$ is infinite dimensional which is denoted by $\mathbb{R}^\infty$.

(*vi*) $C = \{(x) \in \mathbb{R}^\infty \mid \lim\limits_{n \to \infty} x_n \text{ exists}\}$ (the space of convergent sequences) is infinite dimensional.

(*vii*) $C = \{(x) \in \mathbb{R}^\infty \mid \{x_n\} \text{ is bounded}\}$ (the space of bounded sequences) is infinite dimensional.

In future, unless and otherwise stated, we shall always consider only finite dimensional vector spaces.

Definition 3.3.14

A subset $B = \{x_1, x_2, \ldots, x_n\}$ of V is called basis iff $V = Fx_1 \oplus Fx_2 \oplus \ldots \oplus Fx_n$.

Now from the definition it is obvious that $[B] = V$ and $\{x_1, x_2 \ldots x_n\}$ are linearly independent and further if $[B] = V$ and B is linearly independent then we can trivially get $V = Fx_1 \oplus Fx_2 \oplus \ldots \oplus Fx_n$. Therefore, we can define $B = \{x_1, x_2, \ldots, x_n\}$ to be a basis of V iff B is linearly independent and $[B] = V$. Further, from $V = Fx_1 \oplus Fx_2 \oplus \ldots \oplus Fx_n$, it is obvious that every $x \in V$ can be uniquely expressed as $x = \alpha_1 x_1 + \alpha_2 x_2 + \ldots + \alpha_n x_n$ for some $\alpha_i \in F$.

Theorem 3.3.15

If $V \neq 0$ is a vector space over F, then V has a basis.

Proof

Let $T = \{B \subseteq V | B \text{ is linearly independent}\}$, then $T \neq 0$, since for every $x \in V x \neq 0$, $\{x\}$ is linearly independent. We partially order T by inclusion '$\subseteq$' relation and using Zorn's lemma there exists a maximal element A (say), then A is linearly independent. Let $x \in V$ such that $x \notin [A]$ then $A \notin \{x\}$ is linearly independent, which contradicts the maximality of $A \Rightarrow V = \{A\}$ and hence A is a basis of V.

Theorem 3.3.16

Let V be a vector space over F and $B = \{x_1, x_2, \ldots, x_n\}$, where $x_i \in V$, then the following are equivalent:

(*i*) B is a basic

(*ii*) B is a minimal spanning subset of V.

(*iii*) B is maximal linearly independent subset of V.

Proof.

(*i*) $\Rightarrow$ (*ii*). Let $A \subseteq B$ there exists $x_i \in B$ such that $x_i \notin A$ for at least one i. Now suppose $[A] = V$ then $x_i = \alpha_{i1}x_{i1} + \alpha_{i2}x_{i2} + \ldots + \alpha_{ir}x_{ir}$, $x_{ij} \in A$ $\Rightarrow \{x_{i1}, x_{i2} \ldots, x_{ir}, x_i\}$ is linearly dependent but $A \cup \{x_i\} \subseteq B$ and B

is linearly independent and subset of linearly independent is linearly independent. Therefore we reach a contradiction $\Rightarrow$ there does not exist any proper subset A of B such that $[A] = V \Rightarrow B$ is a minimal spanning subset of V.

(ii) $\Rightarrow$ *(iii)*. For any choice of $\alpha_i \in F$, suppose $\Sigma_{i=1}^{n} \alpha_i\, x_i = 0$ and let $\alpha_k \neq 0$ for some k, then

$$x_k = -\alpha_k^{-1}\alpha_1 x_1 - \alpha_k^{-1}\alpha_2 x_2 - \ldots - \alpha_k^{-1}\alpha_{k-1}x_{k-1} - \alpha_k^{-1}\alpha_{k+1}x_{k+1} - \ldots - \alpha_k^{-1}\alpha_n x_n.$$

But B is a spanning subset of $V \Rightarrow$ for every $x \in V$

$$\begin{aligned} x &= \beta_1 x_1 + \beta_2 x_2 + \ldots + \beta_k x_k + \ldots + \beta_n x_n \\ &= \beta_1 x_1 + \beta_2 x_2 + \ldots + \beta_k(-\alpha_k^{-1}\alpha_1 x_1 - \alpha_k^{-1}\alpha_2 x_2 \\ &\quad - \ldots - \alpha_k^{-1}\alpha_{k-1}x_{k-1} - \alpha_k^{-1}\alpha_{k+1}x_{k+1} - \ldots - \alpha_k^{-1}\alpha_n x_n) \\ &\quad + \ldots + \beta_n x_n. \\ &= (\beta_1 - \beta_k\,\alpha_k^{-1}\alpha_1)\,x_1 + (\beta_2 - \beta_k\,\alpha_k^{-1}\alpha_2)\,x_2 \\ &\quad + \ldots + (\beta_{k-1} - \beta_k\,\alpha_k^{-1}\alpha_{k-1})\,x_{k-1} + (\beta_{k+1} - \beta_k\,\alpha_k^{-1}\alpha_{k+1} \\ &\quad x_{k+1} + \ldots + (\beta_n - \beta_k\,\alpha_k^{-1}\alpha_n)\,x_n. \end{aligned}$$

$\Rightarrow \{x_1, x_2, \ldots, x_{k-1}, x_{k+1}, \ldots, x_n\}$ is a spanning set for V, which contradicts the minimality of B. Therefore, $\alpha_k = 0$ for all k. Hence, B is linearly independent. Now we show that for any $x \in V, x \notin B, B \cup \{x\}$ is linearly dependent. Since B is spanning set for $V \Rightarrow x = \lambda_1 x_1 + \lambda_2 x_2 + \ldots. + \lambda_n x_n$ for some $\lambda_i \in F \Rightarrow \{x, x_1, x_2, \ldots, x_n\}$ is linearly dependent $\Rightarrow B$ is maximal linearly independent subset of V.

(iii) $\Rightarrow$ *(i)*. Trivially B is linearly independent (given). Now since B is maximal linearly independent subset of V, therefore for any $x \in V$, $B \notin \{x\}$ is linearly dependent. Hence, x is a linear combination of elements of B. Therefore B is a spanning set for $V \Rightarrow B$ is a basic of V.

Corollary 3.3.17

If B is a linearly independent subset of V, then B can be extended to be a basis of V.

Proof.

Let $T = \{A \subseteq V | A$ is a linearly independent and $B \subseteq A\}$ then $T \neq 0$ as $B \in T$. We partially order T and appeal to Zorn's lemma to get a maximal element in T. Hence, we get a maximal linearly independent subset L of V containing B. Therefore, by Theorem 3.3.16, L is a basis of V.

Corollary 3.3.18

If $\{x_1, x_2, \ldots, x_t\} = B$ is a spanning set for V, then B contains a basis.

Proof.

If B is minimal spanning set for V, then by Theorem 3.3.16, B itself is a basis. Now if B is not minimal spanning set then by deleting process we can get a minimal spanning subset $A \subseteq B$ for V and hence again by the same theorem, A will be a basis for V.

Theorem 3.3.19

If B is a linearly independent subset of V, then for any $x \in V$, $B \cup \{x\}$ is linearly independent iff $x \notin [B]$.

Proof.

Let $B \cup \{x\}$ is linearly independent then we show that $x \in [B]$. Suppose on contrary $x \in [B]$

$$\Rightarrow \qquad x = \sum_{i=1}^{n} \lambda_i x_i,\ x_i \in B,\ \lambda_i \in F$$

$\Rightarrow \{x, x_1, x_2, \ldots, x_n\}$ is linearly dependent.

$\Rightarrow B \cup \{x\}$ is linearly dependent, a contradiction.

$\Rightarrow x \notin [B]$.

Conversely, suppose $x \notin [B]$ and $\alpha_1 x_1 + \alpha_2 x_2 + \alpha_n x_n + \alpha x = 0$ if $\alpha \neq 0$ then $x \in [B]$, a contradiction. Hence $\alpha \neq 0 \Rightarrow \Sigma_{i=1}^{n} \alpha_i x_i = 0$ but B is linearly independent. So $\alpha_I = 0\ \forall_i \Rightarrow B \cup \{x\}$ is linearly independent.

Theorem 3.3.20

For any subset A of V and element $x \in V\ [A] = [A, x]$ iff $x \in [A]$.

Proof.

Let $[A] = [A, x]$. Trivially $x \in [A, x]$. Hence $x \in [A]$. Now if $x \in [A]$, then we show that $[A] = [A, x]$. Trivially $[A] \subseteq [A, x]$. Let $y \in [A, x] \Rightarrow y = \Sigma\alpha_i a_i + ax$, $\alpha \in F$ and $a_i \in A$ but $ax \in [A]$ and $\Sigma\alpha_i a_i \in [A] \Rightarrow y \in [A] \Rightarrow [A, x] \subseteq [A] \Rightarrow [A, x] = [A]$.

Corollary 3.3.21

If A is a basis of V then $[A, x] = [A]$ for all $x \in V$.

Now we have seen that every vector space has a basis. The questions: Is the basis unique? Are the number of elements in any two bases always same? Are very pertinent. Now we proceed for the answers of these questions.

As we know that every singleton set of nonzero elements is always linearly independent, we see that for any $x \in \mathrm{R}$, $x \neq 0$, $\{x\}$ is a basis of

R. Therefore the basis is not unique. The vector space may have infinite number of bases. However, for the answer of the second question we first prove the following:

Theorem 3.3.22.

If $\{x_1, x_2,..., x_t\}$ is a linearly independent subset of V and $\{y_1, y_2,...., y_{t+1}\}$ is a subset of V such that each y_i is a linear combination of x_i's, then $\{y_1, y_2,...., y_{t+1}\}$ is linearly dependent.

Proof. We do it by induction on t. If $t = 1$ and $y_1 = \alpha_1 x_1, y_2 = \alpha_2 x_1$ for some $\alpha_1, \alpha_2 \in F$ then $y_1 = \alpha_1 \alpha_2^{-1} y_2$. Hence $\{y_1, y_2\}$ is linearly dependent. Now suppose the result is true for k. Now if any y_i is zero then $\{y_1, y_2, \ldots, y_{k+1}\}$ is linearly dependent. Suppose no y_i is zero.

Now let

$$\begin{aligned} y_1 &= \lambda_{11}x_1 + \lambda_{12}x_2 + \ldots + \lambda_{1k}x_k \\ y_2 &= \lambda_{21}x_1 + \lambda_{22}x_2 + \ldots + \lambda_{2k}x_k \\ &\vdots \\ y_{k+1} &= \lambda_{k+1,1}x_1 + \lambda_{k+12}x_2 + \ldots + \lambda_{k+1,k}x_k \end{aligned}$$

since $y_1 \neq 0$ we may suppose $\lambda_{11} \neq 0$

$$\Rightarrow \qquad x_1 = \lambda_{11}^{-1}y_1 - \lambda_{11}^{1}\lambda_{12}x_2 - .... - \lambda_{11}^{-1}\lambda_{1k}x_k$$

Now substituting the value of x_1 in the expressions for $y_2, y_3, \ldots, y_{k+1}$, we get $y_2, y_3, \ldots, y_{k+1}$ as a linear combination of $x_2, x_3, \ldots, x_k, y_1$. As we know

$$\begin{aligned} y_2 &= \lambda_{12}x_1 + \lambda_{22}x_2 + \ldots + \lambda_{2k}x_k \\ &= \lambda_{21}(\lambda_{11}^{-1}y_1 - \lambda_{11}^{-1}\lambda_{12}x_2 - .... - \lambda_{11}^{-1}\lambda_{1k}x_k) + \lambda_{22}x_2 + \ldots + \lambda_{2k}x_k \\ &= \lambda_{21}\lambda_{11}^{-1}y_1 - (\lambda_{21}\lambda_{11}^{-1}\lambda_{12} - \lambda_{22})x_2 - \ldots - (\lambda_{21}\lambda_{11}^{-1}\lambda_{1k} - \lambda_{2k})x_k \end{aligned}$$

$$\Rightarrow \quad y_2 - \lambda_{21}\lambda_{11}^{-1}y_1 = -(\lambda_{21}\lambda_{11}^{-1}\lambda_{12} - \lambda_{22})x_2 - \ldots - (\lambda_{21}\lambda_{11}^{-1}\lambda_{1k} - \lambda 2k)x_k$$

$\Rightarrow \quad y_2 - \lambda_{21}\lambda_{11}^{-1}y_1$ is a linear combination of $x_2, x_3, \ldots, x_k$.

Similarly, $y_3 - \lambda_{31}\lambda_{11}^{-1}y_1, \ldots, y_{k+1} - \lambda_{k+1,1}\lambda_{11}^{-1}y_1$ are linear combinations of $x_2, x_3, \ldots, x_k$. Since $\{x_2, x_3, \ldots, x_k\}$ is a linearly independent set therefore by induction $y_2 - \lambda_{21}\lambda_{11}^{-1}y_1$, $y_3 - \lambda_{31}\lambda_{11}^{-1}y_1, \ldots, y_{k+1,1}\lambda_{11}^{-1}y_1$ are linearly dependent vectors. Hence, there exists a set of scalars $(\alpha_2, \alpha_3, \ldots, \alpha_{k+1}$, not all zero such that

$$\alpha_2(y_2 - \lambda_{21}\lambda_{11}^{-1}y_1) + \alpha_3(y_3 - \lambda_{31}\lambda_{11}^{-1}y_1) + \ldots + \alpha_{k+1}(y_{k+1} - \lambda_{k+1}\lambda_{11}^{-1}y_1) = 0$$

$$\Rightarrow \alpha_2 y_2 + \alpha_3 y_3 + \ldots + \alpha_{k+1} y_{k+1} + (-\alpha_2 \lambda_{21} \lambda_{11}^{-1} - \alpha_3 \lambda_{31} \lambda_{11}^{-1} - \ldots$$
$$- \alpha_{k+1} \lambda_{k+1,1} \lambda_{11}^{-1}) y_1 = 0$$

$\Rightarrow \{y_1, y_2, \ldots, y_{k+1}\}$ is linearly dependent.

Corollary 3.3.23

If B is a basis of a vector space V such that $|B| = n$ then every subset A of V containing more than n elements is linearly dependent.

Proof.

Let $|A| = m$ such that $m > n$. Since B is a basis of V, for any $x \in A$, $x = \Sigma_{i=1}^{n} \lambda_i x_i$. Hence by Theorem 3.3.22, A is linearly dependent.

Corollary 3.3.24

Any two bases of a vector space will have the same number of elements.

Proof:

Let B_1 and B_2 be two bases of V such that $|B_1| = n$ and $|B_2| = m$. If $n > m$ then appealing to Corollary 3.3.23, we get B_2 linearly dependent which is not possible. Similarly, if $m > n$ then again appealing to the same corollary we get B_2 to be linearly dependent which is again not possible $\Rightarrow n = m$.

Corollary 3.3.25

Any two bases of a vector space will have either finite number of elements or both will be infinite.

Proof.

Let B_1 and B_2 be two bases of V. If $|B_1| = m$ then by Corollary 3.3.24, $|B_2| = m$. If $|B_1| = \infty$ then $|B_2| = \infty$ and hence $|B_2| = \infty$.

Now on account of above corollaries, in case of finite dimensional vector space we can define:

Definition 3.3.26.

If V is a vector space and V has a basis B with n elements then we say that dimension of $V = n$ and write dim $(V) = n$.

Example 3.3.27

1. In $\mathbb{R}^n$, $\{(1, 0, \ldots, 0), (0, 1, \ldots 0), \ldots, (0, 0, \ldots, 1)\}$ is a basis known as standard basis. It can be seen that $\mathbb{R}^n$ has infinite number of bases.
2. $\{(1, 1)\}$ is linearly independent subset of $\mathbb{R}^2$, so $\{(1, 1), (1, 2)\}$, $\{(1, 1), (1, 3)\}$, ..., $\{(1, 1), (1, m)\}$ are bases of $\mathbb{R}^2$. It can be seen that every nonzero element (a, b) can be extended to a basis $\{(a,$

$b), (c, d)\}$ such that $ad - bc \neq 0$.

3. In $P_2(x)$, $\{1, x, x^2\}$ is standard basis. Further, $\{1, 1-x, 1+x^2\}$. $\{1 + x, 1 - x, x^2\}$, ... are also bases.
4. In M_2 ($\mathbb{R}$), $\left\{\begin{bmatrix}1 & 0\\0 & 0\end{bmatrix}, \begin{bmatrix}0 & 1\\0 & 0\end{bmatrix}, \begin{bmatrix}0 & 0\\1 & 0\end{bmatrix}, \begin{bmatrix}0 & 0\\0 & 1\end{bmatrix}\right\}$

   Is a standard basis. For any $a \neq 0, b \neq 0, c \neq 0, d \neq 0$

   $$\left\{\begin{bmatrix}a & 0\\0 & 0\end{bmatrix}, \begin{bmatrix}0 & b\\0 & 0\end{bmatrix}, \begin{bmatrix}0 & 0\\c & 0\end{bmatrix}, \begin{bmatrix}0 & 0\\0 & d\end{bmatrix}\right\}$$

   Is also a basis of $M_2(\mathbb{R})$.
5. For any vector $V = V_1 \oplus V_2 \oplus \dots \oplus V_n$ if B_i is basis of V_i then $\bigcup_{i=1}^{n} B_i$ is a basis of V.

4 DIMENSIONS

In this section we shall prove some results regarding dimensions of the vector space. First we define.

Definition 3.4.1

Let V and W be two vector spaces over a field F then a mapping $f : V \to W$ is called vector homomorphism if $f(x + y) = f(x + y) = f(x) + f(y)$ and $f(\alpha x) = \alpha f(x)$ for all $x, y \in V$ and $\alpha \in F$.

A vector homomorphism f is called an isomorphism if f is one-one and onto. The vector homomorphism is also called a linear transformation which will be dealt in Chapter 4.

Two vectors space V and W over F are called isomorphic if there exists an isomorphism between V and W and we write $V \cong W$.

Lemma 3.4.2

If V and W are finite dimensional vector spaces over F and $f : V \to W$ is an isomorphism then $\{f(x_1), f(x_2), \dots, f(x_n)\}$ is a basis of W for any basis $\{x_1, x_2, \dots x_n\}$ of V.

Proof:

$\Sigma_{i=1}^{n} \lambda_i f(x_i) = 0 \Rightarrow \left(\Sigma_{i=1}^{n} \lambda_i x_i\right) = 0 \Rightarrow \Sigma_{i=1}^{n} \lambda_i x_i = 0$ as f is one-one. Now

since $\{x_1, x_2, \dots x_n\}$ is a basis.

$\Rightarrow \lambda_I = 0\ \forall i \Rightarrow \{f(x_1), f(x_2) \dots, f(x_n)\}$ is linearly independent.

Let $w \in W$ then as f is onto there exists $v \in V$ such that $f(v) = w$.

But $\{x_1, x_2, \ldots x_n\}$ is a basis of $V \Rightarrow v = \Sigma_{i=1}^{n} \alpha_i x_i$, $\alpha_I \in F$.

Hence $f(v) = \Sigma_{i=1}^{n} \alpha_i f(x_i)$

$\Rightarrow \{f(x_1), f(x_2), \ldots, f(x_n)\}$ spans W

$\Rightarrow \{f(x_1), f(x_2), \ldots, f(x_n)\}$ is a basis of W.

Theorem 3.4.3

If V and W are vector spaces over F, then $V \cong W$ iff dim (V) = dim (W).

Proof.

Let $f: V \rightarrow W$ be an isomorphism and dim $V = n$ then prove that dim $W = n$. If $\{x_1, x_2, \ldots, x_n\}$ is a basis of V then due to Lemma 3.4.2, we get $\{f(x_1), f(x_2), \ldots f(x_n)\}$ to be a basis of W. Hence dim $W = n$.

Conversely, suppose dim V = dim $W = t$, then we prove that $V \cong W$. Let $\{x_1, x_2, \ldots, x_n\}$ be a basis of V then any $x \in V$ we have unique representation as $\Sigma_{i=1}^{t} \lambda_i x_i$ *i.e.*, $x = \Sigma_{i=1}^{t} \lambda_i x_i$. Let $\{w_1, w_2, \ldots, w_t\}$ be a basis of W. Define $f: V \Rightarrow W$ such that $f(x) = \Sigma_{i=1}^{t} \lambda_i w_i$ then trivially f is well defined. Now for $x, y \in V$, $x = \Sigma_{i=1}^{t} \lambda_i x_i$, $y = \Sigma_{i=1}^{t} \beta_i x_i$

$$\Rightarrow f(x + y) f\left(\sum_{i=1}^{t} (\lambda_i + \beta_i) x_i\right)$$

$$= \sum_{i=1}^{t} (\lambda_i + \beta_i) w_i = \sum_{i=1}^{t} \lambda_i w_i + \sum_{i=1}^{t} \beta_i w_i$$

$$= f(x) + f(y).$$

Also
$$f(\alpha x) = f\left(\sum_{i=1}^{t} \alpha \lambda_i w_i\right)$$

$$= \sum_{i=1}^{t} \alpha \lambda_i w_i = \alpha\left(\sum_{i=1}^{t} \lambda_i w_i\right) = \alpha f(x).$$

Further, for any $w \in W$, $w = \Sigma_i \quad \alpha_i \; _i$, Therefore we get $x = \Sigma_{i=1}^{t} \alpha_i x_i$ $\in V$ such that $f(x) = w \Rightarrow f$ is onto.

Also let $f(x) = 0 \Rightarrow \Sigma_{i=1}^{t} \lambda_i w_i = 0$, but $\{w_1\ w_2, \ldots, w_t\}$ is linearly independent $\Rightarrow \lambda_i = 0$ for each $i \Rightarrow x = 0 \Rightarrow f$ is one-one

$\Rightarrow f$ is an isomorphism.

$\Rightarrow V \cong W$.

Corollary 3.4.4. If dim VF $= n$ then $V \cong F^n$.

Proof. Since dim $(F^n) = n$. Hence, by Theorem 3.4.3, $V \cong F^n$.

Theorem 3.4.5. If U is a subspace of a finite dimensional vector space V then dm $U \leq$ dim V and dim $\left(\frac{V}{U}\right)$ = dim (V) dim (U).

Proof. Let $\{x_1, x_2, \ldots, x_r\}$ be a basis of U then $\{x_1, x_2, \ldots, x_r\}$ is a linearly independent subset of V. Hence by Corollary 3.3.17 $\{x_1, x_2, \ldots, x_r\}$ can be extended to a basis $\{x_1, x_2, \ldots, x_r, y_1, y_2, \ldots y_t\}$ of $V \Rightarrow$ dim $U \leq$ dim V. Further for any $x \in V$, $x = \Sigma_{i=1}^{r} \lambda_i x_i + \Sigma_{i=1}^{t} \beta_i y_i$

$$\Rightarrow \quad x + U = \sum_{i=1}^{r} \lambda_i (x_i + U) + \sum_{i=1}^{r} \beta_i (y_i + U) = \sum_{i=1}^{t} \beta_i (y_i + U)$$

$$\Rightarrow \quad \{y_1 + U, y_2 + U, \ldots, y_t + U\} \text{ spans } \frac{V}{U}.$$

Let $\alpha_1(y_1 + U) + \alpha_2(y_2 + U) + \ldots + \alpha_t(y_t + U) = 0 = U \Rightarrow \sum_{i=1}^{t} \alpha_i y_i \in U$

$$\Rightarrow \quad \sum_{i=1}^{t} \alpha_i y_i = \sum_{i=1}^{t} \lambda_i x_i \text{ , as } \{x_1, x_1, \ldots, x_r\} \text{ is a basis of } U.$$

$$\Rightarrow \quad \sum_{i=1}^{t} \lambda_i x_i - \sum_{i=1}^{t} \alpha_i y_i = 0$$

But $\{x_1, x_2, \ldots x_r, y_1, y_2 \ldots, y_t\}$ is linearly independent.

$\Rightarrow \quad \lambda_i = 0, \alpha_i = 0$ for each i.

$$\Rightarrow \quad \{y_1 + U, y_2 + U, \ldots, y_t\} \text{ is a basis of } \frac{V}{U}.$$

$$\Rightarrow \quad \dim \left(\frac{V}{U}\right) = t = r + t - r = \dim V - \dim U.$$

Theorem 3.4.6

If U_1, U_2 are subspaces of V, then

$$\dim (U_1 + U_2) = \dim U_1 + \dim U_2 - \dim (U_1 \cap U_2).$$

Proof.

Let $\{x_1, x_2, \ldots, x_r\}$ be a basis of $U_1 \cap U_2$ is a subspace of U_1 and U_2, $\{x_1, x_2, \ldots, x_r\}$ can be extended to bases $\{x_1, x_2, \ldots, x_r, y_1, y_2, \ldots, y_t\}$ and $\{x_1, x_2, \ldots, x_r, w_1, w_2, \ldots, w_p\}$ of U_2 and U_2, respectively.

Let $z \in U_1 + U_2$, then $z = x + y$, $x \in U_1$, $y \in U_2$.

$$\Rightarrow \quad z = \sum_{i=1}^{r} \lambda_i x_i + \sum_{i=1}^{t} \beta_i y_i + \sum_{i=1}^{r} \alpha_i x_i + \sum_{i=1}^{p} \delta_i w_i$$

$$= \sum_{i=1}^{r} (\lambda_i + \alpha_i) x_i + \sum_{i=1}^{t} \beta_i y_i + \sum_{i=1}^{p} \delta_i w_i$$

$\Rightarrow \quad \{x_1, x_2, \ldots, x_r, y_1, y_2, \ldots, x_t, w_1, w_2, \ldots, w_p,\}$ spans $U_1 + U_2$.

Now let

$$\sum_{i=1}^{r} \alpha_i x_i + \sum_{i=1}^{t} \beta_i y_i + \sum_{i=1}^{p} \lambda_i w_i = 0$$

$$\Rightarrow \quad \sum_{i=1}^{r} \alpha_i x_i + \sum_{i=1}^{p} \lambda_i w_i = -\sum_{i=1}^{t} \beta_i y_i \in U_1 \cap U_2$$

$\Rightarrow \quad \sum_{i=1}^{t} \beta_i y_i = \sum_{i=1}^{t} \delta_i x_i$ as $\{x_1, x_2, \ldots, x_r\}$ is a basis of $U_1 \cap U_2$.

$\Rightarrow \quad \sum_{i=1}^{r} \delta_i x_i + \sum_{i=1}^{t} \beta_i y_i = 0 \Rightarrow \delta_i = 0$ and β_i for all i, as

$\Rightarrow \quad \{x_1, x_2, \ldots, x_r, y_1, y_2, \ldots, y_t\}$ is linearly independent.

$\Rightarrow \sum_{i=1}^{r} \delta_i x_i + \sum_{i=1}^{p} \lambda_i w_i = 0$ but $\{x_1, x_2, \ldots, x_r, w_1, w_2, \ldots, y_p\}$ is linearly independent.

$\Rightarrow \alpha_I = 0$ and $\lambda_i = 0$ for all i.

$\Rightarrow \{x_1, x_2, \ldots, x_r, y_1, y_2, \ldots, y_t, w_1, w_2, .., w_p\}$ is linearly independent.

Hence dim $(U_1 + U_2) = r + t + p = r + t + r + p - r =$ dim U_1 + dim $U_2 - (U_1 \cap U_2)$.

Corollary 3.4.7

If $V = U_1 \oplus U_2$, then dm $(V) =$ dim U_1 + dim U_2.

Proof.

Since $U_1 \oplus U_2 = 0$, therefore the proof follows from the above theorem.

On account of Theorem 3.3.16 the following is trivial.

Corollary 3.4.8

(*i*) If B is a linearly independent subset of V, then $|B| \leq$ dim V.

(*ii*)If B is a spanning set for V, then dim $V \leq |B|$.

Theorem 3.4.9
If U is a subspace of V, then $V = U \oplus U'$ for a subspace U' of V.
Proof.
Let $\{x_1, x_2, \ldots, x_r\}$ be a basis of U then extend $\{x_1, x_2, \ldots, x_r\}$ to a basis $\{x_1, x_2, \ldots, x_r, y_1, y_2, \ldots, y_t\}$ of V. Let $U' = [y_1, y_2, \ldots, y_t]$, then trivially $V = U \oplus U'$.

5 CHANGE OF BASIS

We know that the basis in a vector space is not unique. Let $B = \{(1, 0), (0, 1)\}$ and $B' = \{(2, 3), (3, 2)\}$ be the bases of $\mathbb{R}^2$ then $(2, 3) = 2(1, 0) + 3(0, 1)$ and $(3, 2) = 3(1, 0) + 2(0, 1)$, we get a matrix

$$A = \begin{bmatrix} 2 & 3 \\ 3 & 2 \end{bmatrix}$$

Similarly, $(1, 0) = -\frac{2}{5}(2,3)+\frac{3}{5}(3,2)$ and $(0, 1) = \frac{3}{5}(2,3)-\frac{2}{5}(3,2)$, we get a matrix

$$B = \begin{bmatrix} -\frac{2}{5} & \frac{3}{5} \\ \frac{3}{5} & -\frac{2}{5} \end{bmatrix} = \frac{1}{5}\begin{bmatrix} -2 & 3 \\ 3 & -2 \end{bmatrix}$$

It is trivial to see that $AB = BA = I$. Hence, A and B are inverse of each other.
Now, let V be a vector space with $B = \{x_1, x_2, \ldots, x_n\}$ and $B' = \{y_1, y_2, \ldots, y_n\}$ as bases. Then considering B' as basis, we can write $x_i = \Sigma_{j=1}^{n} \alpha_{ji} y_j, \alpha_{ji} \in F$. Then

$$A = \begin{bmatrix} \alpha_{11} & \alpha_{12} & \cdots & \alpha_{1n} \\ \alpha_{21} & \alpha_{22} & \cdots & \alpha_{2n} \\ \vdots & \vdots & & \vdots \\ \alpha_{n1} & \alpha_{n2} & \cdots & \alpha_{nn} \end{bmatrix}$$

is called transition matrix from B to B'. Similarly considering B to be a basis, we can write $y_j = \Sigma_{k=1}^{n} \beta_{kj} x_k$ and the matrix

$$B = \begin{bmatrix} \beta_{11} & \beta_{12} & \cdots & \beta_{1n} \\ \beta_{21} & \beta_{22} & \cdots & \beta_{2n} \\ \vdots & \vdots & & \vdots \\ \beta_{n1} & \beta_{n2} & \cdots & \beta_{nn} \end{bmatrix}$$

is known as transition matrix from B' to B.

Now as done earlier in the example, we prove that transition matrices from any basis B to B' and from B' to B are inverse of each other.

Theorem 3.5.1

If V is a vector space and $B = \{x_1, x_2, \ldots, x_n\}$, $B' = \{y_1, y_2, \ldots, y_n\}$ are based on V then the transition matrices from B to B' and from B' to B are inverse of each other.

Proof. Let

$$A = \begin{bmatrix} \alpha_{11} & \alpha_{12} & \cdots & \alpha_{1n} \\ \alpha_{21} & \alpha_{22} & \cdots & \alpha_{2n} \\ \vdots & \vdots & & \vdots \\ \alpha_{n1} & \alpha_{n2} & \cdots & \alpha_{nn} \end{bmatrix}$$

and
$$B = \begin{bmatrix} \beta_{11} & \beta_{12} & \cdots & \beta_{1n} \\ \beta_{21} & \beta_{22} & \cdots & \beta_{2n} \\ \vdots & \vdots & & \vdots \\ \beta_{n1} & \beta_{n2} & \cdots & \beta_{nn} \end{bmatrix}$$

be the transition matrices from B to B' and from B' to B, respectively. Hence by definition, $x_i = \Sigma_{j=1}^{n} \alpha_{ji} y_j$ and $y_i = \Sigma_{k=1}^{n} \beta_{kj} x_k$. So

$$\begin{aligned} x_i &= \sum_{j=1}^{n} \alpha_{ji} \left(\sum_{k=1}^{n} \alpha_{kj} x_k \right) \\ &= \sum_{k=1}^{n} \left(\sum_{j=1}^{n} \alpha_{ji} \beta_{kj} \right) x_k \\ &= \sum_{k=1}^{n} c_{ki} x_k \end{aligned}$$

Trivially c_{ki} is the $(k, i)^{\text{th}}$ entry of BA. Since $\{x_1, x_2, \ldots, x_n\}$ is linearly independent $\Rightarrow c_{ki} = 0$ for all $k \neq i$ and $c_{ii} = 1$ for all $i \Rightarrow BA = I$.

Similarly, we can show that $AB = I \Rightarrow AB = BA = I \Rightarrow A = B^{-1}$.

Now if B is a known basis of a vector space and the transition matrix A from B to some basis B′ is known, then we can find the basis B', since $B = \{x_1, x_2, \ldots, x_n\}$ is a known basis. Let $B' = \{y_1, y_2, \ldots, y_n\}$ be the other basis, then $X = AY$, where

$$X = \begin{bmatrix} x_1 \\ x_2 \\ \vdots \\ x_n \end{bmatrix} \text{ and } Y = \begin{bmatrix} y_1 \\ y_2 \\ \vdots \\ y_n \end{bmatrix}$$

Then, $A^{-1} X = Y$ and we can calculate Y.

For example, consider R^2 and $B = \{(1, 1), (1, 2)\}$ as a basis and $\begin{bmatrix} 2 & 1 \\ 3 & 2 \end{bmatrix}$ be the transition matrix from B to some basis B', then

$$\begin{bmatrix} 2 & -1 \\ -3 & 2 \end{bmatrix}\begin{bmatrix} (1,1) \\ (1,2) \end{bmatrix} = \begin{bmatrix} y_1 \\ y_2 \end{bmatrix}$$

$\Rightarrow \quad 2(1, 1) - 1(1, 2) = y_1$

$\Rightarrow \quad y_1 = (1, 0)$

$\Rightarrow \quad -3(1, 1) + 2(1, 2) = y_2$

$\Rightarrow \quad y_2 = (-1, 1)$

and we see that $\{(1, 0), (-1, 1)\}$ is a basis of $\mathbb{R}^2$.

Now towards the end we discuss one more example.

Consider $\mathbb{R}^3$ and $\{(1, 1, 0), (0, 1, 1), (-1, 0, 1)\} = B$ and $\{(1, 0, 0),\}, (0, 1, 0), (0, 0, 1) = B'$ as bases of $\mathbb{R}^3$, then

$$(1, 1, 0) = 1(1, 0, 0) + 1(0, 1, 0) + 0(0, 0, 1)$$

$$(0, 1, 1) = 0(1, 0, 0) + 1(0, 1, 0) + 1(0, 0, 1)$$

$$(-1, 0, -1) = -1\,(1, 0, 0) + 0\,(0, 1, 0) + (-1)\,(0, 0, 1)$$

and we get

$$\begin{bmatrix} 1 & 0 & -1 \\ 1 & 1 & 0 \\ 1 & 1 & -1 \end{bmatrix}$$

As the transition matrix from B to B'. Sometimes it is known as transition matrix of B with respect to B'.

6 SOME SOLVED EXAMPLES

1. If $W = \{(x, y, z) \mid 2x + y - z = 0\}$ then trivially W is a subspace of $\mathbb{R}^3$. Putting $x = 0$, $y = 1$ we get $z = 1$. Further, take $y = 0$, $x = 1$ then $z = 2$. We see that $\{(0, 1, 1), (1, 0, 2)\}$ is linearly independent and

$2x + y - z = 0$ is an equation of a plane $\Rightarrow$ dim $W = 2$ and $\dfrac{R^3}{W}$ is spanned by one element.

2. If $W_1 = \{x, y, z\} \mid 2x + y - z = 0\}$ and $W_2 = \{(x, y, z) \in \mathbb{R}^3 \mid x + 2y + 2z = 0\}$ then find a basis of $W_1 \cap W_2$. Solving $2x + y - z = 0$ and $x + 2y + 2z = 0$, we get $x = -4r$, $y = 5r$, $z = -3r$ for $r \neq 0$ and $\{(-4, 5, -3)\}$ as a basis of $W_1 \cap W_2$. Hence, dim $(W_1 \cap W_2) = 1$.
3. If $W_1 = \{x, y, z\} \in \mathbb{R}^3 \mid x + y + z = 0\}$, $W_2 = \{(x, y, z) \in \mathbb{R}^3 \mid 3x + y - 2z = 0\}$ and $W_3 = \{(x, y, z) \in \mathbb{R}^3 \mid x - 7y + 3z = 0\}$ then find a basis of $W_1 \cap W_2 \cap W_3$.

   Solving the three equations, we see that $x = 1$, $y = 1$, $z = 2$. Hence $\{(1, 1, 2)\}$ is a basis of $W_1 \cap W_2 \cap W_3$ and dim $(W_1 \cap W_2 \cap W_3) = 1$

   $$\begin{aligned}\dim (W_1 + W_2) &= \dim W_1 + \dim W_2 - \dim (W_1 \cap W_2)\\ &= 2 + 2 - 1 = 3\end{aligned}$$

   $$\begin{aligned}\text{and } \dim (W_1 + W_3) &= \dim W_1 + \dim W_3 - \dim (W_1 \cap W_3)\\ &= 2 + 2 - 1 = 3\end{aligned}$$

   Thus, we get $\mathbb{R}^3 = W_1 \oplus W_2$, $\mathbb{R}^3 = W_1 \oplus W_3$, $\mathbb{R}^3 = W_2 \oplus W_3$. So complement is not unique.
4. For any subspace $W \subseteq V$ if dim W = dim V, then $W = V$. Trivially any basis of W can be extended to a basis of V but dim W = dim V. Hence, any basis of W is also a basis of $V \Rightarrow W = V$.
5. f $U = \{(x, y) \in \mathbb{R}^2 \mid y = mx\}$ and $W = \{(x, y) \in \mathbb{R}^2 \mid y = tx, t \neq m\}$ then U and W are subspaces of $\mathbb{R}^2$ (indeed they are lines passing through the origin).

   Now dim $(U + W)$ = dim U + dim W – dim $(U \cap W) = 1 + 1 - 0 = 2$.

   Hence $\mathbb{R}^2 = U \oplus W$.
6. If $W_1 = \{(x, y, z) \in \mathbb{R}^3 \mid x + 2y - z = 0\}$, $W_2 = \{(x, y, z) \in \mathbb{R}^3 \mid 3x + y - 2z = 0\}$ and $W_3 = \{(x, y, z) \in \mathbb{R}^3 \mid x - 7y + 3z = 0\}$ then the basis of $W_1 \cap W_2 \cap W_3$ is Ø as $W_1 \cap W_2 \cap W_3 = 0$.
7. In $P_2(x)$, we can check that $\{1 + x^2, 1 - x^2\}$ is linearly independent. Hence the space U generated by $1 + x^2$, $1 - x^2$ is of dimension 2.

   Trivially $x \neq \alpha(1 + x^2) + \beta(1 - x^2)$ for any α, β.

   $\Rightarrow \{x, 1 - x^2, 1 + x^2\}$ is a basis of $P_2(x)$ and $P_2(x) = U \oplus W$, where $W = [x]$.

   Similarly $\{1 + x, 1 - x\}$ is linearly independent and $\{1 + x, 1 - x,$

$x^2\}$ is a basis of $P_2(x)$ and $P_2(x) = U \oplus U'$ where $U = [1 + x, 1 - x]$ and $U' = [x^2]$.

8. If $S_n = \{A \in M_n(\mathbb{R}) \mid A^t = A\}$ and $K_n = \{A \in M_n(\mathbb{R}) \mid A_t = -A\}$, then S_n and K_n are subspaces of the vector space $M_n(\mathbb{R})$.

Let for $A, B \in S_n$

$$\Rightarrow \quad A^t = A, B^t = B$$
$$\Rightarrow \quad (A + B)^t = A^t + B^t = A + B$$
$$\text{and} \quad (\alpha A)^t = \alpha A^t = \alpha A$$
$$\Rightarrow \quad A + B \in S_n \text{ and } \alpha A \in S_n.$$

Similarly for $A, B \in K_n$, $(A - B)^t = A^t - B^t$

$$= -A + B = -(A - B)$$
$$\text{and} \quad (\alpha A)^t = \alpha A^t = -\alpha A$$
$$\Rightarrow \quad A - B \in K_n, \alpha A \in K_n.$$

The space S_n is the set of symmetric and K_n is the space of skew symmetric matrices. Further, since for any

$$A \in M_n(\mathbb{R}), A = \frac{A + A^t}{2} + \frac{A - A^t}{2}$$

$$\text{and} \quad \frac{A + A^t}{2} \in S_n$$

$$\text{and} \quad \frac{A - A^t}{2} \in K_n$$

$$\Rightarrow \quad M_n(\mathbb{R}) = S_n + K_n.$$

Further

$$S_n \cap K_n = 0$$
$$\Rightarrow \quad M_n(\mathbb{R}) = S_n \oplus K_n$$

$$\left[\text{Indeed dim } S_n = \frac{n(n+1)}{2} \text{ and dim } K_n = \frac{n(n-1)}{2}\right]$$

9. Find a basis of S_2 and K_2. It is easy to see that $\begin{bmatrix} 1 & 0 \\ 0 & 0 \end{bmatrix}, \begin{bmatrix} 0 & 0 \\ 0 & 1 \end{bmatrix}$,

$\begin{bmatrix} 0 & 1 \\ 1 & 0 \end{bmatrix}$ are symmetric and are linearly independent. Further, any

symmetric matrix of order 2 will be of the form $\begin{bmatrix} a & b \\ b & d \end{bmatrix}$ and

$$\begin{bmatrix} a & b \\ b & d \end{bmatrix} = a\begin{bmatrix} 1 & 0 \\ 0 & 0 \end{bmatrix} + b\begin{bmatrix} 0 & 1 \\ 1 & 0 \end{bmatrix} + d\begin{bmatrix} 0 & 0 \\ 0 & 1 \end{bmatrix}$$

$\Rightarrow$ dim $S_2 = 3$.

Hence dim $M_2(\mathbb{R})$ = dim S_2 + dim $K_2 \Rightarrow 4 = 3$ + dim $K_2 \Rightarrow$ dim $K_2 = 1$.

10. In the vector space $M_n(\mathbb{R})$ the matrices

$$E_{ij} = \begin{bmatrix} 0 & 0 & \cdots & \cdots & \cdots & 0 \\ \vdots & \vdots & \vdots & \vdots & \vdots & \vdots \\ 0 & 0 & \cdots & 1 & 0\cdots & 0 \\ \vdots & \vdots & \vdots & \vdots & \vdots & \vdots \\ 0 & 0 & \cdots & \cdots & \cdots & 0 \end{bmatrix} \rightarrow i^{\text{th}} \text{ row}$$

$\uparrow j^{th}$ column

Where 1 occurs at i^{th} row and j^{th} column. It is easy to check that $\{E_{ij}\}_1 \leq i \leq n$ form a basis of $M_n(\mathbb{R}) \Rightarrow \dim(M_n(\mathbb{R})) = n^2$.

Now take $A \in S_n$, then by definition of symmetric matrix we know that $a_{ij} = a_{ji}$ for $1 \leq i \leq n$ and $j = 1, 2, \ldots, n$. Hence there are n entries in the main diagonal, not depending upon any other entries. In first row there are $(n-1)$ entries which are respectively the same in first column. Therefore, for constructing a basis of S_n we can choose $(n-1)$ different entries exhausting the first row and first column. Continuing this process to get different entries in the basis elements of S_n, we get

$$n + (n-1) + (n-2) + \ldots + 3 + 2 + 1 = \frac{n(n+1)}{2} \text{ entries}$$

Hence any basis of S_n will have $\frac{n(n+1)}{2}$ elements.

$$\Rightarrow \dim S_n = \frac{n(n+1)}{2}$$

$$\Rightarrow \dim K_n = \dim M_n(\mathbb{R}) - \dim Sn = n^2 - \frac{n(n+1)}{2}$$

$$= \frac{2n^2 - n^2 - n}{2} = \frac{n(n-1)}{2}$$

As seen in Example 9.

$$B = \left\{\begin{bmatrix} 1 & 0 \\ 0 & 0 \end{bmatrix}, \begin{bmatrix} 0 & 0 \\ 0 & 1 \end{bmatrix}, \begin{bmatrix} 0 & 1 \\ 1 & 0 \end{bmatrix}\right\}$$

is a basis of S_2 and $B' = \left\{\begin{bmatrix} 0 & -1 \\ 1 & 0 \end{bmatrix}\right\}$ is a basis of K_2. $B \cup B'$ is a basis of $M_2(R)$.

11. In $P_2(x)$, consider $U = R(x^2 + x + 1) + Rx$ and $W = R(x^2 + 1) + Rx$ then it is easy to check that $\{x^2 + x + 1, x\}$ are linearly independent and $U \subseteq W$ but $\dim U = \dim W \Rightarrow U = W$.

12. Let $S = \{ax^2 + bx + c | a + b + c = 0\}$ then it is easy to see that S is a subspace of $P_2(x)$ and $\{x - x^2, 1 - x^2)$ is a basis of S.

13. Let $B = \{(1, -1, 2), (1, 3, 1), (-1, 2, 1)\}$ be a subset of $\mathbb{R}^3$ and $\alpha\{(1, -1, 2) + \beta (1, 3, 1) + \gamma (-1, 2, 1) = 0$

$$\Rightarrow \quad \alpha + \beta - \gamma = 0, -\alpha + 3\beta + 2\gamma = 0, 2\alpha + \beta + \gamma = 0$$

$$\Rightarrow \quad \begin{bmatrix} 1 & 1 & -1 \\ -1 & 3 & 2 \\ 2 & 1 & 1 \end{bmatrix}\begin{bmatrix} \alpha \\ \beta \\ \gamma \end{bmatrix} = \begin{bmatrix} 0 \\ 0 \\ 0 \end{bmatrix}$$

$$\Rightarrow \quad AX = O.$$

Now
$$\det(A) = 1(3 - 2) - 1(-1 - 4) - 1(1 - 1 - 6)$$
$$= 1 + 5 + 7 = 13 \neq 0$$

$$\Rightarrow A^{-1} \text{ exists} \Rightarrow \begin{bmatrix} \alpha \\ \beta \\ \gamma \end{bmatrix} = \begin{bmatrix} 0 \\ 0 \\ 0 \end{bmatrix}$$

$\Rightarrow \alpha = 0, \beta = 0, \gamma = 0$

$\Rightarrow B$ is linearly independent and since $\dim R^3 = 3 \Rightarrow B$ is a basis of R^3.

14. If

$$S = \left\{\begin{bmatrix} \lambda_1 & 0 & 0 \\ 0 & \lambda_2 & 0 \\ 0 & 0 & \lambda_3 \end{bmatrix} \lambda_i \in R\right\}$$

then S is a subspace of $M_3(\mathbb{R})$ and $\dim S = 3$. If

$$W = \text{span}\left\{\begin{bmatrix} 0 & 1 & 0 \\ 0 & 0 & 0 \\ 0 & 0 & 0 \end{bmatrix}, \begin{bmatrix} 0 & 0 & 1 \\ 0 & 0 & 0 \\ 0 & 0 & 0 \end{bmatrix}, \begin{bmatrix} 0 & 0 & 0 \\ 1 & 0 & 0 \\ 0 & 0 & 0 \end{bmatrix}\right\}$$

$$\left\{\begin{bmatrix} 0 & 0 & 0 \\ 0 & 0 & 1 \\ 0 & 0 & 0 \end{bmatrix}, \begin{bmatrix} 0 & 0 & 0 \\ 0 & 0 & 0 \\ 1 & 0 & 0 \end{bmatrix}, \begin{bmatrix} 0 & 0 & 0 \\ 0 & 0 & 0 \\ 0 & 1 & 0 \end{bmatrix}\right\}$$

then dim $W = 6$ and $M_3(\mathbb{R}) = S \oplus W$.

15. Let $S = \{(1, 2, -2, 1), (-3, 0, -4, 3), (1, 3, 1, 1), (-1, 4, -2, 1), (2, 4, 1, 3)\}$ be a subset of $\mathbb{R}^4$. As we know that dim $\mathbb{R}^4 = 4 \Rightarrow S$ cannot be a basis of $\mathbb{R}^4$. Let $W = [S]$ then we can calculate dim W. This is done by deleting process which amounts same as under. Let

$$A = \begin{bmatrix} 1 & -3 & 1 & -1 & 2 \\ 2 & 0 & 3 & 4 & 4 \\ -2 & -4 & 1 & -2 & 1 \\ 1 & 3 & 1 & 1 & 3 \end{bmatrix}$$

Now we change A into reduced row echelon form

$$A \sim \begin{bmatrix} 1 & -3 & 1 & -1 & 2 \\ 0 & 6 & 1 & 6 & 0 \\ -2 & -4 & 1 & -2 & 1 \\ 1 & 3 & 1 & 1 & 3 \end{bmatrix}$$

$$R_2 \to R_2 + R_1(-1)$$
$$R_3 \to R_3 + R_1(2)$$
$$R_4 \to R_4 + R_1(-1)$$

$$\sim \begin{bmatrix} 1 & -3 & 1 & -1 & 2 \\ 0 & 1 & \frac{1}{6} & 1 & 0 \\ 0 & -10 & 3 & -4 & 5 \\ 0 & 6 & 0 & 2 & 1 \end{bmatrix} \quad R_2 \to R_2\left(\frac{1}{6}\right)$$

$$\sim \begin{bmatrix} 1 & 0 & \frac{3}{2} & -2 & 2 \\ 0 & 1 & \frac{1}{6} & 1 & 0 \\ 0 & 0 & \frac{14}{3} & 6 & 5 \\ 0 & 0 & 1 & 4 & -1 \end{bmatrix}$$

$$\sim \begin{bmatrix} 1 & 0 & 0 & -8 & \frac{7}{2} \\ 0 & 1 & 0 & \frac{1}{3} & \frac{1}{6} \\ 0 & 0 & 0 & -\frac{38}{3} & \frac{29}{3} \\ 0 & 0 & 1 & 4 & -1 \end{bmatrix}$$

$$\sim \begin{bmatrix} 1 & 0 & 0 & -8 & \frac{7}{2} \\ 0 & 1 & 0 & \frac{1}{3} & \frac{1}{6} \\ 0 & 0 & 1 & 4 & -1 \\ 0 & 0 & 0 & -\frac{38}{3} & \frac{29}{3} \end{bmatrix}$$

Now the last row's first entry can be made 1 and then the matrix can be changed into row reduced echelon form $\Rightarrow$ dim $W = 4 \Rightarrow W = \mathbb{R}^4$.

16. If S = {(1, 2, 2), (3, 2, 1), (11, 10, 7), (7, 6, 4)} $\subseteq \mathbb{R}^3$ then S cannot be a basis of $\mathbb{R}^3$. Let $W = [S]$. Now we find dim (W).

$$A = \begin{bmatrix} 1 & 3 & 11 & 7 \\ 2 & 2 & 10 & 6 \\ 2 & 1 & 7 & 4 \end{bmatrix} \sim \begin{bmatrix} 1 & 3 & 11 & 7 \\ 0 & -4 & -12 & -8 \\ 0 & -5 & -15 & -10 \end{bmatrix}$$

$$\sim \begin{bmatrix} 1 & 3 & 11 & 7 \\ 0 & 1 & 3 & 2 \\ 2 & 1 & 7 & 4 \end{bmatrix} \sim \begin{bmatrix} 1 & 3 & 11 & 7 \\ 0 & 1 & 3 & 2 \\ 0 & 0 & 0 & 0 \end{bmatrix}$$

$$\sim \begin{bmatrix} 1 & 0 & 2 & 1 \\ 0 & 1 & 3 & 2 \\ 0 & 0 & 0 & 0 \end{bmatrix}$$

$\Rightarrow$ {(1, 2, 2), (3, 2, 1)} spans W and is linearly independent.

$\Rightarrow$ dim $W = 2$.

17. If $S = \{t^3 + t^2 + -2t + 1, t^2 + 1, t^3 - 2t, 2t^3 + 3t^2 - 4t + 3\} \subseteq P_3(t)$. Now we consider $W = [S]$ and find a basis of W.

$$A = \begin{bmatrix} 1 & 0 & 1 & 2 \\ 1 & 1 & 0 & 3 \\ -2 & 0 & -2 & -4 \\ 1 & 1 & 0 & 3 \end{bmatrix} \sim \begin{bmatrix} 1 & 0 & 1 & 2 \\ 0 & 1 & -1 & 1 \\ 0 & 0 & 0 & 0 \\ 0 & 1 & -1 & 1 \end{bmatrix}$$

$$\sim \begin{bmatrix} 1 & 0 & 1 & 2 \\ 0 & 1 & -1 & 1 \\ 0 & 0 & 0 & 0 \\ 0 & 0 & 0 & 0 \end{bmatrix}$$

$\Rightarrow$ dim $W = 2$ and $\{t^3 + t^2 - 2t + 1, t^2 + 1\}$ is a basis of W.

18. If $S = \{1, 2, -1\}, (2, \lambda, 1), (1, 4, 3)\}$ is linearly dependent, then we find λ. Put

$$\begin{vmatrix} 1 & 2 & 1 \\ 2 & \lambda & 4 \\ -1 & 1 & 3 \end{vmatrix} = 0$$

$$\Rightarrow 1(3\lambda - 4) - 2(4 + 6) + 1(2 + \lambda) = 0$$

$$\Rightarrow 3\lambda - 4 - 20 + 2 + \lambda = 0$$

$$\Rightarrow \lambda = \frac{11}{2}$$

EXERCISE

1. Prove that $\left\{\begin{bmatrix} 1 & 0 \\ 0 & 1 \end{bmatrix}, \begin{bmatrix} 0 & 0 \\ 0 & 0 \end{bmatrix}, \begin{bmatrix} 1 & 1 \\ 1 & 0 \end{bmatrix}, \begin{bmatrix} 0 & 1 \\ 1 & 1 \end{bmatrix}\right\}$ is a field under usual matrix addition and multiplication, where $1 + 1 = 0$.
2. If S is a subset of a vector space then prove that [[S]] = [S].
3. Prove that if B is a linearly independent subset of a vector space V, then $|B| \leq \dim V$ and $\dim V \leq |S|$ for any spanning set S of V.
4. Show that $W = \{(a_1, a_2, \ldots, a_n) \mid a_1 + a_2 + \ldots + a_n = 0\}$ is a subspace of $\mathbb{R}^n$. Find the dim (W).
5. Which of the following are subspaces:

   (*a*) $\{(x, y, z) \mid x = 5\} \subseteq \mathbb{R}^3$

   (*b*) $\{(x, y, z) \mid x^2 - y^2 = 0\} \subseteq \mathbb{R}^3$

   (*c*) $\{(x, y, z) \mid x = z\} \subseteq \mathbb{R}^3$

   (*d*) $\{(x, y) \mid x^2 + y^2 = 4\} \subseteq \mathbb{R}^2$

6. Find the value of λ such that $\{(-1, 2, 3), (\lambda, 1, 5), (3, 7, 2)\}$ is linearly dependent.
7. If $S = \{1, x-2, (x-2)^2\} \subseteq P_2(x)$, then find $[S]$.
8. If $\{e_1, e_2, e_3\}$ is a linearly independent subset of a vector space V, then prove that $\{e_1, e_2 + e_2, e_1 + e_2, e_3\}$ is also linearly independent.
9. If $B = \{v_1, v_2, \ldots, v_r\}$ is a linearly independent subset of a vector space V, then show that elementary row operations on B do not change the linear independence.
10. Which of the following are bases of $\mathbb{R}^3$:
    (*a*) $\{(1, 0, 1), (-1, 1, 0), (1, 1, 1)\}$
    (*b*) $\{(1, 2, 0), (2, 0, 1), (3, 2, 1)\}$
    (*c*) $\{(2, 1, 3), (1, 3, 2), (3, 2, 1)\}$
    (*d*) $\{(1, -1, 1), (-2, 1, 2), (-1, 2, -2)\}$
11. If $S = \{(1, 0, 1), (-1, 0, 2)\}$, $B = \{(0, 1, 2), (1, 0, 2)\}$ and $C = \{(1, 1, 1)\}$, then find the complementary summands of $[S]$, $[B]$ and $[C]$ in $\mathbb{R}^3$. Further extend S, B and C to bases of $\mathbb{R}^3$.
12. If V is a vector space and U_1, U_2, U_3 are subspaces of V, then prove that $\dim(U_1 + U_2 + U_3) = \dim U_1 + \dim U_2 + \dim U_3 - \dim(U_1 \cap U_2) - \dim(U_1 \cap U_3) - \dim(U_2 \cap U_3) + \dim(U_1 \cap U_2 \cap U_3)$
13. Which of the following span $P_2(x)$.
    (*a*) $\{1 + x, 2x + x^2, x^2 + 1\}$
    (*b*) $\{x^2 + 2, 2x^2 + x - 2, x + 3\}$
    (*c*) $\{x^2 + 1, x^2 - 1, x + 1\}$
14. Find the basis of S_3 and K_3, the space of symmetric 3×3 matrices and 3×3 skew symmetric matrices.
15. Show that $\{x - 1, 2x + 2, 1 + x^2\}$ is a basis of $P_2(x)$.
16. If $S = \{(1, 0, 1, 0), (0, 1, 2, 1), (1, 1, 3, 1), (1, 2, 5, 2), (1, 1, 0, 7)\}$, then find the basis of $[S]$ and its dimension.
17. If $S = \{x_1, x_2, \ldots, x_n\}$ is set of nonzero vectors in a vector space such that $x \in V$, can be uniquely written as $\Sigma_{i=1}^{n} \lambda_i x_i$, then prove that S is a basis of V.
18. If U and W are subspaces of V, then prove that

$$\frac{U + W}{U} \cong \frac{W}{U \cap W}$$

Apply it to show that

$$\dim(U + W) = \dim U + \dim W - \dim U \cap W.$$

19. If $f : V \to U$ is a vector homomorphism of vector spaces V and U, then show $V/\ker f \cong f(V)$.

20. If $W = \left\{ \begin{bmatrix} a & b \\ c & -1 \end{bmatrix} \middle| a, b, c \in \mathbb{R} \right\}$

    Then prove that W is a subspace of $M_2(R)$. Find dim W and a subspace $W' \subseteq M_2(R)$ such that $M_2(R) = W \oplus W'$.

21. If S and A are subsets of a vector space V, then show that $[S \cup A] = [S] + [A]$. Hence or otherwise show that for any two subspaces U and W of V $U + W$ is the smallest subspace of V containing U and W.

4

Linear Transformations

We have seen in Chapter 2 that any two sets can be associated by certain rules known as mappings. The maps can be restricted either by saying it is one-one, onto or many to one, but for any sets A and B if they possess some structures i.e. some operations on A and B are defined,then we can put some extra conditions to be satisfied by the maps. The vector spaces have two operations and if the maps preserve these operations, then we give them separate names like linear transformations. It is an important concept for studying vector spaces.

1 DEFINITION AND EXAMPLES

Definition 4.1.1. Let V and U be vector spaces over a field F, then a map $T : V \to U$ is called *linear transformation* if

$$T(x + y) = T(x) + T(y) \quad \ldots(i)$$

and

$$T(\lambda x) = \lambda T(x) \quad \ldots(ii) \text{ for } x, y \in V, \lambda \in F.$$

The linear transformations are also called *vector space homomorphisms.* The vector spaces V and U are taken over the same field so that $T(\lambda x) = \lambda T(x)$ is meaningful. However, if V is a vector space over F and U is vector space over K such that $F \subseteq K$, then both the conditions are in order. Therefore, now onwards, linear transformations of vector spaces over different fields will not be considered.

Examples 4.1.2

(*i*) $T : \mathbb{R}^3 \to \mathbb{R}^2$ such that $T(x, y, z) = (x - y, y - z)$ is a linear transformation.

(*ii*) $T : \mathbb{R}^2 \to \mathbb{R}^3$ such that $T(x, y) = (x, y, 0)$; $S : \mathbb{R}^3 \to \mathbb{R}^2$ such that $S(x, y, z) = (x, y)$ are linear transformations.

(*iii*) $T : \mathbb{R}^n \to \mathbb{R}^m$, where $n > m$, such that

$$T(x_1, x_2, \ldots, x_n) = (x_1, x_2, \ldots, x_m)$$

is a linear transformation.

(*iv*) $T : \mathbb{R}^n \to \mathbb{R}^m$, where $m > n$, such that

$T(x_1, x_2, \ldots, x_n) = (x_1, x_2, \ldots, x_m, 0, \ldots, 0)$

is a linear transformation.

All the above can be trivially verified.

(*v*) $T : \mathbb{R}^3 \to \mathbb{R}^2$ such that $T(x, y, z) = (x + 1, y)$ is not a linear transformation.

(*vi*) $T : \mathbb{R}^3 \to \mathbb{R}^3$ such that $T(x, y, z) = (\lambda x, \lambda y, \lambda z)$, $\lambda > 1$ is a linear transformation called *dilation* and if $0 < \lambda < 1$, then it is called *contraction.*

(*vii*) $T : \mathbb{R}^2 \to \mathbb{R}^2$ such that $T(x, y) = (x, - y)$ is linear transformation known as *reflexion.*

(*viii*) $D : P_3(x) \to P_2(x)$ such that $D(f(x)) = \dfrac{d}{dx}(f(x)) = f'(x)$ is a linear transformation.

(*ix*) $T : P_2(x) \to P_2(x)$ such that $T(f(x)) = \int f(x)\, dx$ is not a well defined map as the constant of the integration may differ but if $T : P_2(x) \to \mathbb{R}$ such that $T(f(x)) = \int_0^1 f(x)\, dx$, then T is a linear transformation.

(*x*) $T : M_{nm}(\mathbb{R}) \to M_{mn}(\mathbb{R})$ such that $T(A) = A^t$, where A^t is transpose of A is a linear transformation.

(*xi*) $T : M_{nn}(\mathbb{R}) \to \mathbb{R}$ such that $T(A) = \text{trace}(A)$ is a linear transformation.

Now we prove some elementary results, throughout V and U will be considered vector spaces over the same field F.

Theorem 4.1.3. A mapping $T : V \to U$ is a linear transformation iff

$$T(\alpha x + \beta y) = \alpha T(x) + \beta T(y) \text{ for all } x, y \in V, \alpha, \beta \in F.$$

Proof.

$T(\alpha x + \beta y) = T(\alpha x) + T(\beta y)$ by condition (*i*) of Definition 4.1.1.

$= \alpha T(x) + \beta T(y)$ by condition (*i*) of Definition 4.1.1.

Taking $\alpha = \beta = 1$, we have

$$T(1 \cdot x + 1 \cdot y) = 1 \cdot T(x) + 1 \cdot (y)$$
$$= T(x) + T(y)$$
$$\Rightarrow \quad T(x + y) = T(x) + T(y)$$

Also taking $\beta = 0$, we have

$$T(\alpha x + 0 \cdot y) = \alpha T(x) + 0 \cdot T(y)$$
$$= \alpha T(x) + 0$$
$$= \alpha T(x)$$
$$\Rightarrow \quad T(\alpha x) = \alpha T(x)$$

Theorem 4.1.4. If $T : V \to U$ is a linear transformation, then

$$T(0) = 0,\ T(-x) = -T(x)$$

Proof. We know that

$$T(\alpha x) = \alpha T(x) \text{ for all } \alpha \in F, x \in V$$

$\Rightarrow$ taking $\alpha = 0$ we have

$$T(0) = 0 \cdot T(x) = 0.$$

We know that

$$\begin{aligned} x + (-x) &= 0 \\ \Rightarrow \quad T(x + (-x)) &= T(0) = 0 \\ \Rightarrow \quad T(x) + T(-x) &= 0 \\ &= T(x) + (-T(x)). \end{aligned}$$

Hence, by left cancellation law

$$T(-x) = -T(x).$$

Remark 1. If $T(0) \neq 0$, then T cannot be a linear transformation. Since if it is a linear transformation, then

$$T(\alpha x) = \alpha T(x) \text{ for all } \alpha \in F, x \in V.$$

Hence, taking $\alpha = 0$, we have $T(0) = 0 \cdot T(x) = 0$ which is a contradiction. Therefore, for linear transformations it is necessary condition that $T(0) = 0$, but is not sufficient. For example $T : \mathbb{R} \to \mathbb{R}$ given as $T(x) = x^2$, $T(0) = 0$, but it is not a linear transformation.

Remark 2. If $T : V \to U$ is a linear transformation, then T maps linearly dependent subsets of V into linear dependent subsets of U. For if $\{x_1, x_2, \ldots, x_n\}$ is a linearly dependent subset of V, then we show that $\{T(x_1), T(x_2), \ldots, T(x_n)\}$ is linearly dependent. Since $\{x_1, x_2, \ldots, x_n\}$ is linearly dependent $\Rightarrow$ there exists at least one $\alpha_i \neq 0$, such that

$$\alpha_1 x_1 + \alpha_2 x_2 + \ldots + \alpha_n x_n = 0$$

$$\Rightarrow T(\alpha_1 x_1 + \alpha_2 x_2 + \ldots + \alpha_i x_i + \ldots + \alpha_n x_n) = T(0) = 0$$

$$\Rightarrow \alpha_1 T(x_1) + \alpha_2 T(x_2) + \ldots + \alpha_i T(x_i) + \ldots + \alpha_n T(x_n) = 0$$

$\Rightarrow \{T(x_1), T(x_2), \ldots, T(x_n)\}$ is linearly dependent.

Remark 3. If $T : V \to U$ is a linear transformation, then T may not map linearly independent subsets of V into linearly independent subsets of U. For example, $T : \mathbb{R}^2 \to \mathbb{R}^2$, such that $T(x, y) = (x, 0)$ is a linear transformation and $\{(1, 0), (0, 1)\}$ is linearly independent subset of $\mathbb{R}^2$, but $\{T(1, 0), T(0, 1)\} = \{(1, 0), (0, 0)\}$ is not linearly independent. We shall later that if T is one-one then it maps linear independent subsets into linearly independent.

Remark 4. If $T : V \to U$ is a map and $\{x_1, x_2, \ldots, x_n\}$ is a basis of V, then for any $x \in V$ there exist unique $\lambda_1, \lambda_2, \ldots, \lambda_n$, such that

$$x = \sum_{i=1}^{n} \lambda_i x_i .$$

Now define

$$T(x) = \sum_{i=1}^{n} \lambda_i T(x_i) .$$

Trivially, T is a well defined map and it can be easily verified that T is a linear transformation. So every linear transformation $T : V \to U$ is uniquely determined by the values $T(x_1), T(x_2), \ldots, T(x_n)$ of the basis $\{x_1, x_2, \ldots, x_n\}$ of V.

Now, we consider and example for this. Let $T : \mathbb{R}^2 \to \mathbb{R}^2$ be a map such that $T(1, 0) = (2, 1)$ and $T(0, 1) = (1, 2)$, we know that, for any $(x, y) \in^2$

$$(x, y) = x(1, 0) + y(1, 0)$$

$$\Rightarrow \quad T(x, y) = xT(1, 0) + yT(0, 1)$$
$$= x(2, 1) + y(1, 2)$$
$$= (2x + y, x + 2y).$$

We consider another example of different nature.

Let $T : P_1(x) \to P_2(x)$ such that

$$T(1 + x) = x + x^2, T(1 - x) = 1 - x^2,$$

then let us find $T(5x + 3)$. Trivially $\{1 + x, 1 - x\}$ is linearly independent as

$$\alpha(1 + x) + \beta(1 - x) = 0$$

$$\Rightarrow \quad \alpha + \beta = 0 \text{ and } \alpha - \beta = 0$$

$$\Rightarrow \quad \alpha = \beta = 0,$$

but $P_1(x)$ is vector space of dimension 2, therefore $\{1 + x, 1 - x\}$ is a basis of $P_1(x)$.

Now, let $$5x + 3 = a(1 + x) + b(1 - x)$$

$$\Rightarrow \quad a + b = 3 \text{ and } a - b = 5$$

$$\Rightarrow \quad a = 4 \text{ and } b = -1$$

$$\Rightarrow \quad T(5x + 3) = 4T(1 + x) + (-1)T(1 - x)$$
$$= 4(x + x^2) - (1 - x^2)$$
$$= 4x + 4x^2 - 1 + x^2$$

$$= 5x^2 + 4x - 1.$$

Now, let $f(x) \in P_1(x)$ and let $f(x) = \alpha + \beta x$, then

$$a + b = \alpha,\ a - b = \beta$$

$$\Rightarrow \qquad a = \frac{\alpha+\beta}{2} \text{ and } b = \frac{\beta-\alpha}{2}$$

$$T(f(x)) = \frac{\alpha+\beta}{2}(x + x^2) + \frac{\beta-\alpha}{2}(1 - x^2)$$

$$= \left(\frac{\alpha+\beta}{2} - \frac{\beta-\alpha}{2}\right)x^2 + \frac{\alpha+\beta}{2}x + \frac{\beta-\alpha}{2}$$

$$= \alpha x^2 + \frac{\alpha+\beta}{2}x + \frac{\beta-\alpha}{2}.$$

A linear transformation can be easily obtained, provided the images of the basis elements are known.

2 RANGE AND NULL SPACE

First, we prove the following:

Theorem 4.2.1. If $T : V \to U$ is a linear transformation, then

$$TV = \{T(x) \mid x \in V\} \text{ and } S = \{x \in V \mid T(x) = 0\}$$

and subspaces of U and V, respectively.

Proof. Let $z_1, z_2 \in TV$, then $z_1 = T(x_1)$, $z_2 = T(x_2)$ for some $x_1, x_2 \in V$

$$\Rightarrow \qquad \alpha z_1 + \beta z_2 = \alpha T(x_1) + \beta T(x_2)$$

$$= T(\alpha x_1) + T(\beta x_2) = T(\alpha x_2 + \beta x_2),$$

as $\qquad \alpha x_1 + \beta x_2 \in V$

$$\Rightarrow \qquad \alpha z_1 + \beta z_2 \in TV \text{ for all } \alpha, \beta \in F$$

$\Rightarrow$ TV is a subspace of U.

Let $x, y \in S$

$$\Rightarrow \qquad T(x) = 0 = T(y)$$

$$\Rightarrow \qquad T(\alpha x + \beta y) = \alpha T(x) + \beta T(y) = 0$$

$$\Rightarrow \qquad \alpha x + \beta y \in S$$

$\Rightarrow$ S is a subspace of V.

Definition 4.2.2. If $T : V \to U$ is a linear transformation, then TV is called range space of T and S denoted as $N(T)$ or ker T is called null space of T or kernel of T. Further, dim TV denoted as $r(T)$ and dim $N(T)$ denoted as $n(T)$ are called rank and nullity of T, respectively.

Example 4.2.3. If $T : \mathbb{R}^2 \to \mathbb{R}^2$ such that $T(x, y) = (x - y, y)$, then $N(T) = 0$; for if $(x, y) \in N(T)$, then $x - y = 0, y = 0 \Rightarrow x = 0$.

Further, $T : \mathbb{R}^2 = \mathbb{R}^2$; for it $(x, y) \in \mathbb{R}^2$, then

$$T(x + y, y) = (x, y)$$

$$\Rightarrow \qquad r(T) = 2, n(T) = 0.$$

Notes:

1. If $I : V \to U$ is the identity transformation, i.e., $I(x) = x$ for all $x \in V$, then $n(I) = 0$, $r(I) = \dim V$. Also, if $O : V \to U$ is zero transformation, then $n(O) = \dim V$ and $r(O) = 0$.
2. Let $D : P_2(x) \to P_2(x)$, such that $D = \frac{d}{dx}$, then $n(D) = 1$ and $r(D) = 2$. For $\frac{d}{dx}(\alpha) = 0$ for all $\alpha \in F$ and $D(\alpha + \beta x + \gamma x^2) = \beta + 2\gamma x$. Also, for any $a + bx \in P_1(x)$, $D\left(\alpha + ax + \frac{bx^2}{2}\right) = a + bx \Rightarrow$ $D : P_2(x) \to P_1(x)$ is onto.
3. Let $T : P_2(x) \to P_2(x)$ such that
   $T(a + bx + cx^2) = (a + b) + (a - c)x$,
   then trivially T is a linear transformation. Let $a + bx + cx^2 \in N(T)$

   $$\Rightarrow \qquad (a + b) + (a - c)x = 0$$

   $$\Rightarrow \qquad a + b = 0 \text{ and } a - c = 0$$

   $$\Rightarrow \qquad a = c = -b$$

   $$\Rightarrow \qquad a + bx + cx^2 = a - ax + ax^2 = a(1 - x + x^2)$$

   $\Rightarrow \{1 - x + x^2\}$ is a basis of $N(T)$

   $\Rightarrow n(T) = 1$ and $TP_2(x)$ is generated by $\{1, x\}$ which is linearly independent $\Rightarrow r(T) = 2$.

The following Theorem, known as Sylvester's law of nullity, gives a technique to calculate the rank and nullity of a linear transformation.

Theorem 4.2.4. Let $T : V \to U$ be linear transformation, then

$$\dim V = r(T) + n(T).$$

Proof. As done in Theorem 4.2.1, $N(T)$ is a subspace of V. Let $\{x_1, x_2, \ldots, x_r\}$ be a basis of $N(T)$, then $\{x_1, x_2, \ldots, x_r\}$ can be extended to a basis $\{x_1, x_2, \ldots, x_r, x_{r+1}, \ldots, x_n\}$ of V. For any $x \in V$ there exist unique $\lambda_1, \lambda_2, \ldots, \lambda_n \in F$ such that

$$x = \sum_{i=1}^{n} \lambda_i x_i$$

$$\Rightarrow \quad T(x) = \sum_{i=1}^{n} \lambda_i T(x_i), \text{ as } T(x_i) = 0 \text{ for } i = 1, 2, \ldots, r.$$

Hence, $\{T(x_{r+1}), T(x_{r+2}), \ldots, T(x_n)\}$ spans TV. Now, let

$$\sum_{j=r+1}^{n} \beta_j T(x_j) = 0, \text{ then } T\left(\sum_{j=r+1}^{n} \beta_j x_j\right) = 0 \text{ gives } \sum_{j=r+1}^{n} \beta_j x_j \in N(T)$$

$$\Rightarrow \quad \sum_{j=r+1}^{n} \beta_j x_j = \sum_{i=1}^{r} \alpha_i x_i$$

$$\Rightarrow \quad \sum_{i=1}^{r} \alpha_i x_i - \sum_{j=r+1}^{n} \beta_j x_j = 0.$$

But $\{x_1, x_2, \ldots, x_r, \ldots, x_n\}$ is linearly independent.
$\Rightarrow \alpha_i = 0, \beta_j = 0$ for $i = 1, 2, \ldots, r; j = r + 1, \ldots, n$.
Hence, $\{T(x_{r+1}), \ldots, T(x_n)\}$ is a basis of TV

$$\Rightarrow \quad r(T) = n - r = n - n(T)$$

$$\Rightarrow \quad \dim V = r(T) + n(T).$$

Example 4.2.5

1. Let $T : M_2(\mathbb{R}) \to \mathbb{R}$ be such that

$$T\left(\begin{bmatrix} a & b \\ c & d \end{bmatrix}\right) = a + b$$

then it is easy to see that T is a linear transformation. Now, for any $\alpha \in \mathbb{R}$, there is

$$\begin{bmatrix} \alpha & 0 \\ c & d \end{bmatrix} \in M_2(\mathbb{R}) \text{ such that } T\left(\begin{bmatrix} \alpha & 0 \\ c & d \end{bmatrix}\right) = \alpha$$

$$\Rightarrow \quad T(M_2(\mathbb{R})) = \mathbb{R}$$

$$\Rightarrow \quad r(T) = 1 \text{ and } n(T) = 3.$$

Further, it is easy to see that

$$\left\{\begin{bmatrix} 1 & -1 \\ 1 & 0 \end{bmatrix}, \begin{bmatrix} 0 & 0 \\ 0 & 1 \end{bmatrix}, \begin{bmatrix} 0 & 0 \\ 1 & 0 \end{bmatrix}\right\}$$

is linearly independent subset of $N(T)$. Also, for any

$$\begin{bmatrix} a & -a \\ b & c \end{bmatrix} \in N(T)$$

$$\begin{bmatrix} a & -a \\ b & c \end{bmatrix} = a\begin{bmatrix} 1 & -1 \\ 1 & 0 \end{bmatrix} + c\begin{bmatrix} 0 & 0 \\ 0 & 1 \end{bmatrix} + (b-a)\begin{bmatrix} 0 & 0 \\ 1 & 0 \end{bmatrix}$$

$$\Rightarrow \left\{\begin{bmatrix} 1 & -1 \\ 1 & 0 \end{bmatrix}, \begin{bmatrix} 0 & 0 \\ 0 & 1 \end{bmatrix}, \begin{bmatrix} 0 & 0 \\ 1 & 0 \end{bmatrix}\right\}$$

is a basis of $N(T)$.

2. If $T : P_2(x) \to \mathbb{R}$ such that $T(f(x)) = \int_0^1 f(x)\, dx$, then trivially T is a linear transformation. Now, we calculate bases of $N(T)$, $TP_2(x)$ and $n(T)$, $r(T)$.

   Since for any $\alpha \in \mathbb{R}$,

$$\alpha = \int_0^1 \alpha\, dx$$

$$\Rightarrow \quad r(T) = 1 \text{ and } n(T) = 2$$

   by above Theorem. So it becomes sometimes very easy to find $r(T)$ and $n(T)$ by using the Sylvester's law of nullity. Even if we define

   $T(f(x)) = \int_0^\lambda f(x)\, dx$ for any $\lambda \in \mathbb{R}, \lambda \neq 0, \lambda > 0$

   then we get same result as

$$\alpha = \int_0^\lambda \frac{\alpha}{\lambda}\, dx.$$

   Now, let us calculate the basis of range space and null space. Trivially every nonzero real number is a basis for range space. Let

$$f(x) = a + bx + cx^2 \in N(T)$$

$$\Rightarrow \quad \int_0^1 (a + bx + cx^2)\, dx = 0$$

$$\Rightarrow \quad a + \frac{b}{2} + \frac{c}{3} = 0$$

$$\Rightarrow \quad 6a + 3b + 2c = 0$$

$$\Rightarrow \quad c = \frac{6a + 3b}{-2}$$

$$\Rightarrow \quad a + bx + cx^2 = a + bx + \frac{6a + 3b}{-2} x^2$$

$$= a(1 - 3x^2) + b\left(x - \frac{3}{2}x^2\right)$$

$$\Rightarrow \quad n(T) = 2$$

$$\Rightarrow \left\{1 - 3x^2, x\left(x - \frac{3}{2}x\right)\right\} \text{ is a basis of } N(T).$$

Theorem 4.2.6. Let $T : V \to U$ be a linear transformation, then the following hold:

(*i*) T is onto iff $r(T) = \dim U$.

(*ii*) T is one-one iff $n(T) = 0$.

Proof. (*i*) Let T be onto

$\Rightarrow$ for every $u \in U\, \exists\, v \in V$ such that

$$T(v) = u$$

$$\Rightarrow \quad TV = U$$

$$\Rightarrow \quad \dim TV = \dim U$$

$$\Rightarrow \quad r(T) = \dim U.$$

Conversely, $TV \subseteq U$ and $\dim TV = r(T) = \dim U$

$$\Rightarrow \quad TV = U$$

$\Rightarrow$ T is onto.

(*ii*) Let T be one-one and $x \in N(T)$

$$\Rightarrow \quad T(x) = 0 = T(0) \text{ but } T \text{ is one-one}$$

$$\Rightarrow \quad x = 0 \Rightarrow N(T) = 0$$

$$\Rightarrow \quad n(T) = 0.$$

Conversely, let $T(x) = T(y)$

$$\Rightarrow \quad T(x - y) = 0$$

$$\Rightarrow \quad x - y \in N(T), \text{ but } n(T) = 0$$

$$\Rightarrow \quad N(T) = 0$$

$$\Rightarrow \quad x - y = 0$$

$$\Rightarrow \quad x = y$$

$\Rightarrow$ T is one-one.

Corollary 4.2.7. A linear transformation $T : V \to U$ is one-one iff T maps every linearly independent subset of V into a linearly independent subset of U.

Proof. Let T be one-one and $\{x_1, x_2, \ldots, x_n\}$ be a linearly independent subset of V, then we show that $\{T(x_1), T(x_2), \ldots, T(x_n)\}$ is a linearly independent subset of U. Let

$$\sum_{i=1}^{n} \lambda_i T(x_i) = 0$$

$$\Rightarrow \qquad T\left(\sum_{i=1}^{n} \lambda_i x_i\right) = 0,$$

but T is one-one $\Rightarrow \sum_{i=1}^{n} \lambda_i x_i = 0$, but $\{x_1, x_2, \ldots, x_n\}$ is linearly independent $\Rightarrow \lambda_i = 0 \ \forall\ i$. Hence $\{T(x_1), T(x_2), \ldots, T(x_n)\}$ is linearly independent.

Conversely, let $x \in N(T)$ and suppose $x \neq 0$, then $\{x\}$ is linearly independent $\Rightarrow$ it can be extended to a basis $A = \{x, x_2, \ldots, x_r\}$ of V for some $r = \dim V$.

Hence, by hypothesis $\{T(x_1), T(x_2), \ldots, T(x_r)\} = \{0, T(x_2), \ldots, T(x_r)\}$ is linearly independent, which is not possible $\Rightarrow x = 0 \Rightarrow N(T) = 0$. Hence, T is one-one.

Corollary 4.2.8. Let $T : V \to U$ be a linear transformation. If any two of the following conditions hold, then the third is also true:

(*i*) $\dim V = \dim U$

(*ii*) $n(T) = 0$

(*iii*) $r(T) = \dim U$.

Proof. Appealing to

$$\dim V = r(T) + n(T)$$

we get the result.

As defined in Chapter 3, we define:

Definition 4.2.9. A linear transformation $T : V \to U$ be such that T is one-one and onto, then T is called isomorphism and we write $V \cong U$ (i.e., V is isomorphic to U).

Theorem 4.2.10. Any two vector spaces over the same field are isomorphic iff they have the same dimension.

Proof. Let V and U be vector spaces over F such that $V \cong U \Rightarrow$ there exists a linear transformation $T : V \to U$, such that T is one-one and onto. Hence, by the above corollary $\dim V = \dim U$.

Conversely, let $\{v_1, v_2, \ldots, v_n\}$ and $\{u_1, u_2, \ldots, u_n\}$ be bases of V and U, respectively, and define $T : V \to U$ such that

$$T\left(\sum_{i=1}^{n} \lambda_i v_i\right) = \sum_{i=1}^{n} \lambda_i u_i$$

Then, trivially, T is a well defined map. Also, for any $x, y \in V$, we have $x = \sum_{i=1}^{n} \lambda_i v_i$, $y = \sum_{i=1}^{n} \alpha_i v_i$ in a unique way

$$\Rightarrow \qquad T(x+y) = T\left(\sum_{i=1}^{n} (\lambda_i + \alpha_i)\, v_i\right)$$

$$= \sum_{i=1}^{n} (\lambda_i + \alpha_i)\, u_i$$

$$= \sum_{i=1}^{n} \lambda_i u_i + \sum_{i=1}^{n} \lambda_i u_i$$

$$= T(x) + T(y).$$

Similarly, $T(\lambda x) = \lambda T(x)$. Further, if

$$T(x) = 0$$

$$\Rightarrow \qquad \sum_{i=1}^{n} \lambda_i u_i = 0$$

$$\Rightarrow \qquad \lambda_i = 0 \text{ as } \{u_1, u_2, \ldots, u_n\}$$

is linearly independent $\Rightarrow x = 0 \Rightarrow T$ is one-one. Obviously, T is onto $\Rightarrow$ T is an isomorphism of V onto U.

Corollary 4.2.11. If V is an n-dimensional vector space over F, then $V \cong F^n$.

3 MATRIX REPRESENTATION OF A LINEAR TRANSFORMATION

We discussed linear transformation and different examples in Sections 1 and 2.

In this section, we first consider an example.

Let $T : \mathbb{R}^2 \to \mathbb{R}^2$ be given as

$T(x, y) = (x + 2y, y - 3x, 2x + 5y)$.

It is trivial to see that T is a linear transformation and we can rewrite same as

$$T(x, y) = \begin{bmatrix} 1 & 2 \\ -3 & 1 \\ 2 & 5 \end{bmatrix} \begin{bmatrix} x \\ y \end{bmatrix}.$$

So we can see that for any linear transformation we can get a matrix. Now, the question: What is the matrix representing a linear transformation? For this purpose, let V and U be finite dimensional vector spaces over the same field F and dim $V = n$, dim $U = m$ and $T : V \to U$ be a linear transformation. Let $B = \{v_1, v_2, \ldots, v_n\}$ and $B' = \{u_1, u_2, \ldots, u_m\}$ be bases of V and U, respectively. Now $T(v_i) \in U$, so $T(v_i) = \sum_{j=1}^{m} a_{ij}u_j$ for unique $\alpha_{ij} \in F$. The matrix $A = [\alpha_{ij}]^t$ is called the matrix representation of the linear transformation T with respect to the given bases. A is of order $m \times n$ denoted by $m(T)_{B,B'}$.

Remarks:

1. The matrix of the linear transformation is different if the positions of the basis vectors are changed. Therefore, to find the matrix of a linear transformation, one considers the ordered basis (*i.e.,* once a basis is chosen the positions of the elements in the basis are fixed).
2. The matrix of the linear transformation changes according to the change of basis. However, the matrix of zero transformation is always zero for every choice of basis.
3. Whenever the basis is not given, one is free to choose any basis to calculate the matrix of the linear transformation. However, in case of vector spaces $\mathbb{R}^n$ for different n, if the basis is not given, it is practice and customary to choose the standard bases.
4. Since the basis is not unique, the matrix representation always depends upon the choice of the basis. The matrix of identity transformation is identity matrix with respect to the same basis, otherwise it is different from the identity matrix.

Example 4.3.1

1. Let $T : \mathbb{R}^2 \to \mathbb{R}^2$, such that

$$T(x, y) = (x - y, x + y) = \begin{bmatrix} 1 & -1 \\ 1 & 1 \end{bmatrix}\begin{bmatrix} x \\ y \end{bmatrix}.$$

If we take standard basis $\{(1, 0), (0, 1)\}$, then it is easy to see that

$$m(T) = \begin{bmatrix} 1 & -1 \\ 1 & 1 \end{bmatrix}$$

which follows from the definition itself. Now, suppose we want to calculate $m(T)$ with respect to the basis $\{(1, 2), (2, 1)\}$ for both the vector spaces (which are the same), then

$$T(1, 2) = (-1, 3) = \frac{7}{3}(1, 2) + \left(-\frac{5}{3}(2, 1)\right)$$

and

$$T(2, 1) = (1, 3) = \frac{5}{3}(1, 2) + \left(-\frac{1}{3}(2, 1)\right)$$

$$\Rightarrow \qquad m(T) = \frac{1}{3}\begin{bmatrix} 7 & 5 \\ -5 & -1 \end{bmatrix}.$$

2. Let $T : \mathbb{R}^3 \to \mathbb{R}^3$, such that

$$T(x, y, z) = (2x - y + z, y - z, x + 2z)$$
$$= \begin{bmatrix} 2 & -1 & 1 \\ 0 & 1 & -1 \\ 1 & 0 & 2 \end{bmatrix}\begin{bmatrix} x \\ y \\ z \end{bmatrix},$$

then with respect to the standard basis of $\mathbb{R}^3$, the matrix of T,

$$m(T) = \begin{bmatrix} 2 & -1 & 1 \\ 0 & 1 & -1 \\ 1 & 0 & 2 \end{bmatrix}.$$

However, consider $\{\underbrace{(1, 1, 0)}_{f_1}, \underbrace{(1, 0, -1)}_{f_2}, \underbrace{(1, 1, 1)}_{f_3}\}$ as a basis of ^{3},

then

$$\begin{aligned} T(1, 1, 0) &= (1, 1, 1) = 0f_1 + 0f_2 + 1f_3 \\ T(1, 0, -1) &= (1, 1, -1) = 2f_1 + 0f_2 + (-1)f_3 \\ T(1, 1, 1) &= (2, 0, 3) = -5f_1 + 2f_2 + 5f_3 \end{aligned}$$

$$\Rightarrow \qquad m(T) = \begin{bmatrix} 0 & 2 & -5 \\ 0 & 0 & 2 \\ 1 & -1 & 5 \end{bmatrix}.$$

3. Let $T : \mathbb{R}^3 \to \mathbb{R}^2$, such that

$$T(x, y, z) = (x - z, x + y)$$
$$= \begin{bmatrix} 1 & 0 & -1 \\ 1 & 1 & 0 \end{bmatrix}\begin{bmatrix} x \\ y \\ z \end{bmatrix},$$

then with respect to the standard basis $\mathbb{R}^3$ and $\mathbb{R}^2$, trivially the matrix of T,

$$m(T) = \begin{bmatrix} 1 & 0 & -1 \\ 1 & 1 & 0 \end{bmatrix}.$$

Now, let us consider a new set of bases for both the vector spaces. Let $\{f_1 = (1, 1, 0), f_2 = (1, 0, -1), f_3 = (1, 1, 1)\}$ and $\{e_1 = (2, 1)\}$ be basis of $\mathbb{R}^3$ and $\mathbb{R}^2$, respectively, then

$$T(f_1) = (1, 2) = 1e_1 + 0e_2$$
$$T(f_2) = (2, 1) = 0e_1 + 1e_2$$
$$T(f_3) = (1, 2) = \frac{4}{3}(e_1) + \left(-\frac{2}{3}\right)e_2$$

$$\Rightarrow \qquad m(T) = \begin{bmatrix} 1 & 0 & \frac{4}{3} \\ 0 & 1 & -\frac{2}{3} \end{bmatrix} = \frac{1}{3}\begin{bmatrix} 3 & 0 & 4 \\ 0 & 3 & -2 \end{bmatrix}.$$

4. Let $T : P_2(x) \to P_2(x)$, such that $T(f(x)) = \frac{d}{dx}(f(x))$, then considering $\{1, x, x^2\}$ as a basis of $P_2(x)$, trivially

$$m(T) = \begin{bmatrix} 0 & 1 & 0 \\ 0 & 0 & 2 \\ 0 & 0 & 0 \end{bmatrix}.$$

If we consider $\{1 - x, 1 + x, 1 + x^2\}$ as a basis of $P_2(x)$, then $m(T)$ with respect to this basis is $\frac{1}{2}\begin{bmatrix} -1 & 1 & -2 \\ -1 & 1 & 2 \\ 0 & 0 & 0 \end{bmatrix}$.

Therefore, we have seen that whenever the bases are changed, the matrix representation of a linear transformation is different.

Now we discuss the situation when the bases are changed and see that the change of bases makes the corresponding matrices are similar.

Theorem 4.3.2. If $T : V \to V$ is a linear transformation and $\{x_1, x_2, \ldots, x_n\}$ and $\{y_1, y_2, \ldots, y_n\}$ are bases of V if $A = (\alpha_{ij})$ and $B = (\beta_{ij})$ are matrices of T with respect to $\{x_1, x_2, \ldots, x_n\}$ and $\{y_1, y_2, \ldots, y_n\}$, respectively, then A and B are similar matrices.

Proof. Let $T(x_i) = \sum_{j=1}^{n} \alpha_{ji} x_i$ and $T(y_i) = \sum_{j=1}^{n} \beta_{ji} y_i$. Let $P = (p_{ij})$ be the transition matrix from $\{x_1, x_2, \ldots, x_n\}$ to $\{y_1, y_2, \ldots, y_n\}$, then

$$x_j = \sum_{k=1}^{n} p_{kj} y_k$$

$$\Rightarrow \qquad T(x_i) = \sum_{j=1}^{n} p_{ji}\, T(y_j)$$

$$= \sum_{j=1}^{n} p_{ji} \sum_{k=1}^{n} \beta_{kj} y_k$$

$$= \sum_{k=1}^{n} \left(\sum_{j=1}^{n} \beta_{kj} p_{ji} \right) y_k \qquad \ldots(1)$$

Also since

$$T(x_i) = \sum_{j=1}^{n} \alpha_{ji} x_j$$

$$= \sum_{j=1}^{n} \alpha_{ji} \sum_{k=1}^{n} p_{kj} y_k$$

$$= \sum_{k=1}^{n} \left(\sum_{j=1}^{n} p_{kj} \alpha_{ji} \right) y_k \qquad \ldots(2)$$

Since $\{y_1, y_2, \ldots, y_n\}$ is linearly independent, so from (1) and (2), we get

$$\sum_{j=1}^{n} \beta_{kj} p_{ji} = \sum \quad {}_{kj} \quad {}_{ji}, \qquad \ldots(3)$$

but left hand side of (3) is $(k, i)^{\text{th}}$ element of PA and since it is true for all $k = 1, 2, \ldots, n$, $i = 1, 2, \ldots, n$

$$\Rightarrow \qquad BP = PA$$

$$\Rightarrow \qquad B = PAP^{-1}.$$

Since we know that the transition matrices are invertible

$\Rightarrow A$ and B are similar matrices.

Now, taking different vector spaces V and U, we do the following theorem which is on similar lines as Theorem 4.3.2 and show that the matrices of the linear transformation with the choice of two sets of bases are equivalent.

Theorem 4.3.3. If V and U are vector spaces of dimensions n and m respectively and $T : V \to U$ is a linear transformation; if $\{x_1, x_2, \ldots, x_n\}$ and $\{x\ _1, x\ _2, \ldots, x\ _n\}$ are bases of V and $\{y_1, y_2, \ldots, y_m\}$ and $\{y\ _1, y\ _2, \ldots, y\ _m\}$ are bases of U, then the matices of T with respect to $\{x_1, x_2, \ldots, x_n\}$, $\{y_1, y_2, \ldots, y_m\}$, $\{x\ _1, x\ _2, \ldots, x\ _n\}$ are equivalent.

Proof. Let $A = (\alpha_{ij})$ be the matrix of T with respect to (x_i) and (y_i) bases and $B = (\beta_{ij})$ be the matrix of T with respect to $(x\ _i)$ and $(y\ _i)$. Then by definition

$$T(x_i) = \sum_{j=1}^{m} \alpha_{ji} y_j \text{ and } T(x'_i) = \sum_{k=1}^{m} \beta_{kj} y'_k .$$

Let P and Q be the transition matrices from (x_i) to (x'_i) and (y_i) to (y'_i), respectively. Then by definition of transition matrices

$$x_i = \sum_{j=1}^{n} p_{ji} x'_j \text{ and } y_j = \sum_{k=1}^{n} q_{kj} y'_k$$

$$\Rightarrow \quad T(x_i) = \sum_{j=1}^{n} p_{ji} T(x'_j)$$

$$= \sum_{j=1}^{n} p_{ji} \sum_{k=1}^{m} \beta_{kj} y'_k$$

$$= \sum_{k=1}^{n} \left(\sum_{j=1}^{n} \beta_{kj} p_{ji} \right) y'_k \qquad \ldots(1)$$

Also

$$T(x_i) = \sum_{j=1}^{m} \alpha_{ji} \left(\sum_{k=1}^{m} q_{kj} y'_k \right)$$

$$= \sum_{k=1}^{m} \left(\sum_{j=1}^{m} q_{kj} \alpha_{ji} y'_k \right). \qquad \ldots(2)$$

Since $\{y'_1, y'_2, \ldots, y'_m\}$ is linearly independent, therefore, appealing to (1) and (2), we get

$(k, i)^{\text{th}}$ entry of $BP = (k, i)^{\text{th}}$ entry of QA

$$\Rightarrow \quad BP = QA$$

$$\Rightarrow \quad B = QAP^{-1}.$$

Since transition matrix P is non-singular

$\Rightarrow A$ and B are equivalent.

Now, we discuss some examples elaborating Theorems 4.3.2 and 4.3.3.

Examples 4.3.4

1. Let $T : \mathbb{R}^2 \to \mathbb{R}^2$, such that $T(x, y) = (x + 3y, x - y)$ and let $B = \{(1, 0), (0, 1)\}$ and $B' = \{(1, 2), (2, 1)\}$ be bases of $\mathbb{R}^2$, then with respect to B,

$$m(T) = \begin{bmatrix} 1 & 3 \\ 1 & -1 \end{bmatrix}$$

and with respect to B',

$$m(T) = \begin{bmatrix} -3 & -1 \\ 5 & 3 \end{bmatrix}.$$

Now, the transition matrix from B' to B is

$$\begin{bmatrix} 1 & 2 \\ 2 & 1 \end{bmatrix} = P.$$

Then we can see that

$$P\begin{bmatrix} -3 & -1 \\ 5 & 3 \end{bmatrix} = \begin{bmatrix} 1 & 3 \\ 1 & -1 \end{bmatrix} P$$

$$\Rightarrow \begin{bmatrix} 1 & 3 \\ 1 & -1 \end{bmatrix} \text{ and } \begin{bmatrix} -3 & -1 \\ 5 & 3 \end{bmatrix} \text{ are similar matrices.}$$

2. Let $T : \mathbb{R}^2 \to \mathbb{R}^2$, such that $T(x, y) = (x \cos\theta + y \sin\theta, -x \sin\theta + y \cos\theta)$, then $m(T)$ with respect to the standard basis is

$$\begin{bmatrix} \cos\theta & \sin\theta \\ -\sin\theta & \cos\theta \end{bmatrix}$$

and its inverse is

$$\begin{bmatrix} \cos\theta & -\sin\theta \\ \sin\theta & \cos\theta \end{bmatrix}$$

$\Rightarrow T^{-1}(x, y) = (x \cos\theta - y \sin\theta, x \sin\theta + y \cos\theta).$

3. Let $T : \mathbb{R}^3 \to \mathbb{R}^2$, such that $T(x, y, z) = (x + z, y + z)$ and $\{(1, 0, 0), (0, 1, 0), (0, 0, 1), (0, 0, 1)\}$ and $\{(1, 0), (0, 1)\}$ be the standard bases of $\mathbb{R}^3$ and $\mathbb{R}^2$ respectively, then

$$m(T) = \begin{bmatrix} 1 & 0 & 1 \\ 0 & 1 & 1 \end{bmatrix} = A \text{ (say)},$$

Now suppose $\{(1, 0, 1), (1, 0, 0), (1, 1, 1)\}$ and $\{(1, 1), (1, -1)\}$ are other bases of $\mathbb{R}^3$ and $\mathbb{R}^2$, respectively, then

$$T(1, 0, 1) = (2, 1) = \frac{3}{2}(1, 1) + \frac{1}{2}(1, -1)$$

$$T(1, 0, 0) = (1, 0) = \frac{1}{2}(1, 1) + \frac{1}{2}(1, -1)$$

$$T(1, 1, 1) = (2, 2) = 2(1, 1) + 0(1, -1)$$

$$\Rightarrow \qquad m(T) = \begin{bmatrix} \frac{3}{2} & \frac{1}{2} & 2 \\ \frac{1}{2} & \frac{1}{2} & 0 \end{bmatrix}$$

$$= \frac{1}{2}\begin{bmatrix} 3 & 1 & 4 \\ 1 & 1 & 0 \end{bmatrix} = B \text{ (say)}.$$

Now, transition matrix from {(1, 0, 1), (1, 0, 0), (1, 1, 1)} to {(1, 0, 0), (0, 1, 0), (0, 0, 1)} is

$$\begin{bmatrix} 1 & 1 & 1 \\ 0 & 0 & 1 \\ 1 & 0 & 1 \end{bmatrix} = P \text{ (say)}.$$

Also, the transition matrix from {(1, 1), (1, – 1)} to {(1, 0), (0, 1)} is

$$\begin{bmatrix} 1 & 1 \\ 1 & -1 \end{bmatrix} = Q \text{ (say)}.$$

Now, we can see that

$$\begin{bmatrix} 1 & 0 & 1 \\ 0 & 1 & 1 \end{bmatrix} P = QB \text{ i.e., } AP = QB$$

$$\Rightarrow \qquad B = Q^{-1}AP,$$

where P and Q are non-singular

$\Rightarrow A$ and B are equivalent.

4. Let $T : C_{\mathbb{R}}^2 \to C_{\mathbb{R}}^2$ given as $T(a + ib, c + id) = T(a - ib, c - id)$, then trivially T is a linear transformation. Now choose a basis $\{(1, 0), (0, 1), (i, 0), (0, i)\}$, then

$$T(1, 0) = 1(1, 0) + 0(0, 1) + 0(i, 0) + 0(0, i)$$
$$T(0, 1) = 0(1, 0) + 1(0, 1) + 0(i, 0) + 0(0, i)$$
$$T(i, 0) = 0(1, 0) + 0(0, 1) + (-1)(i, 0) + 0(0, i)$$
$$T(0, i) = 0(1, 0) + 0(0, 1) + 0(i, 0) + (-1)(0, i)$$

$$\Rightarrow \qquad m(T) = \begin{bmatrix} 1 & 0 & 0 & 0 \\ 0 & 1 & 0 & 0 \\ 0 & 0 & -1 & 0 \\ 0 & 0 & 0 & -1 \end{bmatrix}$$

5. Let $T : M_2(\mathbb{R}) \to P_2(x)$, such that

$$T\left(\begin{bmatrix} a & b \\ c & d \end{bmatrix}\right) = (a + c) + (a - d)x + (b + c)x^2.$$

Then, trivially T is a linear transformation. Choosing

$$\left\{\begin{bmatrix} 1 & 0 \\ 0 & 0 \end{bmatrix}, \begin{bmatrix} 0 & 0 \\ 1 & 0 \end{bmatrix}, \begin{bmatrix} 0 & 1 \\ 0 & 0 \end{bmatrix}, \begin{bmatrix} 0 & 0 \\ 0 & 1 \end{bmatrix}\right\}$$

and $\{1, x, x^2\}$ as bases of $M_2(\mathbb{R})$ and $P_2(x)$, we have

$$T\left(\begin{bmatrix} 1 & 0 \\ 0 & 0 \end{bmatrix}\right) = 1 + x = 1 \cdot 1 + 1 \cdot x + 0 \cdot x^2$$

$$T\left(\begin{bmatrix} 0 & 0 \\ 1 & 0 \end{bmatrix}\right) = 1 + x^2 = 1 \cdot 1 + 0 \cdot x + 1 \cdot x^2$$

$$T\left(\begin{bmatrix} 0 & 1 \\ 0 & 0 \end{bmatrix}\right) = x^2 = 0 \cdot 1 + 0 \cdot x + 1 \cdot x^2$$

$$T\left(\begin{bmatrix} 0 & 0 \\ 0 & 1 \end{bmatrix}\right) = -x = 0 \cdot 1 + (-1) \cdot x + 0 \cdot x^2$$

$$\Rightarrow \qquad m(T) = \begin{bmatrix} 1 & 1 & 0 & 0 \\ 1 & 0 & 0 & -1 \\ 0 & 1 & 1 & 0 \end{bmatrix}$$

6. Let $T : P_2(x) \to P_3(x)$, such that

$$T(f(x)) = f(x) + 5\int_0^x f(t)\, dt$$

then, trivially T is a linear transformation. Choosing $\{1, 1 + x, 1 - x^2\}$ and $\{1, x, x^2, x^3\}$ as bases of $P_2(x)$ and $P_3(x)$, respectively we calculate $m(T)$

$$T(1) = 1 + 5x,$$

$$T(1 + x) = 1 + x + 5\left(x + \frac{x^2}{2}\right)$$

$$= 1 + 6x + \frac{x^2}{2}.$$

$$T(1 - x^2) = 1 - x^2 + 5\left(x - \frac{x^3}{3}\right)$$

$$= 1 + 5x + (-1)x^2 - \frac{5}{3}x^3$$

so

$$m(T) = \begin{bmatrix} 1 & 1 & 1 \\ 5 & 6 & 5 \\ 0 & \frac{1}{2} & -1 \\ 0 & 0 & -\frac{5}{3} \end{bmatrix}$$

As seen in above examples, given by linear transformation we can obtain a matrix representing the linear transformation. The natural question: "Given any matrix, can we find a linear transformation representing the matrix?" will be exhibited in the rest of this section. Let $A = [a_{ij}]$ be an $m \times n$ matrix and V and U be vector spaces of dimensions n and m, respectively. Let $\{v_1, v_2, \ldots, v_n\}$, $\{u_1, u_2, \ldots, u_m\}$ be bass of V and U, respectively. Define

$$T(v_i) = \sum_{j=1}^{m} a_{ji} u_j \text{ and } T(x) = \sum_{i=1}^{n} \lambda_i \, T(v_i),$$

where $x \in V$ and $x = \sum_{i=1}^{n} \lambda_i v_i$, then T is a linear transformation (verify) corresponding to the matrix A and with respect to the chosen bases.

In case no specific basis is given then we always because the standard basis.

Example 4.3.5

1. If

$$A = \begin{bmatrix} 1 & -1 & 2 \\ 0 & 3 & 1 \\ 1 & -2 & 5 \end{bmatrix},$$

then with respect to standard basis of $\mathbb{R}^2$ we have

$$T(x, y, z) = (x - y + 2z, 3y + z, x - 2y + 5z)$$

as desired linear transformation.

If we have to find the linear transformation with respect to the basis $\{(0, 1, 1), (1, 1, 0), (1, 0, 1)\}$, then for any $(x, y, z) \in {}^3$, we have

$$(x, y, z) = \frac{y-x+z}{2}(0,1,1) + \frac{x+y-z}{2}(1,1,0) + \frac{z-y+x}{2}(1,0,1)$$

$$\Rightarrow \quad T(x, y, z) = \frac{y-x+z}{2}T(0,1,1) + \frac{x+y-z}{2}T(1,1,0) + \frac{z-y+z}{2}T(1,0,1).$$

As given, we also have

$$T(0, 1, 1) = 1(0, 1, 1) + 0(1, 1, 0) + 1(1, 0, 1) = (1, 1, 2)$$
$$T(1, 1, 0) = -1(0, 1, 1) + 3(1, 1, 0) + 2(1, 0, 1) = (1, 2, -3)$$
$$T(1, 0, 1) = 2(0, 1, 1) + 1(1, 1, 0) + 5(1, 0, 1) = (6, 3, 7)$$

$$\Rightarrow T(x, y, z) = \frac{y-x+z}{2}(1,1,2) + \frac{x+y-z}{2}(1,2,-3) + \frac{z-y+x}{2}(6,3,7)$$

$$= \left(\frac{-4y+6x+6z}{2}, \frac{4x+2z}{2}, \frac{2x-8y+12z}{2}\right)$$

$$= (3x - 2y + 3z, 2x + z, x - 4y + 6z)$$

2. If

$$A = \begin{bmatrix} 5 & -2 \\ 1 & 3 \end{bmatrix}$$

is a matrix representation of a linear transformation $T : \mathbb{C}^2 \to \mathbb{C}^2$, where $\mathbb{C}^2$ is a vector space over $\mathbb{C}$, with respect to basis $\{(1, 1), (1, 0)\}$, then let us find T. Now

$$T(1, 1) = 5(1, 1) + 1(1, 0) = (6, 5)$$
$$T(1, 0) = -2(1, 1) + 3(1, 0) = (1, -2)$$

Also

$$(x, y) = y(1, 1) + (x - y)(1, 0)$$
$$\Rightarrow \quad T(x, y) = y(6, 5) + (x - y)(1, -2)$$
$$= (x + 5y, 7y - 2x)$$

3. If

$$A = \begin{bmatrix} 1 & -1 & 2 \\ -2 & 1 & -1 \\ 1 & 2 & 3 \end{bmatrix}$$

is the matrix representation of a linear transformation $T : P_2(x) \to P_2(x)$ with respect to the bases $\{1 - x, x(1 - x), x(1 + x)\}$ and $\{1, 1 + x, 1 + x^2\}$, then let us find T. Now

$$T(1 - x) = 1 - 2(1 + x) + 1 + x^2 = -2x + x^2$$
$$T(x(1 - x)) = -1 + 1(1 + x) + 2(1 + x^2) = 2 + x + 2x^2$$
$$T(x(1 + x)) = 2 - 1(1 + x) + 3(1 + x^2) = 4 - x + 3x^2$$

Let $f(x) \in P_2(x)$, such that $f(x) = a + bx + cx^2$, then

$$f(x) = a(1 - x) + \frac{a + b - c}{2}(x - x^2) + \frac{a + b + c}{2}(x + x^2)$$

$$\Rightarrow T(f(x)) = a(-2x + x^2) + \frac{a + b - c}{2}(2 + x + 2x^2)$$

$$+ \frac{a + b + c}{2}(4 - x + 3x^2)$$

$$= (3a + 3b + 3c) + \left(-2a + \frac{a + b - c}{2} - \frac{a + b + c}{2}\right)x$$

$$+ \left(a + a + b - c + \frac{3a + 3b + 3c}{2}\right)x^2$$

$$= (3a + 3b + c) + \left(\frac{-4a - 2c}{2}\right)x + \left(\frac{7a + 5b + c}{2}\right)x^2$$

$$= (3a + 3b + c) + (-2a - c)x + \left(\frac{7a + 5b + c}{2}\right)x^2$$

4. If

$$A = \begin{bmatrix} 2 & 1 & 3 \\ -1 & 0 & 4 \\ -2 & 1 & -1 \end{bmatrix}$$

is a matrix of a linear transformation $T : \mathbb{R}^3 \to \mathbb{R}^3$ with respect to the standard basis, then let us find the matrix of T with respect to the basis $\{(1, 1, -1), (-1, 1, 1), (1, -1, 0)\}$. Trivially with respect to the standard basis

$$T(x, y, z) = (2x + y + 3z, -x + 4z, -2x + y - z).$$

Now,

$$T(1, 1, -1) = (0, -5, 0)$$
$$= -\frac{5}{2}(1, 1, -1) - \frac{5}{2}(-1, 1, 1) + 0(1, -1, 0)$$

$$\begin{aligned} T(-1, 1, 1) &= (2, 5, 2) \\ &= \frac{7}{2}(1, 1, -1) + \frac{11}{2}(-1, 1, 1) + 4(1, -1, 0) \\ T(1, 1, -1) &= (1, -1, -3) \\ &= 0(1, 1, -1) + (-3)(-1, 1, 1) + (-2)(1, -1, 0) \end{aligned}$$

$$\Rightarrow \qquad m(T) = \begin{bmatrix} -\frac{5}{2} & \frac{7}{2} & 0 \\ -\frac{5}{2} & \frac{11}{2} & 0 \\ 0 & 4 & -2 \end{bmatrix}$$

$$= \frac{1}{2}\begin{bmatrix} -5 & 7 & 0 \\ -5 & 11 & -6 \\ 0 & 8 & -4 \end{bmatrix}$$

4 SOME WORKED OUT EXAMPLES

1. Find a linear transformation $T : \mathbb{R}^3 \to \mathbb{R}^3$ such that $T(e_1) = e_1 - e_2$, $T(e_2) = e_1 + e_3$, $T(e_3) = e_2 - e_3$, where e_1, e_2, e_3 are the standard bases of $\mathbb{R}^3$.

$$\begin{aligned} T(x, y, z) &= xT(1, 0, 0) + yT(0, 1, 0) + zT(0, 0, 1) \\ &= x(1, -1, 0) + y(1, 0, 1) + z(0, 1, -1) \\ &= (x + y, z - x, y - z) \end{aligned}$$

2. Find a linear transformation $T : \mathbb{R}^3 \to \mathbb{R}^4$ which maps (1, 2, 1), (1, 1, 1), (1, 0, 3) to (1, 0, 1, 1), (2, 1, 1, 0) and (0, 2, 1, 0), respectively. Trivially {(1, 2, 1), (1, 1, 1),(1, 0, 3)} is a basis of $\mathbb{R}^3$ (verify). Let $(x, y, z) \in {}^3$ such that $(x, y, z) = a(1, 2, 1) + b(1, 1, 1) + c(1, 0, 3)$, then $x = a + b + c$, $y = 2a + b$, $z = a + b + 3c$, after solving, we get

$$(x, y, z) = \frac{2y + z - 3x}{2}(1, 2, 1) + (3x - y - z)(1, 1, 1) + \frac{z - x}{2}(1, 0, 3)$$

$$\Rightarrow T(x, y, z) = \frac{2y + z - 3x}{2}T(1, 2, 1) + (3x - y - z)T(1, 1, 1) + \frac{z - x}{2}T(1, 0, 3)$$

$$= \frac{2y+z-3x}{2}(1,0,1,1) + (3x-y-z)(2,1,1,0) + \frac{z-x}{2}(0,2,1,0)$$

$$= \left(\frac{-9x-2y-3z}{2}, 2x-y, \frac{2y+z-3x}{2}\right)$$

$$= \frac{1}{2}(-9x-2y-3z, 4x-2y, 0, 2y+z-3x)$$

3. Find a linear transformation $T : \mathbb{R}^2 \to \mathbb{R}^3$, such that
   (*a*) T maps $y = mx$ onto origin.
   (*b*) T maps $y = mx$ onto itself, $m \neq 0$.
   (*i*) Define $T : \mathbb{R}^2 \to \mathbb{R}^2$, such that $T(x, y) = (y - mx, 0)$, then if (x, y) lies on $y = mx$, we have $T(x, y) = (0, 0)$.
   (*ii*) $(1, m)$ lies on the line $y = mx$. Trivially $\{(1, m), (1, 0)\}$ is a basis of ^{2}. Define $T : \mathbb{R}^2 \to \mathbb{R}^2$ such that $T(1, m) = (1, m)$ and $T(1, 0) = (0, 0)$, then $T(x, y) = \left(\frac{y}{m}, y\right)$ is a linear transformation which maps $y = mx$ onto itself.
4. Find a linear transformation $T : \mathbb{R}^3 \to \mathbb{R}^3$, such that
   (*a*) it maps the plane $x + 3y - z = 0$ onto origin.
   (*b*) it maps $x + 3y - z = 0$ onto itself.
   (*i*) Define $T : \mathbb{R}^3 \to \mathbb{R}^3$ such that $T(x, y, z) = (x + 3y - z, 0, 0)$, then trivially T maps $W = \{(x, y, z) \mid x + 3y - z = 0\}$ onto origin.
   (*ii*) Choose $\{(1, 0, 1), (-3, 1, 0)\}$ as a basis of W. Define $T : \mathbb{R}^3 \to \mathbb{R}^3$ such that $T(1, 0, 1) = (2, 0, 2)$, $T(-3, 1, 0) = (1, 0, 1)$ and $T(1, 0, 0) = (0, 0, 0)$, where $\{(1, 0, 1), (-3, 1, 0), (1, 0, 0)\}$ can be taken to be a basis of $\mathbb{R}^3$. Then

$$\begin{aligned} T(x, y, z) &= zT(1, 0, 1) + yT(-3, 1, 0) \\ &= (2x + y, 0, 2z + y). \end{aligned}$$

   Then T maps W onto W. Or, only seeing the equation of the plane, we can find $T : \mathbb{R}^3 \to \mathbb{R}^3$ such that $T(x, y, z) = (x, 0, x)$ but in both the cases $T : \mathbb{R}^3 \to W \subseteq \mathbb{R}^3$. However if we take $T(1, 0, 0) = (1, 0, 0)$, $T(1, 0, 1) = (-3, 1, 0)$, $T(-3, 1, 0) = (1, 0, 1)$, then

$$(x, y, z) = z(1, 0, 1) + y(-3, 1, 0) + (x + 3y - z)(1, 0, 0)$$

   and

$$\begin{aligned} T(x, y, z) &= z(-3, 1, 0) + y(1, 0, 1) + (x + 3y - z)(1, 0, 0) \\ &= (x + 4y - 4z, z, y). \end{aligned}$$

It is trivial to see that $n(T) = 0$ and $r(T) = 3$ and $TW \subseteq W$.

5. Let $T : P_2(x) \to P_2(x)$, such that $T(ax^2 + bx + c) = (a - c)x + (b - c)$, then trivially T is a linear transformation and
$$N(T) = \{ax^2 + bx + c \mid a = c = b\}$$
$\Rightarrow N(T)$ is spanned by $\{1 + x + x^2\}$
$\Rightarrow n(T) = 1$ and $r(T) = 2$.

6. Let $T : \mathbb{R}^3 \to \mathbb{R}^3$, given as $T(x, y, z) = (x - y, x + 2y, z - x)$, then T is a linear transformation and $n(T) = 0$ and $r(T) = 3$. Check that T maps the plane $W = \{(x, y, z) \mid x + y + z = 0\}$ onto itself.

7. Let $T : \mathbb{R}^4 \to \mathbb{R}^3$ such that $T(x, y, z, w) = (x - y, z + w, x + z)$. Let us find the basis of $N(T)$ and $R(T)$.
$$N(T) = \{(x, y, z, w) \mid x = y = -z = w\}$$
$$\Rightarrow \qquad N(T) = \{(x, x, x, -x) \mid x \in \ \},$$
therefore $\{(1, 1, 1, -1)\}$ is a basis of $N(T) \Rightarrow n(T) = 1$ and $r(T) = 3$
$$\Rightarrow \qquad T\mathbb{R}^4 = \mathbb{R}^3$$
$\Rightarrow$ any basis of $\mathbb{R}^3$ is a basis of $R(T)$.

8. Let $T : \mathbb{R}^2 \to P_2(x)$, such that $T(a, b) = ax^2 + bx + a$. Then it is easy to show that T is a linear transformation.
$$N(T) = \{(a, b) \mid ax^2 + bx + a = 0\}$$
$$\Rightarrow \qquad N(T) = 0$$
$$\Rightarrow \qquad n(T) = 0 \text{ and } r(T) = 2.$$
Further, if we define $S : P_2(x) \to \mathbb{R}^2$ such that $S(ax^2 + bx + c) = (a, c)$, then $N(S) = [x]$
$$\Rightarrow \qquad n(S) = 1 \text{ and } r(S) = 2.$$

9. Let $T : \mathbb{R}^3 \to \mathbb{R}^3$ such that $T(x, y, z) = (x - y, x + z, 2y - z)$, then T is a linear transformation and $N(T) = 0$ and $n(T) = 0$, so $r(T) = 3$.

10. Let $T : \mathbb{R}^4 \to \mathbb{R}^4$, such that
$T(x, y, z, w) = (2x + 2y + z + w, 2x + y - z, x - z, 2x + y - z)$,
then trivially T is a linear transformation and $N(T) = \{(x, -x, x, -x) \mid x \in \mathbb{R}\}$, so $n(T) = 1$ and $\{(1, -1, 1, -1)\}$ is a basis of $N(T)$.

We can rewrite

$T(x, y, z, w) = x(2, 2, 1, 2) + y(2, 1, 0, 1) + z(1, -1, -1, -1) + w(1, 0, 0, 0)$

$\Rightarrow R(T)$ is spanned by $\{(2, 2, 1, 2), (1, -1, -1, -1), (2, 1, 0, 1), (1, 0, 0, 0)\}$.

Now, to determine the basis of $R(T)$ consider the matrix

$$A = \begin{bmatrix} 1 & 0 & 0 & 0 \\ 1 & -1 & -1 & -1 \\ 2 & 1 & 0 & 1 \\ 2 & 2 & 1 & 2 \end{bmatrix} \text{ which can be reduced to } \begin{bmatrix} 1 & 0 & 0 & 0 \\ 0 & 0 & 0 & 0 \\ 0 & 1 & 0 & 1 \\ 0 & 0 & 1 & 0 \end{bmatrix}$$

$\Rightarrow$ {(1, 0, 0, 0), (2, 1, 0, 1), (2, 2, 1, 2)} is a spanning set for $R(T)$. Hence the same is a basis of $R(T)$ as $r(T) = 3$.

11. Let $T : \mathbb{R}^3 \to \mathbb{R}^3$ such that $T(x, y, z) = (x + y, x + z, z - y)$, then trivially T is a linear transformation and

$$\begin{aligned} N(T) &= \{(x, y, z) \mid x + y = 0, x + z = 0, z = y\} \\ &= \{(x, -x, -x) \mid x \in \ \}. \end{aligned}$$

Hence, {(1, – 1, – 1)} can be taken as a basis of $N(T)$

$\Rightarrow$ $n(T) = 1$. We write

$$T(x, y, z) = x(1, 1, 0) + y(1, 0, -1) + z(0, 1, 1)$$

$\Rightarrow$ $R(T) = [(1, 1, 0), (1, 0, -1), (0, 1, 1)]$

Now consider

$$A = \begin{bmatrix} 1 & 1 & 0 \\ 1 & 0 & -1 \\ 0 & 1 & 1 \end{bmatrix} \sim \begin{bmatrix} 1 & 1 & 0 \\ 0 & -1 & -1 \\ 0 & 1 & 1 \end{bmatrix} \sim \begin{bmatrix} 1 & 1 & 0 \\ 0 & -1 & -1 \\ 0 & 0 & 0 \end{bmatrix}$$

Hence, {(1, 1, 0), (1, 0, – 1)} is also spanning set for $R(T)$

$\Rightarrow$ {(1, 1, 0), (1, 0, – 1)} is basis of $R(T)$

$\Rightarrow$ $r(T) = 2$.

EXERCISE

1. Which of the following are linear transformations:

(*a*) $T : \mathbb{R}^3 \to \mathbb{R}^2$ such that $T(x, y, z) = (x - y, z + \pi)$.

(*b*) $T : M_2(\mathbb{R}) \to \mathbb{R}^2$ such that $T\left(\begin{bmatrix} a & b \\ c & d \end{bmatrix}\right) = (a, d)$.

(*c*) $T : M_2(\mathbb{R}) \to \mathbb{R}$ such that $T\left(\begin{bmatrix} a & b \\ c & d \end{bmatrix}\right) = a - d + 2b$.

(*d*) $T : P_2(x) \to \mathbb{R}^2$ such that $T(ax^2 + bx + c) = (a, c)$.

(*e*) $T : \mathbb{C} \to \mathbb{C}$ such that $T(x + iy) = iy$, where $\mathbb{C}$ is the set of complex numbers.

(*f*) $T : M_n(\mathbb{R}) \to M_n(\mathbb{R})$ such that $T(A) = A^t$.

(*g*) $T : P(x) \to P(x)$ such that $T(f(x)) = f(x + 1) - f(x)$, where $P(x)$ is vector space of polynomials over $\mathbb{R}$.

(*h*) $T : P_1(x) \to P_2(x)$ such that $T(a + bx) = x(a + bx) + 2x^2$.

(*i*) $T : P_1(x) \to P_2(x)$ such that $T(f(x)) = xf(x) + 2f(1)$.

(*j*) $T : P_1(x) \to P_2(x)$ such that $T(f(x)) = x + f(1)$.

2. Find rank and nullity of the following linear transformations:

(*a*) $T : \mathbb{R}^4 \to {}^2$ such that $T(x_1, x_2, x_3, x_4) = (x_1 + x_2, x_3 + x_4)$.

(*b*) $T : \mathbb{R}^4 \to {}^2$ such that $T(x_1, x_2, x_3, x_4) = (x_1 - x_2, x_3 - x_4)$.

(*c*) $T : \mathbb{R}^4 \to {}^3$ such that $T(x_1, x_2, x_3, x_4) = (x_1 + x_2 + x_3, x_2 + x_3, x_4)$.

(*d*) $T : {}^3 \to {}^3$ such that $T(x, y, z) = (x - 2z, x + y - z, 2x + y + 4z)$.

(*e*) $T : P_2(x) \to {}^2$ such that $T(ax^2 + bx + c) = (a - c, b + c)$.

(*f*) $T : M_2(\mathbb{R}) \to M_2(\mathbb{R})$ such that $T\left(\begin{bmatrix} a & b \\ c & d \end{bmatrix}\right) = \begin{bmatrix} a-d & b-c \\ b+c & a+d \end{bmatrix}$

(*g*) $T : P_2(x) \to \mathbb{R}$ such that $T(f(x)) = \int_0^1 f(x)\, dx$

(*h*) $T : M_2(\mathbb{R}) \to \mathbb{R}$ such that $T\left(\begin{bmatrix} a & b \\ c & d \end{bmatrix}\right) = a - b + c$.

(*i*) $T : \mathbb{R}^3 \to \mathbb{R}^3$ such that $T(x, y, z) = (x + 2y + 2z, 2x + 3y + z, x + z)$.

(*j*) $T : \mathbb{R}^3 \to \mathbb{R}^3$ such that $T(x, y, z) = (2x + y + 3z, x + z, x + y + 2z)$.

3. Find a linear transformation $T : \mathbb{R}^2 \to \mathbb{R}^3$ such

$$T\,\mathbb{R}^2 = \{(x, y, z) \mid x + y + z = 0\}.$$

4. Find the matrix representation of the following linear transformations with respect to the given basis:

(*a*) $T : \mathbb{R}^2 \to \mathbb{R}^2$ such that $T(x, y) = (x - y, x + 2y, y + 2z)$ with respect to the bases $\{(1, 2), (2, 1)\}$ and $\{(1, 1, 0), (0, 1, 1), (0, 2, 3)\}$.

(*b*) $T : P_2(x) \to P_1(x)$ such that $T(f(x)) = \dfrac{d}{dx}(f(x))$ with respect to $\{1 + x^2, 1 - x, 1 + x\}$ and $\{2 + x, 2 - x\}$ as respective bases.

(*c*) $T : P_1(x) \to P_2(x)$ such that $T(f(x)) = xf(x)$ with respect to $\{1 - x, 1 + x\}$ and $\{2, 2 + x, 2 + x^2\}$ as respective bases.

(*d*) $T : \mathbb{R}^2 \to M_2(\mathbb{R})$ such that $T(x, y) = \begin{bmatrix} x - y & x + 2y \\ x - 3y & x + 4y \end{bmatrix}$ with respect to the standard bases.

(*e*) $T : M_2(\mathbb{R}) \to M_2(\mathbb{R})$ such that $T(A) = A^t$ with respect to

$$\left\{\begin{bmatrix} 1 & 1 \\ 0 & 0 \end{bmatrix}, \begin{bmatrix} 1 & 0 \\ 0 & 0 \end{bmatrix}, \begin{bmatrix} 1 & 1 \\ 0 & 1 \end{bmatrix}, \begin{bmatrix} 0 & 0 \\ -1 & 1 \end{bmatrix}\right\}$$

as basis of $M_2(\mathbb{R})$.

5. If

$$A = \begin{bmatrix} 1 & 2 & -1 \\ 3 & 0 & 5 \\ 2 & -1 & 3 \end{bmatrix}$$

is a matrix of a linear transformation $T : \mathbb{R}^3 \to \mathbb{R}^3$ with respect to the standard basis $\{e_1, e_2, e_3\}$ of $\mathbb{R}^3$, then find the matrix of T with respect to $\{e_1, e_1 + e_2, e_1 + e_2 + e_3\}$ and further find T.

6. If

$$A = \begin{bmatrix} 2 & 0 & 3 \\ 1 & 2 & 4 \\ -3 & 1 & -5 \end{bmatrix}$$

is the matrix of a linear transformation $T : \mathbb{R}^3 \to \mathbb{R}^3$ with respect to $\{e_1, e_1 + e_2, e_1 + e_2 + e_3\}$ as a basis of $\mathbb{R}^3$, then find T and hence find the matrix of T with respect to the standard basis $\{e_1, e_2, e_3\}$ of $\mathbb{R}^3$.

7. If $T : \mathbb{R}^2 \to \mathbb{R}^2$ such that $T(x, y) = (x + 2y, x - 3y)$, then find the matrix of T with respect to

(*a*) $\{(1, -1), (1, 1)\}$ and (*b*) $\{(1, 2), (2, 1)\}$.

Further, find the relation between these matrices.

8. If

$$A = \begin{bmatrix} 1 & 0 & 0 \\ 0 & 1 & 1 \\ 1 & 1 & 0 \end{bmatrix}$$

is the matrix of a linear transformation $T : \mathbb{R}^3 \to \mathbb{R}^3$ with respect to the standard basis $\{e_1, e_2, e_3\}$, then show that T is one-one and onto.

9. If $T : M_2(\mathbb{R}) \to M_2(\mathbb{R})$ is such that $T(A) = \dfrac{A + A^t}{2}$ is T is a linear transformation? If so, find the matrix representation of T with respect to the standard basis of $M_2(\mathbb{R})$.

10. Give an example of a map which maps linearly dependent subset to linearly dependent subset but is not a linear transformation.

11. In $\mathbb{R}^2$ if the axes are rotated through an angle θ in anticlockwise direction, then express this phenomena in terms of linear map.

Suppose after rotation the coordinate of P is (x', y') *i.e.,* $T : \mathbb{R}^2 \to \mathbb{R}^2$ such that $T(x, y) = (x', y')$. Now

$$CA = x \tan \theta$$

$$\Rightarrow \quad PC = y - x \tan \theta$$

$$\Rightarrow \quad PC \cos \theta = y \cos \theta - x \sin \theta$$

$$\Rightarrow \quad y' = y \cos \theta - x \sin \theta$$

Also $x' = OC + CB$ (see figure)

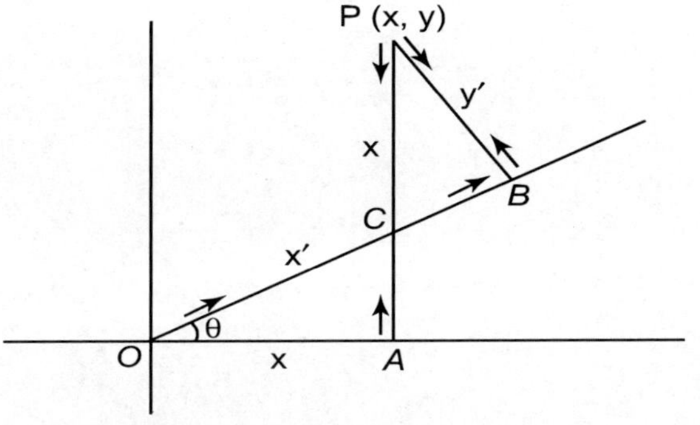

$$\frac{CBN}{PC} = \sin \theta$$

$$\Rightarrow \quad CB = (y - CA) \sin \theta$$

$$= (y - x \tan \theta) \sin \theta$$

$$= y \sin \theta - x \sin \theta \tan \theta$$

Also, $$OC = \frac{x}{\cos \theta}$$

$$\Rightarrow \quad x' = y \sin \theta \tan \theta + x \cos \theta$$

$$= y \sin \theta - x \left[\frac{\sin^2 \theta - 1}{\cos \theta} \right]$$

$$= y \sin \theta + x \left[\frac{1 - \sin^2 \theta}{\cos \theta} \right]$$

$$= y \sin \theta + x \cos \theta$$

$$\Rightarrow \qquad T(x, y) = (x \cos \theta + y \sin \theta, -x \sin \theta + y \cos \theta)$$

$$\Rightarrow \qquad m(T) = \begin{bmatrix} \cos \theta & \sin \theta \\ -\sin \theta & \cos \theta \end{bmatrix}$$

under standard basis $\mathbb{R}^2$.

Now, putting $\theta = \frac{\pi}{4}, \frac{\pi}{2}, \pi,$ and so on, we can find the linear transformation of $\mathbb{R}^2$ in terms of rotation.

5

Algebra of Linear Transformations

We have discussed in Chapter 4 about linear transformations, their ranks and nullity and matrix representation. In this chapter we shall discuss different operations on the set of linear transformations and determine the structures they form. We shall also study different types of linear transformations used in the literature. Throughout, the vector spaces are taken over a field F unless and otherwise stated.

1 OPERATIONS ON LINEAR TRANSFORMATIONS

Definition 5.1.1. If $S, T : V \to U$ are linear transformations, then $S + T$ is defined as $S + T : V \to U$ such that

$$(S + T)(x) = S(x) + T(x) \text{ for all } x \in V.$$

Since

$$\begin{aligned}(S + T)(\alpha x + \beta y) &= S(\alpha x + \beta y) + T(\alpha x + \beta y)\\ &= \alpha S(x) + \beta S(y) + \alpha T(x) + \beta T(y)\\ &= \alpha(S + T)(x) + \beta(S + T)(y) \text{ for all}\\ &\quad \alpha, \beta \in F \text{ and } x, y \in V.\end{aligned}$$

$\Rightarrow S + T$ is again linear transformation.

It is also easy to see that $m(S + T) = m(S) + m(T)$.

Example 5.1.2. Let $S, T : \mathbb{R}^2 \to \mathbb{R}^3$ such that

$$S(x, y) = (x + y, x - y) \text{ and } T(x, y) = (y, x + y, x),$$

then S and T are linear transformations., Now,

$$\begin{aligned}(S + T)(x, y) &= S(x, y) + T(x, y)\\ &= (x + y, x - y) + (y, x + y, x)\\ &= (x + 2y, 2x + y, 2x - y)\end{aligned}$$

which is trivially a linear transformations.

Now, let $\{(1, 0), (0, 1)\}$ and $\{(1, 0, 0), (0, 1, 0), (0, 0, 1)\}$ be bases of $\mathbb{R}^2$ and $\mathbb{R}^3$, respectively. Then

$$m(S) = \begin{bmatrix} 1 & 1 \\ 1 & 0 \\ 1 & -1 \end{bmatrix} \text{ and } m(T) = \begin{bmatrix} 0 & 1 \\ 1 & 1 \\ 1 & 0 \end{bmatrix}.$$

Now

$$m(S) + m(T) = \begin{bmatrix} 1 & 2 \\ 2 & 1 \\ 2 & -1 \end{bmatrix}$$

$$(S + T)(1, 0) = (1, 2, 2)$$

and $$(S + T)(0, 1) = (2, 1, -1)$$

$\Rightarrow$ $$m(S + T) = \begin{bmatrix} 1 & 2 \\ 2 & 1 \\ 2 & -1 \end{bmatrix}$$

$\Rightarrow$ $$m(S + T) = m(S) + m(T).$$

It must be noted that once the bases are chosen they must be taken the same for calculating $m(S)$, $m(T)$ and $m(S + T)$.

Definition 5.1.3. If $T : V \rightarrow U$ is a linear transformation, then for any $\alpha \in F$, define $\alpha T : V \rightarrow U$ such that $(\alpha T)(x) = \alpha T(x)$.

$$\begin{aligned} (\alpha T)(ax + by) &= \alpha T(ax + by) \\ &= \alpha[aT(x) + bT(y)] \\ &= a(\alpha T)(x) + b(\alpha T)(y), \text{ as } a, b, \alpha \in F \end{aligned}$$

$\Rightarrow \alpha T$ is again a linear transformation.

This operation is known as scalar operation. As done in Example 5.1.2 it is trivial to see that $m(\alpha T) = \alpha m(T)$.

Definition 5.1.4. Let $S : V \rightarrow U$ and $T : U \rightarrow W$ be linear transformations, then $T o S : V \rightarrow W$ defined as $(T o S)(x) = T(S(x))$ is called composition of T and S, where $x \in V$ and we shall mostly write $T o S = TS$.

Trivially for any $\alpha, \beta \in F$, $x, y \in V$, then

$$\begin{aligned} (T o S)(\alpha x + \beta y) &= T(S(\alpha x + \beta y)) \\ &= T(\alpha S(x) + \beta S(y)) \\ &= \alpha T(S(x)) + \beta T(S(y)) \\ &= \alpha (T o S)(x) + \beta (T o S)(y) \end{aligned}$$

$\Rightarrow T o S$ is a linear transformation.

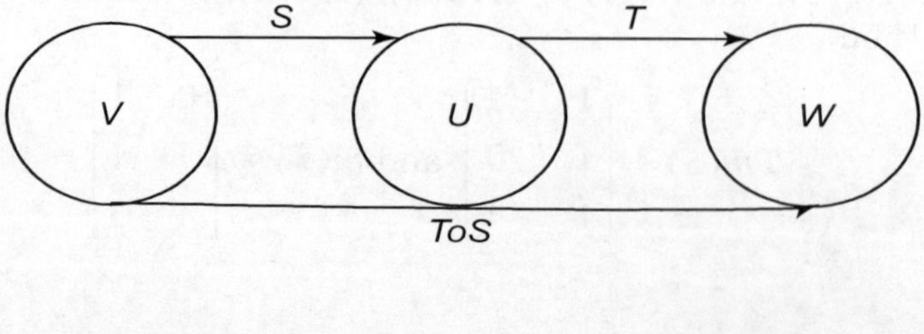

Sometimes $T \circ S$ and $S \circ T$ may not even be defined as in the above case $S \circ T$ (under our notation of taking elements on right side) is not defined. However in case ST and TS both are defined, they may not be the same. For example, if $S : \mathbb{R}^2 \to \mathbb{R}^2$ is such that $S(x, y) = (y, x)$ and $T: \mathbb{R}^2 \to \mathbb{R}^2$ such that $T(x, y) = (x, 0)$, then

$$(ST)(x, y) = S(T(x, y)) = S(x, 0) = (0, x) \text{ and}$$
$$(TS)(x, y) = T(S(x, y)) = T(y, x) = (y, 0), \text{ so } ST \neq TS.$$

Examples 5.1.5

(*i*) $T : \mathbb{R}^2 \to \mathbb{R}^2$ such that $T(x, y) = (x, 0)$ and $S : \mathbb{R}^2 \to \mathbb{R}^2$ such that $S(x, y) = (y, x)$. Let {(1, 0), (0, 1)] be a basis of $\mathbb{R}^2$, then

$$m(T) = \begin{bmatrix} 1 & 0 \\ 0 & 0 \end{bmatrix} \text{ and } m(S) = \begin{bmatrix} 0 & 1 \\ 1 & 0 \end{bmatrix}.$$

$$(ST)(x, y) = (0, x), \text{ so } m(ST) = \begin{bmatrix} 0 & 0 \\ 1 & 0 \end{bmatrix}.$$

But

$$m(S)\, m(T) = \begin{bmatrix} 0 & 1 \\ 1 & 0 \end{bmatrix}\begin{bmatrix} 1 & 0 \\ 0 & 0 \end{bmatrix} = \begin{bmatrix} 0 & 0 \\ 0 & 0 \end{bmatrix}.$$

Therefore, $m(ST) = m(S)\, m(T)$. Further

$$(TS)(x, y) = (y, 0)$$
$$\Rightarrow \quad m(TS) = \begin{bmatrix} 0 & 1 \\ 0 & 0 \end{bmatrix}.$$

Also

$$m(T)\, m(S) = \begin{bmatrix} 1 & 0 \\ 0 & 0 \end{bmatrix}\begin{bmatrix} 0 & 1 \\ 1 & 0 \end{bmatrix} = \begin{bmatrix} 0 & 1 \\ 0 & 0 \end{bmatrix}$$
$$\Rightarrow \quad m(TS) = m(T)\, m(S).$$

(*ii*) Let $T : \mathbb{R}^3 \to \mathbb{R}^2$ such that $T(x, y, z) = (x - z, y + z)$ and $S : \mathbb{R}^2 \to P_2(x)$ such that $S(a + b) = a + bx^2$, then S and T are easily seen to be linear transformations. Now $S \circ T : \mathbb{R}^3 \to P_2(x)$ is a linear transformation and

$$(ST)(a, b, c) = S(a - b, b + c) = a - c + (b + c)x^2.$$

Now for the sake of simplicity choose the standard bases {(1, 0, 0), (0, 1, 0), (0, 0, 1)}, {(1, 0), (0, 1)} and $\{1, x, x^2\}$ of $\mathbb{R}^3$, $\mathbb{R}^2$ and $P_2(x)$, respectively. Now

$$T(1, 0,0) = (1, 0) = 1(1,0) + 0(0, 1)$$

$$T(0, 1, 0) = (0, 1) = 0(1, 0) + 1(0, 1)$$
$$T(0, 0, 1) = (-1, 1) = -1(1, 0) + 1(0, 1)$$

$$\Rightarrow \qquad m(T) = \begin{bmatrix} 1 & 0 & -1 \\ 0 & 1 & 1 \end{bmatrix}.$$

Also

$$S(1, 0) = 1 = 1 \cdot 1 + 0 \cdot x + 0 \cdot x^2$$
$$S(0, 1) = x^2 = 0 \cdot 1 + 0 \cdot x + 1 \cdot x^2$$

$$\Rightarrow \qquad m(S) = \begin{bmatrix} 1 & 0 \\ 0 & 0 \\ 0 & 1 \end{bmatrix}.$$

To find $m(ST)$,

$$(ST)(1, 0, 0) = 1 = 1 \cdot 1 + 0 \cdot x + 0 \cdot x^2$$
$$(ST)(0, 1, 0) = x^2 = 0 \cdot 1 + 0 \cdot x + 1 \cdot x^2$$
$$(ST)(0, 0, 1) = -1 + x^2 = -1 \cdot 1 + 0 \cdot x + 1 \cdot x^2$$

$$\Rightarrow \qquad m(ST) = \begin{bmatrix} 1 & 0 & -1 \\ 0 & 0 & 0 \\ 0 & 1 & 1 \end{bmatrix}.$$

Now

$$m(S)\, m(T) = \begin{bmatrix} 1 & 0 \\ 0 & 0 \\ 0 & 1 \end{bmatrix} \begin{bmatrix} 1 & 0 & -1 \\ 0 & 1 & 1 \end{bmatrix} = \begin{bmatrix} 1 & 0 & -1 \\ 0 & 0 & 0 \\ 0 & 1 & 1 \end{bmatrix}$$

$$\Rightarrow \qquad m(S)\, m(T) = m(ST).$$

Definition 5.1.6. A linear transformation $T : V \to V$ is called *idempotent* if $T^2 = T$.

If we consider $T : \mathbb{R}^2 \to \mathbb{R}^2$ such that $T(x, y) = (y, x)$, then $T^2 \neq T$ but if we also $S(x, y) = (x, 0)$, where $S : \mathbb{R}^2 \to \mathbb{R}^2$, then $S^2 = S$. Therefore, S is idempotent but T is not idempotent.

Trivially maps $I : V \to V$ such that $I(x) = x$ and $w : V \to V$ such that $w(x) = 0$ for all $x \in V$ are linear transformations called identity transformation and zero transformations. Now onward the zero transformation w will be denoted by O.

The identity and zero maps are trivial idempotents. Also if $T : V \to V$ is idempoten, then $I - T$ is also idempotent as

$$(I - T)^2 = I - 2T + T^2 = I - 2T + T^2$$
$$= I - 2T + T = I - T \text{ as } IT = TI = T.$$

Definition 5.1.7. If $T : V \to V$ is a linear transformation, then T is called *nilpotent* if there exists a positive integer k such that $T^k = 0$ and the task k such that $T^k = 0$ is called index of nipotency and will be denoted by $i(T)$. If $i(T) = 1$, then $T = 0$, so $i(T) \geq 2$ for nonzero linear transformation T. Identity transformation is never nilpotent.

We have seen in Chapter 4 that for every linear transformation there is matrix and for every matrix there is a linear transformation, therefore, if the linear transformation is idempotent or nilpotent then the corresponding matrices will also be idempotent or nilpotent. Trivially a nonzero idempontent transformation can never be nilpotent and a nonzero nilpotent can never be idempotent.

Example 5.1.8

1. Let $T : \mathbb{R}^3 \to \mathbb{R}^3$ such that $T(x, y, z) = (y + z, z, 0)$, then
$$T^2(x, y, z) = (z, 0, 0)$$
$$\Rightarrow \quad T^3(x, y, z) = T(z, 0, 0) = (0, 0,0)$$
$$\Rightarrow \quad T^3 = 0$$
$$\Rightarrow \quad T \text{ is nilpotent and } i(T) = 3.$$
2. Let $T : \mathbb{R}^2 \to \mathbb{R}^2$ such that $T(x, y) = (2y, 0)$, then $T^2 = 0$.
3. Let $T : \mathbb{R}^2 \to \mathbb{R}^2$ such that $T(x, y) = (x + y, -x - y)$, then it is easy to check that $T^2(x, y) = (0, 0) \Rightarrow T^2 = 0$.
4. Let $T : \mathbb{R}^3 \to \mathbb{R}^3$ such that $T(x, y, z) = (y + z, 0, 0)$, then $T^2 = 0$.

Definition 5.1.9

A linear transformation $T : V \to V$ is called *involutory* if $T^2 = I$.

If we consider $T : \mathbb{R}^2 \to \mathbb{R}^2$ such that $T(x, y) = (y, x)$, then $T^2(x, y) = (x, y) = I(x, y) \Rightarrow T^2 = I$. It is trivial to see that T is involuntary iff $(1 - T)(1 + T) = 0$.

Example 5.1.10. $T : \mathbb{R}^3 \to \mathbb{R}^3$ such that
$$T(x, y, z) = (4x - y - 4z, 3x - 4z, 3x - y - 3z),$$
then
$$\begin{aligned} T^2(x, y, z) &= (4(4x - y - 4z) - 3x + 4z - 4(3x - y - 3z), \\ &\quad 3(4x - y - 4z) - 4(3x - y - 3z), \\ &\quad 3(4x - y - 4z) - 3x + 4z - 3(3x - y - 3z) \\ &= (16x - 4y - 16z - 3x + 4z - 12x + 4y + 12z, \\ &\quad 12x - 3y - 12z - 12x + 4y + 12z, 12x - 3y \\ &\quad - 12z - 3x + 4z - 9x + 3y + 9z) \end{aligned}$$

$$= (x, y, z)$$

$$\Rightarrow \qquad T^2 = I.$$

The other way is to see if we can find the matrix of T with respect to the standard basis and we have

$$m(T) = \begin{bmatrix} 4 & -1 & -4 \\ 3 & 0 & -4 \\ 3 & -1 & -3 \end{bmatrix}$$

and
$$(m(T))^2 = \begin{bmatrix} 1 & 0 & 0 \\ 0 & 1 & 0 \\ 0 & 0 & 1 \end{bmatrix}$$

$\Rightarrow$ T is involutory.

2 INVERSES OF LINEAR TRANSFORMATIONS

In this section we shall deal with maps which are one-one and onto (as discussed in Chapter 2) and discuss some examples and results needed for future use.

Definition 5.2.1. If $T : V \to V$ is a linear transformation, then T is called non-singular or regular or invertible iff T is one-one and onto.

Since T is a linear map, so we can say that T is non-singular iff T is one-one and onto. For arbitrary sets we have already seen this situation Chapter 2.

Examples 5.2.2.

1. Let $T : \mathbb{R}^2 \to \mathbb{R}^2$ such that $T(x, y) = (y, x)$, then T is trivially one-one and onto and so it is non-singular.
2. Let $T : \mathbb{R}^2 \to \mathbb{R}^2$ such that $T(x, y) = (x, 0)$, then T is neither one-one nor onto, so T is singular.
3. Let $T : \mathbb{R}^2 \to \mathbb{R}^2$ such that $T(x, y) = (x - y, x + y)$, then trivially T is one-one. Also for any $(\alpha, \beta) \in {}^2$,

$$T\left(\frac{\alpha + \beta}{2}, \frac{\beta - \alpha}{2}\right) = (\alpha, \beta),$$

so T is onto $\Rightarrow$ T is non-singular.

Notes

1. As discussed earlier for every linear transformation there is a matrix. Therefore, it is enough to seethe non-singularity of the matrix associated with the linear transformation.

2. If an idempotent linear transformation is non-singular then it is only the identity map.
3. Nipotent linear transformations can never be non-singular.
4. If $T : V \to V$ such that $T^k = 0$, then $I - T$ is always non-singular. (One can check that $(I + T + \ldots + T^{k-1})(I - T) = I$.)

Theorem 5.2.3. A linear transformation $T : V \to V$ is non-singular if and only if $TS = ST = I$ for some $S : V \to V$ (a linear map).

Proof. Let $TS = ST = I$ and $T(x) = T(y)$; $x, y \in V$

$\Rightarrow ST(x) = S(T(y)) \Rightarrow I(x) = I(y) \Rightarrow x = y \Rightarrow T$ is one-one.

Further, for any $y \in V$, $I(y) = y$

$\Rightarrow (TS)(y) = y \Rightarrow T(S(y)) = y \Rightarrow T$ is onto $\Rightarrow T$ is onto-singular.

Conversely, suppose $T : V \to V$ is non-singular. Since T is one-one onto map

$\Rightarrow$ there exists a map $S : V \to V$ such that $TS = ST = I$. Now, for any $\alpha, \beta \in F$ and $x, y \in V$

$$\begin{aligned}(TS)(\alpha x + \beta y) &= \alpha x + \beta y \\ &= \alpha I(x) + \beta I(y) \\ &= \alpha(TS)(x) + \beta(TS)(x) \\ &= \alpha T(S(x)) + \beta T(S(x)) \\ &= T(\alpha S(x) + \beta S(y))\end{aligned}$$

and $(TS)(\alpha x + \beta y) = T(S(\alpha x + \beta y))$. Since T is one-one.

$$\Rightarrow \qquad S(\alpha x + \beta y) = T(S(x)) + \beta(S(y)).$$

$\Rightarrow S$ is a linear transformation.

Now, as we know that if S and T are one-one and onto maps, then ST and TS are also one-one and onto maps. Therefore, due to above theorem we can say that the set of all non-singular linear transformations : $V \to V$ is a group under the composition as defined in this chapter. This amounts the same what we discussed in Chapter 2,that the set of non-singular matrices forms a group under multiplication.

Since we are dealing with only finite dimensional vector spaces, therefore, in defining non-singular linear transformation it is sufficient either to take one-one or onto because it follows from the following:

Theorem 5.2.4. For a finite dimensional vector spaces V, a linear map $T : V \to V$ is one-one iff T is onto.

Proof. Let T be one-one, then $\ker T = 0$, so by Sylvester's law of nullity $\dim V = \dim TV$ but $TV \subseteq V$. Hence, $TV = V \Rightarrow T$ is onto. Also if T is onto then $r(T) = \dim TV$ but $\dim V = r(T) + n(T) \Rightarrow n(T) = 0 \Rightarrow \ker T = 0 \Rightarrow T$ is one-one.

Now, we end this section with the remark that a linear map $T : V \rightarrow V$ is singular iff there exists $v \in V,\ v \neq 0$ such that $T(v) = 0$.

3 PROJECTIONS

As shown in the following figure, we know that if P is a point in the plane and PA and PB are perpendiculars from P on X and Y-axes, respectively, then PA and PB are projections of OP on X and Y-axes.

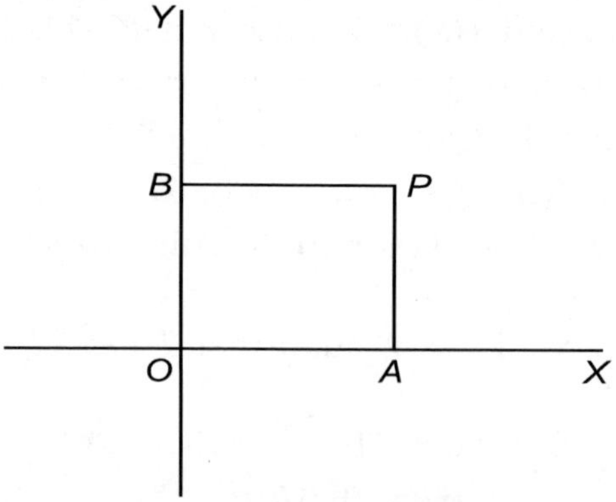

If we consider the coordinate of P as (x, y), then $OA = x$ and $OB = y$ and coordinates of A and B are $(x, 0)$ and $(0, y)$. If we consider the space generated (spanned) by these points $A(x, 0)$ and $B(0, y)$, then we get

$$\mathbb{R}^2 = U \oplus W, \text{ where } U = [(x, 0)] \text{ and } W = [(0, y)].$$

Now, we can define maps T_1 and $T_2 : \mathbb{R}^2 \rightarrow \mathbb{R}^2$ such that $T_1(x, y) = (x, 0)$ and $T_2(0, y)$. It is trivial to see that T_1 and T_2 takes the point $P(x, y)$ to $A(x, 0)$ and $B(0, y)$, respectively. Further, it is trivial to see that T_1 and T_2 are linear maps neither zero nor identity and so they are non-trivial linear maps.

As we have seen in Chapter 3, that for every subspace U of a vector space there exists a complementary summand W (*i.e.*, $V = U \oplus W$). Hence, now we formally define projection.

Definition 5.3.1. Let V be a vector space such that $V = U \oplus W$ for some subspaces U and W, then a map $T : V \rightarrow V$ such that $T(x) = T(u + w) = u$ is called projection of V on U along W, where $x = u + w,\ u \in U,\ w \in W$.

Since $V = U \oplus W$ every $x \in V$ has unique way to write $u + w$, $u \in U$, $w \in W$. It is trivial to see that T is well defined and T is a linear transformation. Further, for every $u \in U$ and $w \in W$, $T(u) = u$, $T(w) = 0$ *i.e.*, $T \mid U = I$ and $T \mid W = 0$. Therefore, we can say that there will be a unique transformation with these properties for the subspaces U and W.

Example 5.3.2

1. Let $\mathbb{R}^2 = U \oplus W$, where $U = [(1, 0)]$ and $W = [(0, 1)]$, then $T : \mathbb{R}^2 \to \mathbb{R}^2$ such that $T(x, y) = (x, 0)$ is a projection of $\mathbb{R}^2$ on U along W.
2. Let $\mathbb{R}^2 = U \oplus W$, where $U = [(1, 1)]$ and $W = [(1, 2)]$, then $T : \mathbb{R}^2 \to \mathbb{R}^2$ such that $T(x, y) = (2x - y, 2x - y)$, then T is a projection on U along W.
3. $P_1(x) = U \oplus W$, where $U = [(1 - x)]$ and $W = [(1 + x)]$, then $T : P_1(x) \to P_1(x)$, such that

$$T(a + bx) = \frac{a - b}{2}(1 - x),$$

then T is a projection on U along W.

Notes

1. From the definition of the projection of $V = U \oplus W$ on U along W, it is obvious that if T is such projection, then trivially $W = \ker T$ and $U = TV$.
2. If $T \mid U$ is identity on U, then every basis element of U must be mapped to itself. Therefore, in order to construct examples one has to keep this point into consideration.
3. Whenever we talk of projections only then it means it is projection on some subspace U along some subspace W.
4. If $V = U \oplus W$ and T is projection on U along W. If $\{x_1, x_2, \ldots, x_r\}$ is a basis of U and $\{y_1, y_2, \ldots, y_t\}$ is a basis of W, then as $T(x_i) = x_i$ and $T(y_i) = 0$, the matrix representation of T will be of the type

$$\begin{bmatrix} I_r & O \\ O & O \end{bmatrix},$$

a matrix of order $r + t = \dim V$.

Now, we state and prove a result which may be more useful in verifying and constructing examples.

Theorem 5.3.3. A linear transformation $T : V \to V$ is projection iff $T^2 = T$.

Proof. Let $V = U \oplus W$ for some subspaces U and W and suppose T is projection on U along W. Let $x \in V$, then $x = u + w$, $u \in U$, $w \in W$

$$\Rightarrow T(x) = u \qquad \Rightarrow T^2(x) = T(u) = u = T(x) \qquad \Rightarrow T^2 = T.$$

Conversely, suppose $T^2 = T$. Trivially TV is a subspace of V and $\ker T$ is also a subspace of $V \Rightarrow TV + \ker T \subseteq V$. Let $x \in V$, then $x = x - T(x) + T(x)$, since

$$T(x - T(x)) = T(x) - T^2(x) = T(x) - T(x) = 0 \Rightarrow x - T(x) \in \ker T.$$

Further $T(x) \in TV \Rightarrow V = TV + \ker T$.

Let $z \in \ker T \cap TV \Rightarrow T(z)=0$ and $z = T(v)$ for some $v \in V$

$$\Rightarrow T(z) = T^2(v) = T(v) = 0 \text{ as } T(z) = 0 \Rightarrow z = 0 \Rightarrow V = TV \oplus \ker T.$$

Now, trivially T is identity map for TV and zero map for $\ker T$

$\Rightarrow T$ is projection on TV along $\ker T$.

Corollary 5.3.4. $T : V \rightarrow V$ is projection iff $I - T$ is also a projection.

It is obvious to see that if T is a projection on U along W, then $I - T$ will be projection on W along U.

Corollary 5.3.5. If $T : V \rightarrow V$ is a linear transformation such that $T^2 = T$, then $TV = \ker (I - T)$ and $\ker T = (I - T)V$.

Definition 5.3.6. Any two projections T_1 and T_2 of V are called orthogonal if $T_1T_2 = T_2T_1 = 0$.

Examples 5.3.7

1. If $T : \mathbb{R}^2 \rightarrow \mathbb{R}^2$ such that $T(x, y) = (x, 0)$ and $S : \mathbb{R}^2 \rightarrow \mathbb{R}^2$ such that $S(x, y) = (0, y)$, then $ST = TS = 0$, so S and T are orthogonal.
2. If T is a projection of V, then $I - T$ is also a projection of V and T and $(I - T)$ are orthogonal.
3. If $T : \mathbb{R}^4 \rightarrow \mathbb{R}^4$ such that $T(x_1, x_2, x_3, x_4) = (x_1, x_2, 0, 0)$ and $S : \mathbb{R}^4 \rightarrow \mathbb{R}^4$ such that $S(x_1, x_2, x_3, x_4) = (0, 0, x_3, 0)$, then S and T are projections but in this case $S \neq I - T$, although S and T are orthogonal.
4. If $T : \mathbb{R}^3 \rightarrow \mathbb{R}^3$ such that $T(x_1, x_2, x_3) = (x_1, 0, x_3)$ and $S : \mathbb{R}^3 \rightarrow \mathbb{R}^3$ such that $S(x_1, x_2, x_3) = (0, x_2, x_3)$, then S and T are projections but neither $ST = 0$ nor $TS = 0$, although $ST = TS$.
5. If V is a vector space such that $\dim V = 1$, then I is the only projection.
6. If $T : \mathbb{R}^n \rightarrow \mathbb{R}^n$ such that $T(x_1, x_2, \ldots, x_n) = (0, \ldots 0, x_i, 0, \ldots, 0)$, then T is projection and is known as projection at i^{th} coordinate. Since $\mathbb{R}^n$ has many subspaces of dimensions $1, 2, \ldots, n - 1$, we can have many projections.

Now, let us consider a vector space V and suppose $\{x_1, x_2, \ldots, x_3\}$ is a basis of V, then we have $V = Fx_1 \oplus \ldots \oplus Fx_n$. Let $x \in \sum_{i=1}^{n} \lambda_i x_i$, $\lambda_i \in F$.

Define $T_i : V \rightarrow V$ such that $T_i(x) = \lambda_i x_i$, then T_i's are projections. It is easy to see that $T_i(x_i) = x_i$ and $T_i(x_j) = 0$ for $i \neq j$. Further

$$T_1 + T_2 + \ldots + T_n = I \quad \text{and } T_iT_j = 0 \text{ for } i \neq j,$$

and

$$T_1V \oplus T_2V \oplus \ldots \oplus T_nV = V.$$

Similarly, if $V = U_1 \oplus U_2 \oplus \ldots \oplus U_t$, then we define $T_i : V \rightarrow V$ such that $T_i(x) = u_i$, where $x = u_1 + u_2 + \ldots + u_t$, $u_i \in U_i$ for $i = 1, 2, \ldots, t$. It is easy to verify that T_i's are projections and $I = T_1 + T_2 + \ldots + T_t$ such that $V = T_1V \oplus T_2V \oplus \ldots \oplus T_tV$ and $T_iT_j = 0$. As in Note 4, if $V = U \oplus W$ and $T : V \rightarrow V$ is a projection then

$$m(T) = \begin{bmatrix} I_r & O \\ O & O \end{bmatrix},$$

where $r = \dim U$. Also $m(I) = I_n$.

$$\Rightarrow \qquad m(I + T) = \begin{bmatrix} 2 & & & & & & & O \\ & 2 & & & & & & \\ & & \cdot & & & & & \\ & & & \cdot & & & & \\ & & & & \cdot & & & \cdot \\ & & & & & 2 & & 1 \\ & & & & & & 1 & \\ & & & & & & & \cdot \\ & & & & & & & \cdot \\ & O & & & & & & \cdot \\ & & & & & & & 1 \end{bmatrix}$$

where 2 occurs r times $2 = 1 + 1$ (denoted) as $1 \in F$ is identity of $F \Rightarrow m(I + T)$ is invertible provided $1 + 1 = 2 \neq 0$ (denoted).

In other words, we can say that $I + T$ is invertible if $1 \in F$ is the identity of F and $1 + 1 \neq 0$.

4 INVARIANT SUBSPACES

We know that if $T : V \rightarrow V$ is a linear transformation, then $TV \subseteq V \Rightarrow T(TV) \subseteq TV \Rightarrow TV$ is mapped to TV by T. The question is whether given a subspace U of V and $T : V \rightarrow V$, is it possible to have $TU \subseteq U$?

For the answer, we proceed as follows:

Definition 5.4.1

Let $T : V \rightarrow V$ be a linear transformation and U be a subspace of V, then U is called invariant under T or T-invariant if $TU \subseteq U$ *i.e.*, for every $x \in U$, $T(x) \in U$.

Trivially all subspaces of V under zero map and identity map are invariant. Now, if U is T-invariant, then $T \mid U$ restriction of T on U is trivially a linear transformation from $U \to U$. Further, it is possible that $T \mid U$ may be onto which we will see in the example. If $\{x_1, x_2, \ldots, x_r\}$ is a basis of U, then $\{x_1, x_2, \ldots, x_n\}$ can be extended to a basis $\{x_1, x_2, \ldots, x_n, y_1, y_2, \ldots, y_n\}$ of V. Now, as we know that every subspace of V is a direct summand $\Rightarrow V = U \oplus W$ for some subspace W of V.

Now, if U is T-invariant then we can easily check that $m(T)$ will be of the type

$$\begin{bmatrix} A & B \\ O & C \end{bmatrix},$$

where $A = m(T \mid U)$ and $r - t = \dim V$. If W is also T-invariant, then

$$m(T) = \begin{bmatrix} A & O \\ O & D \end{bmatrix},$$

where $D = m(T \mid W)$.

Further, $\{O\}$ zero space and whole space V are always invariant under all linear transformations. There may exist some subspaces which are not invariant under some transformations.

Now, we shall first give some examples discussing all these cases.

Example 5.4.2

1. Let $T : \mathbb{R}^2 \to \mathbb{R}^2$ such that $T(x, y) = (x - y, x + y)$, then trivially T is a linear map. Let $U = \{(x, 0) \mid x \in \mathbb{R}\}$, $W = \{(0, y) \mid y \in \mathbb{R}\}$ and $P = \{(x, mx) \mid x \in \mathbb{R}, m \in \mathbb{R}$ and m is fixed$\}$, then U, W, P are subspaces of $\mathbb{R}^2$, then none of these are T-invariant *i.e.*, $TU \not\subseteq U$, $TW \not\subseteq W$, $TP \not\subseteq P$.

2. Let $T : \mathbb{R}^2 \to \mathbb{R}^2$ such that $T(x, y) = (x, 0)$, then the subspaces U and W are T-invariant but P is not T-invariant. It is trivial to see that $T \mid U$ is onto but $T \mid W$ is not onto, where U, W and P are given in Example 5.4.1.

   However, as we know that if T is any projection on V as discussed in Section 5.3, then $V = TV \oplus \ker (I - T)$ and both the subspaces are T-invariant. So, for any projection T on V, we can always get T-invariant subspaces.

3. Let $T : \mathbb{R}^3 \to \mathbb{R}^3$, such that $T(x, y, z) = (x - y, x + y, z)$, then T is a linear map. Let $U = \{(a, b, 0) \mid a, b \in \mathbb{R}\}$ (xy-plane) $W = \{(0, 0, c) \mid c \in \mathbb{R}\}$, then U and W are subspaces of $\mathbb{R}^3$ and $\mathbb{R}^3 = U \oplus W$. Obviously U and W are T-invariant subspaces of $\mathbb{R}^3$. For any

$(a, b, 0) \in U$, $T\left(\dfrac{a+b}{2}, \dfrac{b-a}{2}, 0\right) = 0\ (a, b, 0)$ and for any $(0, 0, c)$ $\in W T(0, 0, c) \Rightarrow T \mid U$ and $T \mid W$ are onto. It is easy to check that under standard basis

$$m(T) = \begin{bmatrix} 1 & -1 & 0 \\ 1 & 1 & 0 \\ 0 & 0 & 1 \end{bmatrix}, \text{ where } m(T \mid U) = \begin{bmatrix} 1 & -1 \\ 1 & 1 \end{bmatrix}$$

and $m(T \mid W) = [1]$.

4. If $T : \mathbb{R}^3 \to \mathbb{R}^3$ such that $T(x, y, z) = (x - y, y - z, y + z)$, then it is easy to check that T is a linear map.
   Now, taking $U = \{(x, 0, 0) \mid x \in \mathbb{R}\}$ and $W = \{(0, y, z) \mid y, z \in \mathbb{R}\}$, then $\mathbb{R}^3 = U \oplus W$. It is easy to see that U is T-invariant but W is not T-invariant. Now under the standard basis it is easy to see that
   $$m(T) = \begin{bmatrix} 1 & -1 & 0 \\ 0 & 1 & -1 \\ 0 & 1 & 1 \end{bmatrix}$$
5. If $T : \mathbb{R}^2 \to \mathbb{R}^2$ such that $T(x, y) = (-y, x)$, then T is a linear map. Since every subspace of $\mathbb{R}^2$ is a straight line passing through origin and T will map the straight line to a perpendicular straight line passing through origin, therefore, there will not exist any proper subspace which is T-invariant.
6. $\dfrac{d}{dx} : P_n(x) \to P_n(x)$ is a linear transformation and every subspace of $P_n(x)$ is invariant under $\dfrac{d}{dx}$.
7. If $T : V \to V$ is a linear transformation, then ker T is always T-invariant.

Theorem 5.4.3. If $V = U_1 \oplus U_2 \oplus \ldots \oplus U_r$ and $T : V \to V$ is a linear map such that U_i's are T-invariant, then

$$m(T) = \begin{bmatrix} A_1 & & & & \\ & A_2 & & O & \\ & & \cdot & & \\ & & & \cdot & \\ & O & & & \cdot \\ & & & & A_r \end{bmatrix},$$

where $A_i = m(T \mid U_i)$ for $i = 1, 2, \ldots, r$.

Proof. Let $B_1 = \{x_1, x_2, \ldots, x_{t_1}\}$, $B_2 = \{y_1, y_2, \ldots, y_{t_2}\}$, ..., $B_r = \{u_1, u_2, \ldots, u_{t_r}\}$ be bases of $U_1, U_2, \ldots, U_r$ respectively, then $B_1 \cup B_2 \cup \ldots \cup B_r$ is a basis of V and so $\dim V = t_1 + t_2 + \ldots + t_r$. Now, since U_i's are T-invariant, so as done in matrix representation, the $m(T)$ with respect to the basis of $B_1 \cup B_2 \cup \ldots \cup B_r$ will be

$$m(T) = \begin{bmatrix} A_1 & & & & \\ & A_2 & & O & \\ & & \cdot & & \\ & & & \cdot & \\ & O & & \cdot & \\ & & & & A_r \end{bmatrix},$$

where $A_i = m(T \mid U_i)$. The detail is easy to verify.

Now, we proceed to construct more invariant spaces as said in the beginning and in Example 5.4.2 (7). Let $T : V \to V$, then $TV \subset V$ and $T^2 V \subset TV$, and so on. So we get a chain of T-invariant subspaces

$$V \supseteq TV \supseteq T^2V \supseteq \ldots \supseteq T^kV \supseteq \ldots \qquad \ldots(1)$$

Similarly, $0 \subseteq \ker T$ and if $x \in \ker T \Rightarrow T(x) = 0 \Rightarrow T^2(x) = 0 \Rightarrow x \in \ker T^2 \Rightarrow \ker T \subseteq \ker T^2$, and so on. Therefore, we get a chain of T-invariant subspaces

$$0 \subseteq \ker T \subseteq \ker T^2 \subseteq \ldots \subseteq \ker T^k \subseteq \ldots \qquad \ldots(2)$$

Since the vector space is finite dimension, therefore, V cannot have infinite descending or ascending chain of subspaces. Hence (1) and (2) must terminate after finite number of steps.

Now, suppose $\qquad T^kV = T^{k+1}V = \ldots$.

Hence by Sylvester's law of nullity.

$$\dim V = \dim T^kV + \dim \ker T^k \text{ and } \dim V = \dim T^{k+1}\text{V} + \dim T^{k+1}$$

$$\Rightarrow \dim \ker T^k = \dim T^{k+1}.$$

But $\ker T^k \subseteq \ker T^{k+1} \Rightarrow \ker T^k = \ker T^{k+1} = \ldots$. Hence, chain (2) will also terminate at the same step. Similarly, the vice versa case follows.

Theorem 5.4.4. If $T : V \to V$ such that $T^kV = T^{k+1}V = \ldots$, then

$$V = T^kV \oplus \ker T^k.$$

Proof. As done above $\dim V = \dim T^kV + \dim \ker T^k$. Since $T^kV \subseteq V$ and $\ker T^k \subseteq V \Rightarrow T^kV + \ker T^k \subseteq V$. So it is sufficient to show that $T^kV \cap \ker T^k = 0$. Let $x \in T^kV \cap \ker T^k \Rightarrow T^k(x) = 0$ and $x = T^kv$ for some $v \in V$

$$\begin{aligned}
\Rightarrow \qquad\qquad T^k(x) &= T^{2k}\, v = 0 \\
\Rightarrow \qquad\qquad v &\in \ker T^k = \ker T^k \\
\Rightarrow \qquad\qquad T^k\, v &= 0 \\
\Rightarrow \qquad\qquad x &= 0 \\
\Rightarrow \qquad T^k\, v \cap \ker T^k &= 0 \\
\Rightarrow \dim\,(T^k\, v + \ker T^k) &= \dim\, T^k v + \dim \ker T^k - \dim 0 \\
&= \dim\, T^k v + \dim \ker T^k = \dim V.
\end{aligned}$$

Hence, $\qquad V = T^k\, V \oplus \ker T^k.$

Theorem 5.4.5. If $T : V \to V$ is a linear transformation and U is T-invariant subspace of V, then there exists a linear transformation $\overline{T} : V/U \to V/U$.

Proof. Define $\overline{T} : V/U \to V/U$ and that $\overline{T}(x+U) = T(x) + U$. If $x + U = U \Rightarrow x \in U \Rightarrow T(x) \in U$ as U is T-invariant $\Rightarrow T(x) + U = U \Rightarrow \overline{T}$ is well defined. Now,

$$\begin{aligned}
\overline{T}(\alpha(x+U) + \beta(y+U)) &= \overline{T}(\alpha x + \beta y + U) \\
&= T(\alpha x + \beta y) + U \\
&= \alpha T(x) + \beta T(y) + U \\
&= \alpha(T(x) + U) + \beta(T(y) + U) \\
&= \alpha\overline{T}\,(x+U) + \beta\overline{T}\,(y+U)
\end{aligned}$$

$\Rightarrow \overline{T}$ is a linear transformation.

Thus for every linear transformation $T : V \to V$ and T-invariant subspace U of V, we can always get $\overline{T} : V/U \to V/U$ such that $\overline{T}(x+U) = T(x) + U$.

Now, the question: If a subspace W/U of V/U is $\overline{T}$-invariant, does it imply that W is T-invariant, is answered in the following:

Theorem 5.4.6. If U is T-invariant where $T : V \to V$ is a linear transformation and $W/U \subseteq V/U$ is a subspace of V/U is invariant under $\overline{T}$ (as given in Theorem 5.4.5), then W is T-invariant.

Proof. Since for all $w \in W$, $w + U \in W/U$ and W/U is $\overline{T}$-invariant

$\Rightarrow \overline{T}\,(w + U) \in W/U$ but $\overline{T}\,(w + U) = T(w) + U$

$\Rightarrow T(w) \in W$ for all $w \in W \Rightarrow W$ is T-invariant.

5 SPACE OF LINEAR MAPS

In this section we shall use the operations on the set of linear maps as defined in Section 1 and investigate the structure formed by the linear maps. Consider V and U to be vector spaces over F, then we denote $L(V, U) = \{T : V \to U \mid T \text{ is a linear map}\}$. It is also denoted by Hom (V, U).

Theorem 5.5.1. If V and U are vector spaces over F, then $L(V, U)$ is a vector space over F.

Since the proof is trivial, therefore, only the sketch is given.

Proof. Let $S, T \in L(V, U)$, then as defined and done Section 5.1 $S + T = T + S \in L(V, U)$. Also $T + O = O + T = T$, where O is zero map. Trivially $(S + T) + T_1 = S + (T + T_1)$, $T, S, T_1 \in L(V, U)$. Also $T + (-T) = (-T) = (-T) + T = 0 \Rightarrow L(V, U)$ is abelian group. As defined in Section 5.1, for every $\lambda \in F$, $T \in L(V, U)$, $\lambda T \in L(V, U)$. It is easy to verify $(\lambda + \mu)T = \lambda T + \mu T$, $\lambda(T_1 + T_2) = \lambda T_1 + \lambda T_2$, $(\lambda\mu)T = \lambda(\mu T)$ and $1 \cdot T = T$ for all $\lambda, \mu \in F$, $T, T_1, T_2 \in L(V, U) \Rightarrow L(V, U)$ is a vector space over F.

Definition 5.5.2. A nonempty set X is called algebra over a field F if X is a vector space over F and X is a ring with $\alpha(xy) = (\alpha x)y = x(\alpha y)$ for all $\alpha \in F$, $x, y \in X$.

Every field is algebra over itself. The set of complex numbers is algebra over $\mathbb{R}$ (field of reals). $M_n(\mathbb{R})$ is an algebra over $\mathbb{R}$.

Theorem 5.5.3. If V is a vector space over F, then $L(V, V)$ is an algebra over F.

Proof. By Theorem 5.5.1, $L(V, V)$ is a vector space over F. As done in Section 1, for any $S, T \in L(V, V)$, $S \circ T = ST$ (denoted) is also an element of $L(V, V)$. It is routine matter to check that

$$(T_1T_2)T_3 = T_1(T_2T_3) \text{ and } T_1(T_2 + T_3) = T_1T_2 + T_1T_3$$

or

$$(T_1 + T_2)T_3 = T_1T_3 + T_2T_3 \text{ for all } T_1, T_2, T_3 \in L(V, V)$$

$\Rightarrow L(V, V)$ is a ring.

Now, let $S, T \in L(V, V)$ and $\alpha \in F$ $x \in V$, then

$$\begin{aligned} \alpha(ST)(x) &= \alpha(ST(x)) \text{ by definition of scalar operation} \\ &= S(\alpha T(x)) \text{ as } S \text{ is a linear map} \\ &= (S(\alpha T))(x) \text{ as } \alpha T \text{ is a linear map} \\ \Rightarrow \quad \alpha(ST) &= S(\alpha T). \end{aligned}$$

Similarly,

$$\begin{aligned} \alpha(ST) &= (\alpha S)T \\ \Rightarrow \quad \alpha(ST) &= (\alpha S)T = S(\alpha T). \end{aligned}$$

$\Rightarrow L(V, V)$ is an algebra over F.

We can further see that $L(V, V)$ has an identity (the identity map $I : V \to V$).

As we are dealing with finite dimensional vector spaces, now we discuss the dimension of $L(V, U)$.

Theorem 5.5.4. dim $(L(V, U))$ = dim V dim U.

Proof. Let $\{x_1, x_2, \ldots, x_n\}$ and $\{y_1, y_2, \ldots, y_n\}$ be a basis of V and U, respectively, then for any $x \in V$, x has representation as $\sum_{i=1}^{n} \lambda_i x_i$. Define $T_{ij} : V \to U$ such that $T_{ij}(x) = \lambda_i y_j$. Now, if $x = 0$, then $\lambda_i = 0 \Rightarrow T_{ij}$ is well defined. Let $y \in V$, then

$$y = \sum_{i=1}^{n} \beta_i x_i$$

$$\Rightarrow \qquad T_{ij}(x + y) = T_{ij}\left(\sum_{i=1}^{n} (\alpha_i + \beta_i) x_i\right) = (\lambda_i + \beta_i) y_j$$

$$= \lambda_i y_j + \beta_i y_j = T_{ij}(x) + T_{ij}(y).$$

Also

$$T_{ij}(\alpha x) = (\alpha\lambda_i) y_j = \alpha(\lambda_i y_j) = \alpha T_{ij}(x).$$

$\Rightarrow T_{ij}$ is a linear transformation for $i = 1, 2, \ldots, n$ and $j = 1, 2, \ldots$, m.

It is clear from the definition that $T_{ij}(x_i) = y_j$ and $T_{ij}(x_k) = 0 \cdot y_j = 0$ *i.e.*, $T_{ij}(x_l) = 0$ if $i \neq l$.

Now, we prove that $\{T_{ij}\}_{\substack{1 \le i \le n \\ 1 \le j \le m}}$ is a basis of $L(V, U)$. Let

$$\sum_{\substack{1 \le i \le n \\ 1 \le j \le m}} \lambda_{ij} T_{ij} = 0$$

$$\Rightarrow \quad \sum_{i=1}^{n} \sum_{j=1}^{m} \lambda_{ij} T_{ij} = 0$$

$$\Rightarrow \quad \sum_{j=1}^{m} (\lambda_{1j} T_{1j} + \lambda_{2j} T_{2j} + \ldots + \lambda_{nj} T_{nj}) = 0$$

$\Rightarrow$ For any x_i, we have

$$\sum_{j=1}^{m} (\lambda_{1j} T_{1j} + \lambda_{2j} T_{2j} + \ldots + \lambda_{nj} T_{nj})(x_i) = 0 \Rightarrow \sum_{j=1}^{m} \lambda_{ij} y_j = 0,$$

applying $T_{ij}(x_i) = y_j$ and $T_{ij}(x_k) = 0$ for $i \neq k$.

Since $\{y_1, y_2, \ldots, y_n\}$ is a basis of U so it is linearly independent. Thus $\lambda_{ij} = 0$ for all i and $j \Rightarrow \{T_{ij}\}$ is linearly independent. Now, we show that

$\{T_{ij}\}$ spans $L(U, V)$. Choose $T \in L(U, V)$. Since $x_1, x_2, \ldots, x_n \in V \Rightarrow T(x_k) \in U$ for every k.

$$\Rightarrow \qquad T(x_k) = \sum_{j=1}^{m} \lambda_{kj} y_j \text{ for all } k, \lambda_{kj} \in F.$$

Now, substituting the action of T_{ij} on x_i, we have

$$T(x_k) = \sum_{j=1}^{m} \lambda_{kj}\, T_{kj}(x_k) = \sum_{i=1}^{n} \sum_{j=1}^{m} \lambda_{ij}\, T_{ij}(x_k).$$

Since T coincides with $\sum_{i=1}^{n} \sum_{j=1}^{m} \lambda_{ij}\, T_{ij}$ on the basis element of V

$$\Rightarrow \qquad T = \sum_{i=1}^{n} \sum_{j=1}^{m} \lambda_{ij}\, T_{ij}$$

$\Rightarrow \{T_{ij}\}$ spans $L(V, U)$

$\Rightarrow \{T_{ij}\}$ is a basis of $L(V, U)$

$\Rightarrow \qquad \dim (L(V, U)) = \dim V \cdot \dim U = n \cdot m = m \cdot n.$

This proves the result.

Corollary 5.5.5. If V is n-dimensional vector space, then the algebra $L(V, V)$ is of dimension n^2.

Since we know that every field over itself is a vector space and is of dimension 1, therefore, appealing to Theorem 5.5.4, we get the following corollary.

Corollary 5.5.6 If V is a vector space over F, then

$$\dim (L(V, F)) = \dim V.$$

The space $L(V, F)$ is called *dual space* and it will be dealt in Chapter 6. However, if we closely see the proof of Theorem 5.5.4, we can coincide that $\{T_{11}, T_{21}, \ldots, T_n\}$ is a basis of $L(V, F)$ and $T_{i1}(x_i) = 1$ and $T_{i1}(x_k) = 0$ for $i \neq k$, where $\{x_1, x_2, \ldots, x_n\}$ is a basis of V. This will be discussed in detail in Chapter 6.

6 SOME WORKED OUT PROBLEMS AND EXAMPLES

1. If $S, T : \mathbb{R}^2 \to \mathbb{R}^2$ such that $S(x, y) = (-y, x)$ and $T(x, y) = (x, -y)$, then find $S + T$, $S \circ T$, $T \circ S$ and $4T - 3S$.

$$\begin{aligned}(S + T)(x, y) &= S(x, y) + T(x, y) \\ &= (x - y, x - y). \\ (S \circ T)(x, y) &= S(T(x, y)) \\ &= S(x, -y) = (y, x). \\ (T \circ S)(x, y) &= T(S(x, y))\end{aligned}$$

$$= T(-y, x) = (-y, -x).$$

$\Rightarrow$ $T \circ S$ is additive inverse of $S \circ T$ and $T \circ S \neq S \circ T$.

$$(4T - 3S)(x, y) = 4(x, -y) - 3(-y, x) = (4x + 3y, -4y - 3x).$$

2. If $T : \mathbb{R}^2 \to \mathbb{R}^2$ such that $T(x, y) = (x, 0)$ and $S : \mathbb{R}^2 \to \mathbb{R}^2$ such that $S(x, y) = (0, y)$, then

$$(S + T)(x, y) = S(x, y) + T(x, y) = (0, y) + (x, 0) = (x, y)$$

$$\Rightarrow \quad S + T = I$$

and

$$(ST)(x, y) = S(T(x, y)) = S(x, 0) = (0, 0)$$

$$(TS)(x, y) = T(S(x, y)) = T(0, x) = (0, 0)$$

$$\Rightarrow \quad TS = ST = 0.$$

Also $\quad S^2 = S,\ T^2 = T.$

3. Let $T : \mathbb{R}^2 \to \mathbb{R}^2$ such that $T(x, y) = (2x + y, x + y)$, then trivially T is one-one and so T is onto. Therefore, T^{-1} exists. The matrix $m(T)$ under the standard basis is

$$\begin{bmatrix} 2 & 1 \\ 1 & 1 \end{bmatrix} \text{ and } (m(T))^{-1} = \begin{bmatrix} 1 & -1 \\ -1 & 2 \end{bmatrix}.$$

Hence, $T^{-1}(x, y) = (x - y, -x + 2y)$. The other method to calculate T^{-1} is as follows. Let

$$T(x, y) = (a, b)$$

$$\Rightarrow \quad T^{-1}(a, b) = (x, y)$$

$$\Rightarrow \quad a = 2x + y \text{ and } b = x + y$$

$$\Rightarrow \quad a - b = x \text{ and } 2b - a = y$$

$$\Rightarrow \quad T^{-1}(a, b) = (a - b, 2b - a)$$

$$\Rightarrow \quad T^{-1}(x, y) = (x - y, 2y - x).$$

4. Let $T : \mathbb{R}^3 \to \mathbb{R}^3$ such that $T(x, y, z) = (x - y, x + y, 2x - y + z)$, then trivially T is a linear map and $\ker T = 0 \Rightarrow T$ is one-one and hence it is onto $\Rightarrow T^{-1}$ exists. Now, under the standard basis

$$m(T) = \begin{bmatrix} 1 & -1 & 0 \\ 1 & 1 & 0 \\ 2 & -1 & 1 \end{bmatrix}, \text{ then } \det(m(T)) = 2 \neq 0.$$

Hence

$$(m(T))^{-1} = \frac{1}{2}\begin{bmatrix} 1 & 1 & 0 \\ -1 & 1 & 0 \\ -3 & -1 & 2 \end{bmatrix}$$

Therefore, $T^{-1}(x, y, z) = \frac{1}{2}(x + y, y - x, -3x - y + 2z)$ .Now, let us calculate T^{-1} by the second method. Let

$$\begin{aligned}
& T(x, y, z) &=& (a, b, c) \\
\Rightarrow\quad & T^{-1}(a, b, c) &=& (x, y, z) \\
\Rightarrow\quad & x - y &=& a, x + y = b, 2x - y + z = c \\
\Rightarrow\quad & x &=& \frac{1}{2}(a + b), y = \frac{1}{2}(b - a) \\
\Rightarrow\quad & z &=& c + \frac{1}{2}(b - a) - (a - b) \\
& &=& \frac{2c + b - a - 2a - 2b}{2} \\
& &=& \frac{1}{2}(-3a - b + 2c) \\
\Rightarrow\quad & T^{-1}(a, b, c) &=& \frac{1}{2}(a + b, b - a, -3a - b + 2c) \\
\text{i.e.,}\quad & T^{-1}(x, y, z) &=& \frac{1}{2}(x + y, y - x, -3x - y + 2z).
\end{aligned}$$

5. Let $T : \mathbb{R}^3 \to \mathbb{R}^3$ such that $T(x, y, z) = (x + y + z, 2x + 3z, 5x + 5y + z)$, then, under the standard basis

$$m(T) = \begin{bmatrix} 1 & 1 & 1 \\ 0 & 2 & 3 \\ 5 & 5 & 1 \end{bmatrix}, \text{ then } \det(m(T)) = -8.$$

Hence

$$(m(T))^{-1} = \begin{bmatrix} \frac{13}{8} & -\frac{1}{2} & -\frac{1}{8} \\ -\frac{15}{8} & \frac{1}{2} & \frac{3}{8} \\ \frac{5}{4} & 0 & -\frac{1}{4} \end{bmatrix}$$

$$= \frac{1}{8}\begin{bmatrix} 13 & -4 & -1 \\ -15 & 4 & 3 \\ 10 & 0 & -2 \end{bmatrix} \text{ (verify)}$$

$$\Rightarrow T^{-1}(x, y, z) = \frac{1}{8}(13x - 4y - x, -15x + 4y + 3z, 10z - 2z).$$

As done in Example 4, we can also calculate $T^{-1}(x, y, z)$ to be the same as above.

6. If $S, T \in L(V, V)$ and S, T are invertible (non-singular), then

$$(ST)^{-1} = T^{-1}S^{-1}$$

$$\Rightarrow \quad (ST)(T^{-1}S^{-1}) = S(TT^{-1})S^{-1}$$

$$= (ST)S^{-1} = SS^{-1} = I.$$

$$\Rightarrow \quad (ST)^{-1} = T^{-1}S^{-1}.$$

7. If $T : \mathbb{R}^2 \to \mathbb{R}^2$ such that $T(x, y) = (2x, y - x)$ and $S : \mathbb{R}^2 \to \mathbb{R}^2$ such that $S(x, y) = (bx, (a - b)x + ay)$, where $a, b \in \mathbb{R}$, then

$$T(S(x, y)) = T(bx, (a - b)x + ay)$$

$$= (2bx, (a - b)x + ay - bx)$$

$$= (2bx, (a - 2b)x + ay).$$

$$S(T(x, y)) = S(2x, y - x)$$

$$= (2bx, (a - b)(2x) + a(y - x))$$

$$= (2bx, (a - 2b)x + 2y).$$

$$\Rightarrow \quad ST = TS.$$

Hence, choosing as many values of a and b as we like, we can get infinite number of linear transformations S such that $ST = TS$.

8. A linear map $T : V \to V$ is singular iff there exists $v \in V$, such that $v \neq 0$ and $T(v) = 0$.

If T is singular $\Rightarrow$ T is not one-one $\Rightarrow$ ker $T \neq 0$

$\Rightarrow$ there exists $x \in V$, such that $x \neq 0$ and $x \in$ ker T

$\Rightarrow T(x) = 0$. Also if for some $x \in V, x \neq 0$ if $T(x) = 0$, then $T(x) = T(0)$. If T is one-one then $x = 0$ which is a contradiction $\Rightarrow T$ is not one-one $\Rightarrow T$ is singular.

9. $T \in L(V, V)$ is singular iff there exists $S \in L(V, V)$, such that $S \neq 0$ and $TS = 0$.

If T is singular $\Rightarrow$ ker $T \neq 0 \Rightarrow V =$ ker $T \oplus U$ for some subspaces U of V. Let S be a projection on ker T along $U \Rightarrow S(x) = x$ for all $x \in$ ker T. Let

$v \in V \Rightarrow v = x + y, x \in \ker T$ and $y \in U$
$\Rightarrow S(v) = S(x) + S(y) = x$
$\Rightarrow T(S(v)) = T(x) = 0 \Rightarrow TS = 0.$
Also if $TS = 0 \Rightarrow (TS)V = 0$
$\Rightarrow SV \subseteq \ker T \Rightarrow \ker T \neq 0$
$\Rightarrow T$ is not one-one $\Rightarrow T$ is singular.

10. If $T \in L(V, V)$, such that $TV = \ker T$, then T is nilpotent and dim V is even. Trivially $T^2V = 0 \Rightarrow T$ is nilpotent. Also
dim $V = r(T) + n(T) = 2r(T)$ as $n(T) = r(T)$
$\Rightarrow$ dim V is even.

11. If $T \in L(V, V)$ and $TV = \ker(T^2)$, then T is nilpotent and $T^2V = \ker(T)$.
It is easy to see that $T^3V = 0 \Rightarrow T^3 = 0 \Rightarrow T$ is nilpotent. Also $T^2V = \{T(x) \mid x \in \ker T^2$ *i.e.*, $T(T(x)) = 0\} = \ker T.$

12. If V and W are the sets of real functions $y = f(x)$ such that

$$\frac{d^2y}{dx^2} + 4y = 0 \qquad \ldots(1)$$

and

$$\frac{d^3y}{dx^3} - 6\frac{d^2y}{dx^2} + 11\frac{dy}{dx} - 6y = 0 \qquad \ldots(2)$$

respectively, then V and W can be decomposed of subspaces which are $\frac{d}{dx}$-invariant.

Solving equation (1), we get two solutions, e^{2x} and e^{-2x}. Trivially $V = [e^{2x}] \oplus [e^{-2x}] = U_1 \oplus U_2$

and U_1, U_2 are $\frac{d}{dx}$-invariant.

Also solving (2), we get e^x, e^{2x} and e^{3x} as solutions. So $W = [e^{2x}] \oplus [e^{2x}] \oplus [e^{3x}] = U_1 \oplus U_2 \oplus U_3.$

We can see that U_i's are $\frac{d}{dx}$-invariant subspace of W.

13. Let $\mathbb{R}^3 = U_1 \oplus W_1 = U_2 \oplus W_2$, where $U_1 = \{(x, y, 0) \mid x, y \in \ \}$, $W_1 = \{(0, 0, z) \mid z \in \ \}$, $U_2 = \{(x, 0, 0) \mid x \in \ \}$ and $W_2 = \{(0, y, z) \mid y, z \in \ \}$. Let T_1 and T_2 be projections on U_1 along W_1 and on U_2 along W_2, respectively. Now,

$$T_1T_2(a, b, c) = T_1(T_2(a, b, c)) = T_1(a, 0, 0)$$
$$= (a, 0, 0) = T_2(a, b, c)$$
$$\Rightarrow \qquad T_1T_2 = T_2.$$

Also

$$T_2T_1(a, b, c) = T_2(T_1(a, b, c))$$
$$= T_2(a, b, 0) = T_2((a, 0, 0) + (0, b, 0))$$
$$= (a, 0, 0) = T_2(a, b, c)$$
$$\Rightarrow \qquad T_2T_1 = T_2 = T_1T_2.$$

14. If $T \in L(V, V)$, then V can be expressed as sum of non-singular linear maps $V \to V$.

If T is non-singular, then $T = T_1 + T_2$, where $T_1 = \lambda T$ and $T_2 = \mu T$ such that $\lambda, \mu \neq 0$ and $\lambda + \mu = 1$, then trivially T_1 and T_2 are non-singular. Now, suppose T is singular, then is neither one-one nor onto $\Rightarrow$ ker $T \neq 0$ and $TV \neq V$. Let $\{x_1, x_2, \ldots, x_n\}$ be a basis of ker T, then $\{x_1, x_2, \ldots, x_n\}$ can be extended to $\{x_1, x_2, \ldots, x_r, y_1, y_2, \ldots, y_t\}$ a basis of V, such that dim $V = r + t$. Consider $\{T(y_1), T(y_2), \ldots, T(y_t)\}$ such that

$$\sum_{i=1}^{t} \lambda_i T(y_i) = 0$$

$$\Rightarrow \qquad T\left(\sum_{i=1}^{t} \lambda_i y_i\right) = 0$$

$$\Rightarrow \qquad \sum_{i=1}^{t} \lambda_i y_i \in \ker T$$

$$\Rightarrow \qquad \sum_{i=1}^{t} \lambda_i y_i = \sum_{i=1}^{r} \beta_i x_i$$

$$\Rightarrow \qquad \lambda_i = 0 = \beta_i$$

as $\{x_1, x_2, \ldots, x_r, y_1, y_2, \ldots, y_t\}$ is a basis of V.

Further, let $z \in T(V) \Rightarrow z = T(x)$, where $x \in V$

$$\Rightarrow \qquad x = \sum_{i=1}^{t} \lambda_i x_i + \sum_{j=1}^{r} \beta_j x_j$$

$$\Rightarrow \qquad z = T(x) = \sum_{j=1}^{t} \beta_j T(y_j)$$

$\Rightarrow \{T(y_1), T(y_2), \ldots, T(y_t)\}$ is a basis of TV.

Let $T(y_i) = v_i$, then $\{v_1, v_2, \ldots, v_t\}$ can be extended to a basis $\{v_1, v_2, \ldots, v_t, v_{t+1}, v_{t+2}, \ldots, v_n\}$, where $n = r + t = \dim V$. Now,

$$
\begin{aligned}
T(y_1) &= v_1 = 1v_1 + 0v_2 + \ldots + 0v_n \\
T(y_2) &= v_2 = 0v_1 + 1v_2 + \ldots + 0v_n \\
&\vdots \\
T(y_t) &= v_t = 0v_1 + 0v_2 + \ldots + 1v_t + 0v_{t+1} + \ldots + 0v_n \\
T(x_1) &= 0 = 0v_1 + 0v_2 + \ldots + 0v_n \\
&\vdots \\
T(x_r) &= 0 = 1v_1 + 0v_2 + \ldots + 0v_n
\end{aligned}
$$

$$
\Rightarrow \quad m(T) = A = \begin{bmatrix} I_t & O \\ O & O \end{bmatrix}
$$

$$
= \begin{bmatrix} \alpha_1 & & & & & & & O \\ & \alpha_2 & & & & & & \\ & & \ddots & & & & & \\ & & & \alpha_t & & & & \\ & & & & 1 & & & \\ & & & & & \ddots & & \\ O & & & & & & & 1 \end{bmatrix}
$$

$$
= \begin{bmatrix} \lambda_1 & & & & & & & O \\ & \lambda_2 & & & & & & \\ & & \ddots & & & & & \\ & & & \lambda_t & & & & \\ & & & & -1 & & & \\ & & & & & \ddots & & \\ O & & & & & & & -1 \end{bmatrix}
$$

where $\alpha_i + \lambda_i = 1$ but $\alpha_i \neq 0$, $\lambda_i \neq 0$. Therefore, $A = A_1 + A_2$. Trivially A_1 and A_2 are non-singular and we can get $T_1, T_2 \in L(V, V)$ corresponding to A_1 and A_2, such that $m(T_1) = A_1$ and $m(T_2) = A_2 \Rightarrow T = T_1 + T_2$, such that T_i's are non-singular.

EXERCISE

1. Give an example of idempotent, involuntary linear transformation.

2. If $T : \mathbb{R}^3 \to \mathbb{R}^3$ such that $T(x, y, z) = (x - y, x + z, y - 2z)$ and $S : \mathbb{R}^3 \to \mathbb{R}^3$ such that $S(x, y, z) = (x + y, z, x)$, then find $S + T$, $S \text{ o } T$, $T \text{ o } S$ and $4S - 3T$ and their matrices.

3. If $x \in V$ such that $x \neq 0$ and $T \in L(V, V)$, then $\{x, T(x), \ldots, T^k(x), \ldots\}$ can be linearly independent as V is infinite dimensional. If k is the least positive integer such that $\{x, T(x), \ldots, T^{k-1}(x)\}$ is linearly independent, then $\{x, T(x), \ldots, T^{k-1}(x)\}$ is called T-cyclic space of T and x. Show that T-cyclic subspaces are T-invariant.

4. If $U = \ker T^k$ such that $T^k V = T^{k+1} V = \ldots$, then prove that $\overline{T} : V/U \to V/U$ such that $\overline{T}(x + U) = T(x) + U$ is non-singular. Also if $W = T^k V$, then $\overline{T} : V/W$ such that $\overline{T}(x + W) = T(x) + W$ is nilpotent.

5. If T_1 and T_2 are projections of V on U_1 along W_1 and on U_2 along W_2, then prove that $T_1T_2 = T_2$ iff $U_2 \subseteq U_1$.

6. Find a basis of $L(\mathbb{R}^2, \mathbb{R}^2)$.

7. If dim $V = 1$, then prove that $L(V, V)$ is commutative.

8. If $S, T \in L(V, V)$, then show that

(*a*) $r(\lambda T) = r(T)$ if $\lambda \neq 0$.

(*b*) $r(S) - r(T) \leq r(S + T) \leq r(S) + r(T)$.

(*c*) $r(S) + r(T) - n \leq r(S \text{ o } T) \leq \min(r(T), r(S))$.

(*d*) $r(ST) \leq r(T)$.

(*e*) $r(ST) = r(TS) = r(T)$ for S non-singular.

6

Dual Spaces

We have seen earlier that for any vector spaces V and U over the same field F, $L(V, U)$ is a vector space and the elements of $L(V, U)$ is a vector space and the elements of $L(V, U)$ are called linear transformations. Since every field over itself is also a vector space, considering F_F as a vector space we have $L(V, F)$. The elements of $L(V, F)$ are also linear transformations, but being scalar valued, they are called *linear functional*. So all linear functionals are linear transformations but not vice versa.

1 LINEAR FUNCTIONAL AND DUAL BASIS

Definition 6.1.1. The vector space $L(V, F)$ is called the *dual* space of V, so every vector space V has its dual space and $L(V, F)$ will be denoted by V^*.

Definition 6.1.2. Given a basis $\{v_1, v_2, \ldots, v_n\}$ of V, a subset $A = \{f_1, f_2, \ldots, f_n\}$ of V^* is called *dual basis* of V^*, with respect to $\{v_1, v_2, \ldots v_n\}$ if A is a basis of V^* and $f_i(v_i) = 1$ and $f_i(v_j) = 0$ for $i \neq j$.

It can be noted that the dual basis with respect to any basis of the vector space is always unique. We shall see later that changing the base of V, we get different dual basis of V^*.

Theorem 6.1.3 For a vector space V_F, V_F^* has a dual basis.

Proof: Let $\{v_1, v_2, \ldots v_n\}$ be a basis of V, then for any $x \in V$, E unique $\lambda_1, \lambda_2, \ldots, \lambda_n, \in F$, such that $x = \sum_{i=1}^{n} \lambda_i v_i$. Define $f_i : V \to F$ as $f_i(x) = \lambda_i$. Trivially it is well defined. Now, we show that $f_i \in V^*$. Let $x, y \in V$, then there exists unique $\lambda_1, \lambda_2, \ldots, \lambda_n$, and $\beta_1, \beta_2, \ldots, \beta_n$, such that

$$x = \sum_{i=1}^{n} \lambda_i v_i \text{ , } y = \sum_{i=1}^{n} \beta_i v_i$$

Hence, $f_i(x+y) = f_i\left(\sum_{i=1}^{n} (\lambda_i + \beta_i) v_i\right) = \lambda_i + \beta_i = f_i(x) + f_i(y)$

and $$f_i(\alpha x) = f_i\left(\sum_{i=1}^{n}(\alpha\lambda_i)v_i\right) = \alpha\lambda_i = \alpha f_i(x)$$

So we get $f_i \in V^*$. Trivially $f_i(v_i) = 1$ and $f_i(v_j) = 0$ for all $i \neq j$. To show that $\{f_1, f_2, \ldots f_n\}$ is a basis, let

$$\sum_{i=1}^{n}\alpha_i f_i = 0 \text{ for } \alpha_i \in F,$$

Then $\left(\sum_{i=1}^{n}\alpha_i f_i\right)(v_j) = \alpha_j = 0$

This implies that $\{f_1, f_2, \ldots f_n\}$ is linearly independent. Since dim (V^*) = dim $(V) = n$. Therefore, $\{f_1, f_2, \ldots f_n\}$ is a basis of V^*.

Example 6.1.4

$\mathbb{R}^2 = \{(x, y) \mid x, y, \in \mathbb{R}\}$ is a vector space over $\mathbb{R}$. Taking $A = \{(1, 0), (0, 1)\}$, $B = \{(1, 2), (2, 1)\}$ as different bases of V, we find the dual bases of A and B, respectively. Define $\phi_1, \phi_2, : \mathbb{R}^2 \to \mathbb{R}$ such that $\phi_1(1, 0) = 1$, $\phi_1(0, 1) = 0$, $\phi_2(1, 0) = 0$ and $\phi_2(0, 1) = 0$. Now, for any $(x, y) \in \mathbb{R}^2$, $(x, y) = x(1, 0) + y(0, 1)$. So $\phi_1(x, y) = x$ and $\phi_2(x, y) = y$ will serve the purpose and $\{\phi_1, \phi_2\}$ will be the dual of $\{(0, 1), (0, 1)\}$. Now, taking $B = \{(1, 2), (2, 1)\}$ as a basis of $\mathbb{R}^2$, we see that

$$(x, y) = \frac{2y - x}{3}(1, 2) + \frac{2x - y}{3}(2, 1)$$

Define $f_1, f_2 : \mathbb{R}^2 \to \mathbb{R}$ such that

$$f_1(x, y) = \frac{2y - x}{3}$$

and $$f_2(x, y) = \frac{2x - y}{3}$$

Trivially $f_1, f_2 \in (\mathbb{R}^2)^*$ and $f_1(1, 2) = 1$, $f_2(2, 1) = 0$, $f_2(1, 2) = 0$ and $f_2(2, 1) = 1$.

As we know that if $T : V \to U$ is a linear transformation, then T can be represented by a matrix of order $m \times n$, where $m = \dim U$ and $n = \dim V$. Now, if $T : V \to F$ is a linear transformation, where dim $V = n$, then T can be represented by an $1 \times n$ matrix $[\alpha_1, \alpha_2, \ldots, \alpha_n]$ (say), so T can be treated as a row vector. Given $\{v_1, v_2 \ldots, v_n\}$ as a basis of V, we express

$$V = Fv_1 \oplus Fv_2 \oplus \ldots \oplus Fv_n \cong F \oplus F \oplus \ldots \oplus F.$$

so T will be a map such that

$$T(\lambda_1, \lambda_2, \ldots, \lambda_n) = (\alpha_1, \alpha_2, \ldots, \alpha_n)\begin{pmatrix}\lambda_1\\ \lambda_2\\ \vdots\\ \lambda_n\end{pmatrix}$$

$$= \alpha_1\lambda_1 + \alpha_2\lambda_2 + \ldots + \alpha_n\lambda_n.$$

However, if we consider $\mathbb{R}^n$. then any map $f : \mathbb{R}^n \to \mathbb{R}$ is given by

$$f(\alpha_1, \alpha_2, \ldots, \alpha_n) = (\alpha_1, \alpha_2, \ldots, \alpha_n)\begin{pmatrix}\alpha_1\\ \alpha_2\\ \vdots\\ \alpha_n\end{pmatrix} = \alpha_1\alpha_1 + \alpha_2\alpha_2 + \ldots + \alpha_n\alpha_n.$$

Example 6.1.5

Considering the basis as given in Example 6.1.4, define $f_1, f_2 : R^2 \to R$ such that

$$f_1(x, y) = ax + by,\ a, b \in \mathbb{R}$$
$$f_2(x, y) = cx + dy,\ c, d \in \mathbb{R}.$$

For dual basis of {(1, 0), (0, 1)}, we need

$$f_1(1, 0) = 1, f_2(1, 0) = 0, f_1(0, 1) = 0, f_2(0, 1) = 1$$

$$\Rightarrow \qquad \alpha = 1, c = 0, b = 0, d = 1,$$

so we get $f_1(x, y) = y$, which are the projections.

Now, taking the basis as {(1, 2), (2, 1)}, we get

$$f_1(1, 2) = a + 2b = 1,$$
$$f_1(2, 1) = 2a + b = 0,$$
$$f_2(1, 2) = c + 2d = 0$$
$$f_2(2, 1) = 2c + d = 1$$

Hence, $a + 2a + b = 0$ gives

$$-3a = 1$$

$$\Rightarrow \qquad a = -\frac{1}{3}, b = \frac{2}{3}$$

$$\Rightarrow \qquad f_1(x, y) = -\frac{x}{3} + \frac{2}{3}y = \frac{2y - x}{3}$$

Similarly, we can solve the other equations and get

$$f_2(x, y) = \frac{2x - y}{3}$$

Now we discuss some examples in the reverse direction.

Example 6.1.6

Define $\phi_1(x, y) = x + 2y$, $\phi_2(x, y) = 2x - y$. It is trivial to see that $\phi_1, \phi_2 \in L(\mathbb{R}^2, \mathbb{R})$. Now we find a basis of $\mathbb{R}^2$ with respect to which $\{\phi_1, \phi_2\}$ is the dual basis. Let $\{(a, b), (c, d)\}$ be a basis of $\mathbb{R}^2$. Then $\phi_1(a, b) = 1 = a + 2b$, $\phi_1(c, d) = 0 = c + 2d$, $\phi_2(a, b) = 0 = 2a - b$ and $\phi_2(c, d) = 1 = 2c - d$. Solving the equations we get $a = \frac{1}{5}$, $b = \frac{2}{5}$, $c = \frac{2}{5}$, $d = -\frac{1}{5}$. Therefore $\left\{\left(\frac{1}{5}, \frac{2}{5}\right), \left(\frac{2}{5}, \frac{1}{5}\right)\right\}$ is the desired basis of $\mathbb{R}^2$.

Example 6.1.7

Let $P_1(x)$ be the vector space of polynomials of degree ≤ 1 choosing $\{1 + x, x - 1\}$ as a basis of $P_1(x)$, we find the dual basis.

For any

$$a + bx \in P_1(x),\ a + bx = \frac{a+b}{2}(1 + x) + \frac{b-a}{2}(x - 1)$$

Now define $\phi_1, \phi_2 : P_1(x) \to \mathbb{R}$ such that

$$\phi_1(a + bx) = \frac{a+b}{2}$$

and

$$\phi_2(a + bx) = \frac{b-a}{2}$$

Trivially $\{\phi_1, \phi_2\}$ is the dual of $\{1 + x, x - 1\}$.

Example 6.1.8

Consider the vector space $P_2(x)$ and the maps $\phi_1, \phi_2, \phi_3 : P_2(x) \to \mathbb{R}$ such that

$$\phi_1(f(x)) = \int_0^1 f(x)dx,$$

$$\phi_2(f(x)) = f'(1),\ \phi_3(f(x)) = f'(0)$$

Now, we find the dual basis of $\{\phi_1, \phi_2, \phi_3\}$.

Let $\left\{\underbrace{a + bx + cx^2}_{e_1}, \underbrace{d + ex + fx^2}_{e_2}, \underbrace{\alpha + \beta x + \gamma x^2}_{e_3}\right\}$ be the required basis of $P_2(x)$, then

$$\phi_1(e_1) = ax + \frac{bx^2}{2} + \frac{cx^3}{3}\Big|_0^1 = a + \frac{b}{2} + \frac{c}{3} = 1$$

$$\phi_1(e_2) = d + \frac{e}{2} + \frac{f}{3} = 0$$

$$\phi_1(e_3) = \alpha + \frac{\beta}{2} + \frac{\gamma}{3} = 0$$

$$\phi_2(e_1) = b + 2c = 0$$

$$\phi_2(e_2) = e + 2f = 1$$

$$\phi_2(e_2) = \beta + 2\gamma = 0$$

$$\phi_3(e_1) = b = 0$$

$$\phi_3(e_2) = e = 0$$

$$\phi_3(e_3) = \beta = 1$$

$$b = 0, c = 0, a = 1, e = 0, f = \frac{1}{2}, d = -\frac{1}{6},$$

$$\beta = 1, \gamma = -\frac{1}{2}, \alpha = -\frac{\beta}{2} - \frac{\gamma}{3}$$

$$= -\frac{1}{2} + \frac{1}{6} = -\frac{1}{3}$$

Therefore $\left\{1, -\frac{1}{6} + \frac{x^2}{2}, -\frac{1}{3} + x - \frac{1}{2}x^2\right\}$

is the required basis of $P_2(x)$.

2 ANNIHILATOR

Definition 6.2.1

Let V_F be a vector space and W be a subset of V, then annihilator of W is defined as $A(W) = \{f \in V^* \mid f(w) = 0 \text{ for all } w \in W\}$. Trivially zero linear function always belongs to $A(W)$. Now, if $f_1, f_2 \in A(W)$, then $f_1(w) = 0$, $f_2(w) = 0$ for all $w \in W$.

Hence $(f_1 - f_2)(w) = 0$ for all $w \in W \Rightarrow f_1\ f_2 \in A(W)$.

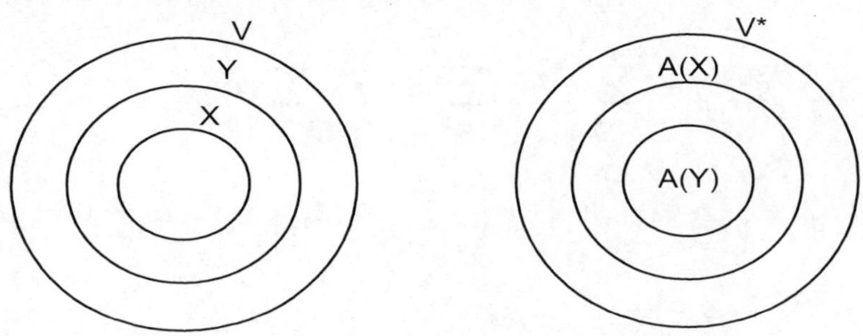

Also for any $\alpha \in F$, $(\alpha f)(w) = \alpha (f(w))$. Hence, if $f \in A(W)$, then $\alpha f \in A(W) \Rightarrow A(W)$ is a subspace of V^*.

It is easy to see that for any two subsets X and Y of V, if $X \subseteq Y$, then $A(X) \supseteq A(Y)$.

Example 6.2.2

Consider projection maps $\pi_i : \mathbb{R}^n \to \mathbb{R}$ such that $\pi_i (x_1, x_2, \ldots, x_i, x_{i+1} \ldots, x_n) = x_i$, then π_i are linear functional. If $W_1 = \{(x_1, 0, \ldots, 0)\}$, $W_2 = \{x_1, x_2, 0, \ldots, 0\} \ldots$, then all the elements of $W_1, W_2, \ldots W_{i-1}$ will be killed by π_i. Although the annihilators are defined for any subset X of V, but our interest lies when X is taken as a subspace of V. Now, first we prove the following lemma.

Lemma 6.2.3

If V is a vector space over F and $x \in V, x \neq 0$, then there exists an element $f \in V^*$ such that $f(x) \neq 0$.

Proof

Since $x \neq 0$, $\{x\}$ can be extended to a basis $\{x, x_2, x_3, \ldots, x_n\}$ of V. Therefore, as done earlier, there will exist a dual basis $\{f, f_2, \ldots f_n\}$ such that $f(x) = 1 \Rightarrow \exists F \in V^*$ such that $f(x) \neq 0$.

Theorem 6.2.4

If U is a subspace of V, then

$$\dim U + \dim A(U) = \dim V = \dim V^*.$$

Proof

Let $\{x_1, x_2, \ldots, x_r\}$ be a basis of U, then it can be extended to a basis $\{x_1, x_2, \ldots, x_r, x_{r+1}, \ldots, x_n\} = A$ of V. Let $\{f_1, f_2, \ldots, f_r, f_{r+1}, \ldots, f_n\}$ be the dual basis of A. Then, by definition $f_j(x_i) = 0$, for all $j = r + 1, \ldots, n$ and $i = 2, \ldots, r \Rightarrow f_j(x) = 0$ for every $x \in U$ and $j = r + 1, \ldots, n \Rightarrow f_{r+1}, f_{r+2}, \ldots, f_n \in A(U)$.

Now we prove that $\{f_{r+1}, f_{r+2}, \ldots, f_n\}$ is a basis of $A(U)$. Let $f \in A(U)$, then $f(x) = 0$ for all $x \in U$. Hence $f(x_i) = 0$ for all $i = 1, 2, \ldots, r$. Since $f \in V^*$.

$$\Rightarrow \qquad f = \sum_{i=1}^{n} \alpha_i f_i = \sum_{i=1}^{n} \alpha_i f_i + \sum_{j=r+1}^{n} \alpha_j f_j$$

Now, $0 = f(x_1) = \left(\sum_{i=1}^{n} \alpha_i f_i\right)(x_1) + \left(\sum_{j=r+1}^{n} \alpha_j f_j\right)(x_1) = \alpha_1 + 0 \Rightarrow \alpha_1 = 0.$

Similarly as $f(x_2) = f(x_3) = \ldots = f(x_r) = 0$, we get $\alpha_2 = \alpha_3 = \ldots = \alpha_r = 0$.

Therefore

$$f = \sum_{j=r+1}^{n} \alpha_j f_j \Rightarrow \{f_{r+1}, f_{r+2}, \ldots f_n\}$$

spans $A(U)$. Also being a subset of a linearly independent set it is linearly independent.

$\Rightarrow \{f_{r+1}, f_{r+2}, \ldots f_n\}$ is a basis of $A(U)$

$\Rightarrow \quad \dim (A(U)) = n - r = \dim V - \dim U.$

Now we, prove another result from which Theorem 6.2.4 follows immediately.

Theorem 6.2.5

For any subspace U of V, $U^* \cong \dfrac{V^*}{A(U)}$

Proof: Let $f \in V^*$ and $g = f \mid U$ (restriction of f on U), then it is easy to see that $g \in U^*$. Now we define a map $\eta : V^* \to U^*$ such that $\eta(f) = f \mid U$. Trivially η is well defined

$$\begin{aligned}\eta(\alpha f + \beta h) &= \alpha f + \beta h|_U, \text{ where } f, h \in V^*, \alpha, \beta \in F\\ &= \alpha (f|_U) + \beta (h|_U) = \alpha\eta(f) + \beta\eta(h).\end{aligned}$$

Hence η is a vector space homomorphism. As done in Theorem 6.2.4, let $\{x_1, x_2, \ldots, x_r\}$ be a basis of U, then $\{x_1, x_2, \ldots, x_r\}$ can be extended to a basis $\{x_1, x_2, \ldots, x_r, x_{r+1}, \ldots, x_n\}$ of V. So we can write $V = U \oplus W$, where W is spanned by $\{x_{r+1}, x_{r+2}, \ldots, x_n\}$. Now for any $f \in U^*$, define a map $\phi : V \to F$ such that for any $x \in V$, $x = u + w$ and $\phi(x) = f(u)$, then ϕ is well defined as if $x = 0$, then $u = 0$, $w = 0 \Rightarrow f(u) = 0$. It is easy to see that ϕ is a linear functional such that $\phi|_U = f$. Hence η is onto. Thus, by Fundamental theorem of vector homomorphism

$$\frac{V^*}{\ker \eta} \cong U^*$$

Let $g \in \ker \eta \Rightarrow \eta(g) = 0 \Rightarrow g \mid_U = 0$

$\Rightarrow g(u) = 0$ for all $u \in U$.

$\Rightarrow g \in A(U)$. Also if $f \in A(U)$ then $f(u) = 0$ for all $u \in U$.

Hence, $f|_U = 0 \Rightarrow f \in \ker \eta \Rightarrow \dfrac{V^*}{\ker \eta} = \dfrac{V^*}{A(U)} \cong U^*$

3 SECOND DUAL SPACE

As seen for any vector space V, we can get V^*, the dual of V. Now, taking V^* as a vector space, we get $(V^*)^*$, the dual of V^* *i.e.*, V^{**} is the set of all linear functional on V^* and is called second dual of the vector space V. As we have noted earlier, dim V^* = dim V. Similarly

$$\dim V^{**} = \dim V^* \cdot \dim F = \dim V^* = \dim V.$$

Now, on account of the following result, we can see that onwards no distinct duals can be obtained.

Theorem 6.3.1

For any vector space V, $V \cong V^{**}$

Proof: For any $x \in V$, define a map $\phi_x : V^* \to F$ such that $\phi_x(f) = f(x)$. Trivially ϕ_x is well defined and is a linear functional on V^*. Now, define $\eta : V \to V^{**}$, such that $\eta(x) = \phi_z$, then

$$\begin{aligned}\eta(\alpha x + \beta y) &= \phi_{\alpha x + \beta y}\\ &= \alpha\phi_x(f) + \beta\phi_y(f) = (\alpha\phi_x + \alpha\phi_y)(f).\end{aligned}$$

Also
$$\begin{aligned}\phi_{\alpha x + \beta y}(f) &= f(\alpha x + \beta y) = \alpha f(x) + \beta(y)\\ &= \alpha\phi_x(f) + \beta\phi_y(f) = (\alpha\phi_x + \beta\phi_y)(f),\end{aligned}$$

for all $f \in V^*$, $x, y, \alpha, \beta \in F$. Hence, η is a linear transformation.

Let $x \in \ker \eta \Rightarrow \eta(x) = 0 \Rightarrow \phi_x = 0$

$\Rightarrow \phi_x(f) = 0$ for all $f \in V^*$

$\Rightarrow f(x) = 0$ for all $f \in V^*$

$\Rightarrow x = 0$ as V^* has a dual basis $\Rightarrow \eta$ is one-one

$\Rightarrow \quad \eta(V) \subseteq V^{**}$ and $V \cong \eta(V)$

$\Rightarrow \quad \dim V = \dim \eta(V)$ but $\dim V = \dim V^{**}$

$\Rightarrow \quad \eta(V) = V^{**} \Rightarrow \eta$ is onto $\Rightarrow V \cong V^{**}$.

Note 1: The isomorphism above is called canonical isomorphism or sometimes natural isomorphism as under this $x \leftrightarrow \phi_x$ so we identify V with V^{**}. Hence the elements of V may be treated as linear functional on V^*. The isomorphism never depends upon the choice of basis.

Note 2: However, V and V^* being of same dimensions are isomorphic but their isomorphism is not of that nature.

Theorem 6.3.2

If W is a subspace of V, then $A(A(W)) = W$.

Proof: Let $x \in W \Rightarrow f(x) = 0$ for all $f \in A(W)$. Therefore under identification of V and V^{**} we can say that x kills all $f \in A(W) \Rightarrow x \in A(A(W)) \Rightarrow W \subseteq A(A(W))$ but by Theorem 6.2.4

$$\dim A(W) = \dim V - \dim W$$

and
$$\begin{aligned}\dim(A(A(W))) &= \dim V^* - \dim A(W)\\ &= \dim V - (\dim V - \dim W)\end{aligned}$$

$$= \dim(W)$$

$$\Rightarrow \qquad W = A(A(W))$$

Theorem 6.3.3

If U and W are subspaces of V, then

$$A(U + W) = A(U) \cap A(W)$$

and
$$A(U \cap W) = A(U) + A(W)$$

Proof: Since $U, W \subseteq U + W \Rightarrow A(U), A(W)$ both contain $A(U + W)$

$$\Rightarrow \qquad A(U + W) \subseteq A(U) \cap A(W).$$

Let $f \in A(U) \cap A(W) \Rightarrow f(x) = 0$ and $f(y) = 0$ for all $x \in U$ and $y \in W \Rightarrow f(x + y) = 0 \Rightarrow f \in A(U + W) \Rightarrow A(U + W) = A(U) \cap A(W)$, using this, we have

$$A(A(U) + A(W)) = A(A(U)) \cap A(A(W)) = U \cap W,$$

by Theorem 6.3.2 Hence

$$A(A(A(U) + A(W))) = A(U \cap W) \Rightarrow A(U) + A(W) = A(U \cap W).$$

Hence the result is proved.

4 DUAL TRANSFORMATIONS

For any vector space V, we can always obtain the dual of V which is denoted by V^*. It is natural to see that given any $T \in L(V, U)$ for any vector spaces V and U, is it possible to obtain a map $S \in L(U^*, V^*)$? The purpose of this section is to check whether we can get such maps and if such maps are possible then how they are related to the original maps and what are their relative theories.

Definition 6.4.1

If V and U are vector spaces over F, then for any $f \in U^*$ and $T \in L^* V, U)$, the map $S = f \text{o} T : V \to F$ is the dual map corresponding to f and T. Trivially

$$\begin{aligned} S(\alpha x + \beta y) &= f(T(\alpha x + \beta y)) \\ &= f(\alpha T(x) + \beta T(y)) \\ &= \alpha f(T(x) + \beta f(T(y)) \\ &= \alpha S(x) + \beta S(y) \text{ for all } x, y \in V, \alpha, \beta \in F \end{aligned}$$

$$\Rightarrow \qquad S \in V^*.$$

Definition 6.4.2

If $T \in L(V, U)$ and $f \in U^*$, then a map $T' : U^* \to V^*$ given as $T'(f) = f o T$ is called transpose of T or dual of T and is denoted by T^t.

For any $f_1, f_2 \in U^*$, $\alpha, \beta \in F$ and $x \in V$.

$$
\begin{aligned}
(T^t(\alpha f_1 + \beta f_2))(x) &= ((\alpha f_1 + \beta f_2) oT)(x) \\
&= (\alpha f_1 + \beta f_2)(T)(x)) \\
&= \alpha f_1 + (T)(x)) + \beta f_2)(T)(x)) \\
&= \alpha(f_1\, oT)(x)) + \beta(f_2 oT)(x) \\
&= (\alpha(f_1\, oT)(x)) + \beta(f_2 oT))(x) \\
\Rightarrow \quad T^t(\alpha f_1 + \beta f_2) &= \alpha T^t(f_1 + \beta T^t(f_2) \\
\Rightarrow \quad T^t &\in L(U^*, V^*).
\end{aligned}
$$

As we know that for every linear transformation there exists a matrix, so T^t will have a matrix representation. Now we show that $m(T^t) = (m(T))^t$.

Theorem 6.4.3

Let V and U be vector spaces of dimensions n and m respectively and $T \in L(V, U)$ then $m(T^t) = (m(T^t))$ with some choice of basis of V, U, V^* and U^*.

Proof: Let $\{x_1, x_2, \ldots, x_n\}$ and $\{y_1, y_2, \ldots, y_m\}$ be a basis of V and U, respectively. Further, suppose that $\{f_1, f_2, \ldots, f_n\}$ and $\{\phi_1, \phi_2, \ldots, \phi_m\}$ be the respective dual bases of V^* and U^*. Let

$$
m(T) = (a_{ij}) = \begin{bmatrix} a_{11} & a_{12} & \ldots & a_{1n} \\ a_{21} & a_{22} & \ldots & a_{2n} \\ \vdots & \vdots & \vdots & \vdots \\ a_{m1} & a_{m2} & \ldots & a_{mn} \end{bmatrix},
$$

where

$$
T(x_i) = \sum_{j=1}^{m} a_{ji} y_j
$$

$$
(m(T))^t = \begin{bmatrix} a_{11} & a_{21} & \ldots & a_{m1} \\ a_{12} & a_{22} & \ldots & a_{m2} \\ \vdots & \vdots & \vdots & \vdots \\ a_{1n} & a_{2n} & \ldots & a_{mn} \end{bmatrix}
$$

Now we show that

$$
T^t(\phi_1) = \begin{matrix} T^t(\phi_1) & = & a_{11} f_1 & + & a_{12} f_2 & + & \ldots & + & a_{1n} f_n \\ T^t(\phi_2) & = & a_{21} f_1 & + & a_{22} f_2 & + & \ldots & + & a_{2n} f_n \\ \vdots & \vdots & \vdots & \vdots & \vdots & \vdots & \vdots & \vdots & \vdots \\ T^t(\phi_m) & = & a_{m1} f_1 & + & a_{m2} f_2 & + & \ldots & + & a_{mn} f_n \end{matrix}
$$

In general, we show that $T^t(\phi_k) = ak_1 f_1 + ak_2 f_2 + \ldots + ak_n f_n$. Now for $i = 1, 2, \ldots, n$,

$$(T^t(\phi_k))\,(x_i) = \phi_k(T(x_i))$$
$$= \phi_k(a_{1i}\,y_1 + a_{2i}\,y_2 + \ldots + \phi_{ki}\,y_k + \ldots + a_{mi}\,y_m$$
$$= a_{ki} \text{ as } \phi_k(y_j) = \delta_{kj} = \begin{cases} 1 \text{ if } k = j \\ 0 \text{ if } k \neq j \end{cases}$$

Also $(\, ak_1 f_1 + ak_2 f_2 + \ldots + ak_n f_n)\,(x_i)$
$$= (\, ak_1 f_1 + \ldots + ak_i f_i + \ldots + ak_n f_n)\,(x_i)$$
$$= a_{ki} \text{ as } f_j\,(x_i) = \delta_{ij} = \begin{cases} 1 \text{ if } i = j \\ 0 \text{ if } i \neq j \end{cases}$$

$\Rightarrow T^t(\phi_k)$ and $ak_1 f_1 + ak_2 f_2 + \ldots + ak_n f_n)$ coincide on the basis elements

$\Rightarrow \qquad T^t(\phi_k) = ak_1 f_1 + ak_2 f_2 + \ldots + ak_n f_n$

$\Rightarrow \qquad (m(T))^t = m(T^t).$

Now suppose $T = \mathbb{R}^2 \to \mathbb{R}^2$ such that $T(x, y) = (x - y, x + y, x)$, then $T \in L(\mathbb{R}^2, \mathbb{R}^3)$, and

$$m(T) = \begin{bmatrix} 1 & -1 \\ 1 & 1 \\ 1 & 0 \end{bmatrix}$$

with respect to the standard bases of $\mathbb{R}^2$ and $\mathbb{R}^3$.

Therefore,

$$m(T^t) = \begin{bmatrix} 1 & 1 & 1 \\ -1 & 1 & 0 \end{bmatrix}$$

with respect to the standard bases of $(\mathbb{R}^3)$ and $(\mathbb{R}^2)$. The detail of this will follow from the following example.

Let $T : \mathbb{R}^2 \to \mathbb{R}^3$ be given as $T(x, y) = (x + 2y, x\ y, x + 3y)$, then $T^t : (\mathbb{R}^3)^* \to (\mathbb{R}^2)^*$ is given as $T^t(\phi) = \phi\ o\ T$, where $\phi \in (\mathbb{R}^3)^*$ choose $\{(1, 0), (0, 1)\}$ and $\{(1, 0, 0), (0, 1, 0), (0, 0, 1)\}$ as the standard bases of $\mathbb{R}^2$ and $\mathbb{R}^3$, respectively, then

$$m(T) = \begin{bmatrix} 1 & 2 \\ 1 & -1 \\ 1 & 3 \end{bmatrix}$$

Now the dual bases of $(\mathbb{R}^2)^*$ and $(\mathbb{R}^3)^*$ induced by the standard bases will be $\{f_1, f_2\}$ and $\{\phi_1, \phi_2, \phi_3\}$ of $(\mathbb{R}^2)^*$ and $(\mathbb{R}^3)^*$ such that $f_1(x, y) = x$, $f_2(x, y) = y$, and $\phi_1 = (x, y, z) = y$, $\phi_2 = (x, y, z) = y$, $\phi_3 = (x, y, z) = z$.

Now $(\phi_1 T)(x, y) = \phi_1(T(x, y)) = \phi_1(x + 2y, x - y, x + 3y)$

$$= x + 2y = f_1(x, y) + 2f_2(x, y)$$

$\Rightarrow$ $\phi_1 T = f_1 + 2f_2.$

Similarly, $\phi_2 T = f_1 - f_2$ and $\phi_3 T = f_1 + 3f_2$

$\Rightarrow$ $$m(T') = \begin{bmatrix} 1 & 1 & 1 \\ 2 & -1 & 3 \end{bmatrix}$$

Theorem 6.4.4

If $T \in L(V, U)$ and $S \in L(U, W)$, then $(S\, o T)^t = T'\, oS'$.

Proof: $(S\, o\, T)^t \in L(W^*, V^*)$ and $T'\, oS' \in L(W^*, V^*)$ are trivial. Now, for any $\phi \in W^*$,

$$(S\, o\, T)^t\, (\phi) = \phi\, o\, (S\, o\, T).$$

For any $x \in V$,

$$(\phi\, o\, S\, oT)(x) = \phi\, (S(T(x))) = \phi\, (S(y)),\ T(x) = y \in U$$
$$= S^t(\phi)(y) = S^t\phi\, (T(x))$$
$$= ((S^t T')\phi)(x)$$

$\Rightarrow$ $(S\, o\, T)^t = T'\, o\, S'$.

Now suppose $V = U = W$, then the following is trivial to prove.

Theorem 6.4.5

If $S, T \in L(V, V)$, then the following holds:

1. $O^t = O$.
2. $I^t = I$.
3. $(S\, oT)^t = T'\, oS'$
4. $(S + T)^t = S' + T'$
5. If T is non-singular, then $(T')^{-1} = (T^{-1})^t$.
6. $(\lambda\, T)^t = \lambda T', \lambda \in F$.
7. $(T')^t = T$.

Now, using Theorem 6.4.5, we can easily establish the following:

Theorem 6.4.6

For any vector space V, $L(V, V)$ and $L(V^*, V^*)$ are canonically isomorphic.

Proof

Define $\eta : L(V, V) \to L(V^*, V^*)$ given as $\eta(T) = T'$, then

$$\eta(\alpha T_1 + \beta T_2) = (\alpha T_1 + \beta\, T_2)^t$$
$$= \alpha T_1^t + \beta T_2^t = \alpha\eta(T_1) + \beta\eta(T_2)$$

$\Rightarrow \eta$ is a linear transformation. Further $\eta(T_1) = \eta(T_2) \Rightarrow T_1 = T_2 \Rightarrow \eta$ is one-one. Also

$$\eta\,(L(V, V)) = L(V^*, V^*).$$

Hence η is an isomorphism. Since η does not depend upon any choice of bases $\Rightarrow \eta$ is a canonical isomorphism.

Theorem 6.4.7

If $T \in L(V, V)$, then $r(T) = r(T')$, $n(T) = n(T')$.

Proof:

$$r(T') = \dim T'V^* \text{ and } r(T) = \dim TV. \text{ Now, let } x \in \ker T$$

$\Rightarrow \quad T(x) = 0 \Rightarrow \phi\,(T(x) = 0 \Rightarrow (\phi T)(x) = 0$

$\Rightarrow \quad (T'\phi)(x) = 0 \Rightarrow T'\phi \in A\,(\ker T)$ for all $\phi \in V^*$.

$\Rightarrow \quad T'V^* \subseteq A\,(\ker T) \Rightarrow r(T') \leq \dim A(\text{ket } T)$,

i.e., $r(T') \leq \dim V - \dim\,(\ker T)$, as $\dim V = \dim W + \dim A(W) = \dim V^*$

$$= n - (n - r(T))$$

by Sylvester's law of nullity, where $\dim V = n$

$\Rightarrow \quad r(T') \leq r(T)$.

Also as $(T')^t = T$, we get

$$r(T')^t \leq r(T)^t \Rightarrow r(T) \leq r(T')$$

$\Rightarrow \quad r(T) = r(T')$

Since $\quad \dim V = r(T) + n(T)$ and $\dim V^* = r(T') + n(T')$

And $\quad \dim V = \dim V^* \Rightarrow n(T) = n(T')$.

Now, we show that if $T \in L(V, V)$ is a projection, then T' is also a projection.

Theorem 6.4.8

If $V = U \oplus W$ and $T \in L(V, V)$ is a projection on U along W, then T' is also a projection on $A(W)$ along $A(U)$.

Proof: $V^* = A(O) = A(U \cap W) = A\,(U) + A(W)$ by Theorem 6.3.3. Also using the same theorem we have $A(U) \cap A(W) = 0$.

$\Rightarrow V^* = A(U) \oplus A(W)$. Trivially $(T')^2 = T' o T' = (T^2)' = T'$.

$\Rightarrow T'$ is a projection of V^*. Now, we show that it is a projection on $A(W)$.

Let $\phi \in A(W)$, then $T'\phi = \phi\, oT$. For any $x \in V$, $x = u + w$.

$\Rightarrow \quad (T'\phi)\,(x) = \phi(T(u + w) = \phi(u) = \phi\,(u + w) = \phi\,(x)$

$\Rightarrow \quad T'\phi = \phi \Rightarrow T'|_{A(W)} = I.$

Let $\sigma \in A(U) \Rightarrow (T'\sigma)(x) = \sigma(T(x)) = \sigma(T(u + w)) = \sigma(u) = 0$

$\Rightarrow \qquad T'|_{A(U)} = 0$

$\Rightarrow T'$ is a projection on $A(W)$ along $A(U)$.

5 HYPER PLANE

Let V be a vector space that dim $V = n$, then we can get many subspaces of V of dimension $n - 1$. For example, in $\mathbb{R}^2$ there are infinite number of straight lines passing through origin and they represent subspaces of $\mathbb{R}^2$ and are of dimensions 1 and they are maximal proper subspaces. So we define.

Definition 6.5.1

A maximal proper subspace of a vector space is called a *hyper plane.*

As discussed above, there may exist many hyper planes. In $\mathbb{R}^2$ hyper planes are straight lines passing through origin, in $\mathbb{R}^3$ hyper planes are planes passing through origin. In $\mathbb{R}$, $\{0\}$ can be considered to be hyper plane provided the hyper planes are not taken to be proper. Therefore, now onwards we will consider only those vector spaces V for which dim $V > 1$. The hyper planes of or complex vector space.

We know that for any $x \in V, x \neq 0$, there exists a nonzero linear functional f such that $f(x) \neq 0$.

Theorem 6.5.2

A subspace W of V is a hyper plane iff $W = \ker f$ for $f \in V^*, f \neq 0$.

Proof: Let W be a hyper plane. Let $x \in V$ such that $x \notin W$, then

$$[W, x] = V$$

$$\Rightarrow \qquad v = w + \lambda x,\ v \in V,\ \lambda \in \mathbb{R}.$$

Define $f : V \to \mathbb{R}$ such that

$$f(v) = f(w + \lambda x) = \lambda\beta \text{ for some } \beta \in \mathbb{R}, \beta \neq 0.$$

then trivially f is well define

$$\begin{aligned} f(av_1 + bv_2) &= f(a(w_1 + \lambda_1 x) + b(w_2 + \lambda_2 x)) \\ &= f(aw_1 + bw_2 + (a\lambda_1 + b\lambda_2)x) \\ &= (a\lambda_1 + b\lambda_2)\beta \\ &= a\lambda_1\beta + b\lambda_2\beta \\ &= af(v_1) + bf(v_2) \end{aligned}$$

$\Rightarrow f$ is linear functional.

Trivially $f \neq 0$. Trivially $W \subseteq \ker f$. Let $y \in \ker f$

$\Rightarrow \qquad f(y) = 0$ then $y = w + \alpha x$ for some $w \in W, \alpha \in \mathbb{R}$

$\Rightarrow \qquad \alpha\beta = 0$ but $\beta \neq 0$

$\Rightarrow \qquad \alpha = 0$

$\Rightarrow \qquad y = w$

$\Rightarrow \qquad y \in W$

$\Rightarrow \qquad \ker f \subseteq W \Rightarrow W = \ker f.$

Conversely, suppose $W = \ker f$. Then W is a subspace of V. To see the maximality of W, let $y \in V$ such that $y \notin W$, then we show that $[W, y] = V$.

Let $x \in V$, then $z = x - \dfrac{f(x)}{f(y)}y$ is an element of W because

$$f(z) = f(x) - \frac{f(x)}{f(y)} f(y) = 0$$

$\Rightarrow \qquad x \in [W, y] \Rightarrow V = [W, y]$

$\Rightarrow W$ is a hyper plane.

Now, if W is a real hyper plane, then iW is also be a real hyper plane.

Theorem 6.5.3 If V is a complex vector space and W is a real hyper plane in V, then $W \cap iW$ is a complex hyper plane.

Proof Let $x \in V$ such that $x \notin W \cap iW$, then we show that $[W \cap iW, x] = V$.

Let $x \notin W$, then $ix \notin iW \Rightarrow [iW, ix] = V \Rightarrow x = w + \alpha ix$, $\alpha \in \mathbb{R}$, $w \in iW$, since iW is real hyper plane. So $(1 + \alpha i)w = (1 + \alpha i)(1 - \alpha i)x = (1 + \alpha^2) x \notin W$ as $x \notin W \Rightarrow w \notin W \Rightarrow [W, w] = V \Rightarrow$ for any $y \in V$, $y = \beta w + h$, for some $\beta \in \mathbb{R}$ and $h \in W$. Since $w \notin W \Rightarrow iw \notin iW \Rightarrow [iW, iw] = V$. As some $h \in W \subseteq V \Rightarrow h = \eta iw + z$ for some $\eta \in \mathbb{R}$, $z \in iW$. We know that $w \in iW$. Hence

$$\begin{aligned} y &= \beta w + \eta iw + z \\ &= (\beta + \eta i)w + z \\ &= (\beta + \eta i)(1 - \alpha i)x + z \\ &= (\beta + \eta\alpha + i(\eta - \alpha\beta))x + z \\ &= (a + bi)x + z. \end{aligned}$$

Since $w \in iW \Rightarrow iw \in W \Rightarrow \eta iw \in W$

$\Rightarrow \qquad z = h - \eta iw \in W$

$\Rightarrow \qquad z \in W \cap iW.$

$\Rightarrow \qquad y = (a + bi)x + z \in [W \cap iW, x]$

$\Rightarrow \qquad V = [W \cap iW, x]$

$\Rightarrow W \cap iW$ is a complex hyper plane.

6 SOME WORKED OUT PROBLEMS AND EXAMPLES

1. In $P_2(x)$ if $\{1, x, x^2\}$, $\{1+x, 1-x, 1+x^2\}$ and $\{x, 2+x, 1-x-x^2\}$ are bases, then find the dual bases of $(P_2(x))^*$ induced by them.

   (*a*) Let $\phi_i(a_1 + a_2x + a_3x^2) = a_i$, then $\{\phi_1, \phi_2, \phi_3\}$ are dual bases of $(P_2(x))^*$ with respect to $\{1, x, x^2\}$, known as standard bases.

   (*b*) $a_1 + a_2x + a_3x^2 = \alpha(1+x) + \beta(1-x) + \gamma(1+x^2)$, then

$$\gamma = a_3,\ \beta = \frac{a_1 - a_2 - a_3}{2},\ \alpha = \frac{a_1 - a_3 + a_2}{2}$$

   Therefore, the maps f_i given as

$$f_1\,(a_1 + a_2x + a_3x^2) = \frac{a_1 - a_3 + a_2}{2}$$

$$f_2\,(a_1 + a_2x + a_3x^2) = \frac{a_1 - a_2 - a_3}{2}$$

$$f_3\,(a_1 + a_2x + a_3x^2) = a_3$$

   are linear functionals and $\{f_1, f_2, f_3\}$ is the dual basis of $(P_2(x))^*$ with respect to $\{1+x, 1-x, 1+x^2\}$.

   (*c*) Let $a + bx + cx^2 = \alpha x + \beta(2+x) + \gamma(1-x-x^2)$, then

$$\alpha = \frac{2b - 3c - a}{2},\ \beta = \frac{a+c}{2},\ \gamma = -c.$$

   Hence the maps $f_i : P_2(x) \to \mathbb{R}$ given as

$$f_1\,(a + bx + cx^2) = \frac{2b - 3c - a}{2}$$

$$f_2\,(a + bx + cx^2) = \frac{a+c}{2}$$

$$f_3\,(a + bx + cx^2) = -c$$

   are linear functionals and $\{f_1, f_2, f_3\}$ are the dual bases of $(P_2(x))^*$ with respect to $\{x, 2+x, 1-x-x^2\}$.

2. If $f_1, f_2, f_3 : \mathbb{R}^3 \to \mathbb{R}$ such that $f_1\,(x, y, z) = x - 2y$, $f_2(x, y, z) = 2x - y$, $f_3(x, y, z) = x + y + z$ and $\{f_1, f_2, f_3\}$ is a dual basis of $(\mathbb{R}^3)^*$, then find a basis of $\mathbb{R}^3$ with respect to which $\{f_1, f_2, f_3\}$ is the dual basis.

   Let $\{(a, b, c), (a', b', c'), (\alpha, \beta, \gamma)\}$ be the desired bases, then $f_1\,(a, b, c) = a - 2b = 1$, $f_2\,(a, b, c) = 2a - b = 0$, $f_3(a, b, c) = a + b + c = 0$.

   Solving these equations we get $a = -\dfrac{1}{3}$, $b = -\dfrac{2}{3}$, $c = 1$.

Similarly using the definition and the process adopted above we can see that

$$a' = \frac{2}{3},\ b' = \frac{1}{3},\ c' = -1 \text{ and } \alpha = \beta = 0, \gamma = 1$$

$$\Rightarrow \left\{\left(-\frac{1}{3}, -\frac{2}{3}, 1\right), \left(\frac{2}{3}, -\frac{1}{3}, -1\right), (0, 0, 1)\right\}$$

is the desired basis.

3. Show that if $\{x_1, x_2, \ldots, x_n\}$ is a basis of a vector space V and $\{f_1, f_2, \ldots, f_n\}$ is the induced dual basis of V^*, then any $x \in V$ can be uniquely written as $\sum_{i=1}^{n} f_i(x)x_i$.

Since $\{x_1, x_2, \ldots, x_n\}$ is a basis of $V \Rightarrow x \in V$ has unique representation as $\sum_{i=1}^{n} \alpha_i x_i$. Now, using the definition, we have

$$\begin{aligned} f_i(x) &= \alpha_1 f_i(x_1) + \alpha_2 f_i(x_2) + \ldots + \alpha_i f_i(x_i) + \ldots + \alpha_n f_i(x_n) \\ &= \alpha_i \end{aligned}$$

$$\Rightarrow \qquad x = \sum_{i=1}^{n} f_i(x)\, x_i .$$

4. Show that if $x, y \in V$ and $f(x) = 0 \Rightarrow f(y) = 0$ for all $f \in V^*$, then $\{x, y\}$ is linearly dependent.

If $\{x, y\}$ is linearly independent, then $\{x, y\}$ can be extended to a basis $\{x, y, x_3, \ldots, x_n\}$ of V. Let $\{f_1, f_2, \ldots, f_n\}$ be the dual basis of V^* with respect to $\{x, y, x_3, \ldots, x_n\}$ of V, then $f_2(x) = 0$ but it does not imply $f_2(y) = 0$ as $f_2(y) = 1 \Rightarrow \{x, y\}$ is linearly dependent.

5. Find $A(W)$ if

   (*a*) W is spanned by $\{(1, 2, 1), (-1, 0, 1)\}$ in $\mathbb{R}^3$.

   (*b*) W is spanned by $\{(2, 1, 3)\ \}$ in $\mathbb{R}^3$.

   (*c*) W is spanned by $\{(1 + x, 1 - x^2\}$ in $P_2(x)$.

   (*a*) Since linear functional f in $\mathbb{R}^3$ is of the form $f(x, y, z) = ax + by + cz$, so if $f \in A(W)$, then

$$f(1, 2, 1) = a + 2b + c = 0,$$

$$f(-1, 0, 1) = -a + c = 0$$

$$\Rightarrow \qquad a = c$$

$$\Rightarrow \qquad b = -a$$

$\Rightarrow$ $f(x, y, z) = x - y + z$ will serve the purpose

$\Rightarrow$ $A(W) = [f]$.

(*b*) Since dim $A(W) = 2 \Rightarrow f_1(2, 1, 3) = 0 \Rightarrow 2a + b + 3c = 0$.

Choosing $a = 1$, $b = 0$ we have $c = -\frac{2}{3}$. Choosing $a = 0$, $b = 1$, we have $c = -\frac{1}{3} \Rightarrow f_1(x, y, z) = x - \frac{2}{3}z$

and $f_2(x, y, z) = y - \frac{1}{3}z$

$\Rightarrow A(W) = [f_1, f_2]$.

In another method we can extend (2, 1, 3) to a basis {(2, 1, 3), (1, 0, 0), (0, 1, 0)} $\Rightarrow f_2(1, 0, 0) = 1, f_3(0, 1, 0) = 1$ and $f_1(2, 1, 3) = 1 \Rightarrow f_2(1, 0, 0) = a, f_3(0, 1, 0) = b, f_1(2, 1, 3) = 2a + b + 3c = 1 \Rightarrow a = 1, b = 1$ and $c = -\frac{2}{3}$.

Similarly $f_2(2, 1, 3) = 0, f_3(2, 1, 3) = 0 \Rightarrow 2a + b + 3c = 0$, then taking respective combinations, we get

$$f_1(x, y, z) = \frac{1}{3}z, f_2(x, y, z) = x - \frac{2}{3}z, f_3(x, y, z) = y - \frac{1}{3}z.$$

$\Rightarrow \{f_1, f_2, f_3\}$ is the dual basis w.r.t. {(2, 1, 3), (1, 0, 0), (0, 1, 0)}

$\Rightarrow A(W) = [f_2, f_3]$ but if we extend (2, 1, 3) to different basis, then we get a different basis of $A(W)$.

(*c*) In general, $f(a + bx + cx^2) = \alpha a + \beta b + \gamma c$

$\Rightarrow$ $f(1 + x) = \alpha + \beta = 0$

and $f(1 - x^2) = \alpha - \gamma = 0$ if $f \in A(W)$.

$\Rightarrow$ $\alpha = \gamma = -\beta$. Now choose $\alpha = 1$, we get

$$f(a + bx + cx^2) = a - b + c$$

$\Rightarrow$ $A(W) = [f]$.

6. If $T \in L(V, U)$ has the matrix representation

$$m(T) = \begin{bmatrix} 2 & 0 & 1 & 7 \\ 3 & 4 & 6 & 8 \\ 0 & 1 & 3 & -1 \end{bmatrix}$$

with respect to some choice of basis of V and U, where dim $U = 3$, dim $V = 4$, then trivially

$$m(T') = \begin{bmatrix} 2 & 3 & 0 \\ 0 & 4 & 1 \\ 1 & 6 & 3 \\ 7 & 8 & -1 \end{bmatrix}$$

7. Show that if W is a subspace of V invariant under $T \in L(V, V)$, then $A(W)$ is invariant under $T' \in L(V^*, V^*)$.

   Let $f \in A(W)$, then we show that $T'(f) \in A(W)$.

   $$(T'f)(w) = (f \circ T)(w) = f(T(w)) = 0.$$

   Since $T(w) \in W$ for all $w \in W \Rightarrow T'f \in A(W)$.

8. If $f \in (\mathbb{R}^2)^*$ such that $f(x, y) = 2x - 5y$ and let $T : \mathbb{R}^3 \to \mathbb{R}^2$ such that $T(x, y, z) = (x - 2y + z, x - z)$, then $T' : (\mathbb{R}^2)^* \to (\mathbb{R}^3)^*$

   $$\begin{aligned}(T'f)(x, y, z) &= f(T(x, y, z)) \\ &= f(x - 2y + z, x - z) \\ &= 2(x - 2y + z) - 5(x - z) \\ &= 2x - 4y + 2z - 5x + 5z \\ &= -3x - 4y + 7z.\end{aligned}$$

9. Show that if $T \in L(V, U)$, then $T'U^* = A(\ker T)$.

   Let $\phi \in T'U^*$

   $$\Rightarrow \qquad \phi = T'f, f \in U^*.$$

   Let $x \in \ker T$

   $$\begin{aligned}\Rightarrow \qquad & (T'f)(x) = f(T(x)) = f(0) = 0 \\ \Rightarrow \qquad & T'f \subseteq A(\ker T) \\ \Rightarrow \qquad & T'U^* \subseteq A(\ker T)\end{aligned}$$

   but we have proved Theorem 6.4.7, that

   $$r(T') = r(T)$$

   $$\begin{aligned}\Rightarrow \qquad \dim T'U^* &= r(T') = r(T) = \dim V - n(T) \\ &= \dim(\ker T) + \dim A(\ker T) - n(T) \\ &= \dim A(\ker T)\end{aligned}$$

   $$\Rightarrow \qquad T'U^* = A(\ker T).$$

EXERCISE

1. Find the dual of each of the following bases in $\mathbb{R}^3$:
   (*a*) $\{(1, 1, 0), (0, 1, 0), (1, 1, 1)\}$
   (*b*) $\{(1, 2, 3), (2, -1, 0), (4, 3, 1)\}$
2. Find the dual of each of the following bases in $P_2(x)$:
   (*a*) $\{x, 1 - x, 1 + x - x^2\}$
   (*b*) $\{2, 1 + x, 1 - x^2\}$
3. In $\mathbb{R}^4$ if W is generated by $\{(1, 2, 0, 3), (-1, -4, 1, 7)\}$, then find a basis of $A(W)$.
4. Find $A(W)$ if
   (*a*) W is generated by $\{(1, -1, 0), (1, 0, -1)\}$ in $\mathbb{R}^3$.
   (*b*) W is generated by $\{(4, 1, 7)\}$ in $\mathbb{R}^3$.
   (*c*) W is generated by $1 + x$ in $P_2(x)$.
   (*d*) W is generated by $\{1 - x + x^2, 2 + x\}$ in $P_2(x)$.
5. For $T \in L(V, U)$, show that $\ker T' = A(TV)$.
6. For $f, g \in V^*$ if $f(x) = 0$ implies $g(x) = 0$ for all $x \in V$, then show that $\{f, g\}$ is linearly dependent.
7. If $f \in (\mathbb{R}^3)^*$ such that $f(x, y, z) = x + 2y - 3z$ and $T : \mathbb{R}^4 \to \mathbb{R}^3$ such that $T(x, y, z, w) = (x + w, x - y - z, y - 4z)$, then find $(T'f)(x, y, z, w)$.
8. Prove that $(T')^t = T$, where $T \in L(V, V)$.
9. For $x, y \in V$ such that $x \neq y$. Show that there exists $f \in V^*$ such that $f(x) \neq f(y)$.

7

Diagonalization

In Chapters 4 and 5, we have studied linear transformations from a vector space V into another vector space U and their matrix representations and other properties. In this chapter we restrict ourselves to the case $V = U$ and the vector spaces are finite dimensional. If V is a vector space such that dim $V = 1$ then we know that every $T \in L(V, V)$ is of the form $T(x) = \alpha x$, α is a scalar and vice versa. Now, in general, we shall consider the cases where an element $x \in V$ is mapped to a scalar multiple of itself. If $T : \mathbb{R}^2 \to \mathbb{R}^2$ such that $T(x, y) = (-x, -y)$ then $T \in L(\mathbb{R}^2, \mathbb{R}^2)$ and $T(x, y) = -1(x, y)$. Here $\alpha = -1$, however if $T : \mathbb{R}^2 \to \mathbb{R}^2$ such that $T(x, y) = (-y, x)$ then there does not exist any real α such that $T(x, y) = \alpha(x, y)$. Hence, it is not guaranteed that if $T \in L(V, V)$ then there will always exist some α such that $T(x) = \alpha x$, $x \in V$.

Throughout this chapter, V denotes the finite dimensional vector space and $T \in L(V, V)$.

1 EIGENVALUES AND EIGENVECTORS

In this section first we define the following:

Definition 7.1.1. Let $T \in L(V, V)$ then a nonzero $x \in V$ is called *eigenvector* (characteristic vector, proper vector, latent vector) of T if $T(x) = \alpha x$ for some scalar α and the scalar α is called *eigenvalue* (characteristic value, proper value, latent value) of T associated with x.

Therefore, a scalar α is a characteristic root of T iff there exists a nonzero $x \in V$ such that $T(x) = \alpha x$. The set of all eigenvalues of T is called spectrum of T.

If α is an eigenvalues of T then $W = \{x \in V : x \neq 0 \text{ and } T(x) = \alpha x\} \cup \{0\}$ is called eigenspace of T associated with α. If $x, y \in S$, then

$$T(x) = \alpha x,\ T(y) = \alpha y$$

$$\Rightarrow \quad T(x - y) = \alpha(x - y)$$

$$\Rightarrow \quad x - y \in S$$

and for any scalar λ.

$$T(\lambda x) = \lambda T(x) = \lambda(\alpha x) = \alpha(\lambda x)$$

$$\Rightarrow \qquad \lambda x \in S$$

$\Rightarrow S$ is a subspace of T.

If T is zero transformation, then 0 is trivially the only eigenvalues of T. If $T = I$, identity transformation, then 1 is the only eigenvalues of T.

Theorem 7.1.2. If $T \in L(V, V)$ then the following hold:

(*i*) If T is a projection then 0 and 1 are the only eigenvalues of iT.

(*ii*) α is an eigenvalue of T iff $T - \alpha I$ is singular.

(*iii*) If T is nilpotent then 0 is the only eigenvalue of T.

(*iv*) If α is an eigenvalue of T then α is an eigenvalue of $S^{-1}\,TS$ for all non-singular $S \in L(V, V)$ and vice versa.

(*v*) If α is an eigenvalue of T then the eigenspace of $T = \ker(T - \alpha I)$.

Proof. (*i*) Let α be given value of $T \Rightarrow$ there exists $x \neq 0$ in V such that

$$T(x) = \alpha x \Rightarrow T^2(x) = T(\alpha x) = \alpha(\alpha x) = \alpha^2 x.$$

$$\Rightarrow \qquad T(x) = \alpha^2 x \Rightarrow \alpha x = \alpha^2 x$$

$$\Rightarrow \qquad \alpha(\alpha - 1)x = 0 \Rightarrow \alpha = 0 \text{ or } \alpha = 1$$

$\Rightarrow$ 0 and 1 are the only eigenvalues of T.

(*ii*) Let α be an eigenvalue of $T \Rightarrow$ there exists a nonzero $x \in V$ such that

$$T(x) = \alpha x$$

$$\Rightarrow \qquad (T - \alpha I)(x) = 0 \Rightarrow T - \alpha I$$

is singular as $x \neq 0$. Now if $T - \alpha I$ is singular then $\ker(T - \alpha I) \neq 0$ $\Rightarrow$ there exists $x \neq 0$ such that $x \in \ker(T - \alpha I) \Rightarrow (T - \alpha I)(x) = 0$ $\Rightarrow T(x) = \alpha x \Rightarrow \alpha$ is an eigenvalue of T.

(*iii*) If T is nilpotent then there exists a positive integer t such that $T^t = 0$.

Let α be an eigenvalue of T then $T(x) = \alpha x$ for some $x \neq 0$

$$T^t(x) = \alpha^t x \Rightarrow \alpha^t x = 0 \Rightarrow \alpha = 0$$

(*iv*) Let β be an eigenvalue of $S^{-1}TS$, then there exists $y \in V, y \neq 0$ and $S^{-1}(TS(y)) = \beta y \Rightarrow TS(y) = \beta(S(y))$. Since S is non-singular $\Rightarrow S(y) \neq 0 \Rightarrow \beta$ is an eigenvalue of T. Now let α be an eigenvalue of T, then there exists $x \neq 0$ such that

$$T(x) = \alpha x$$

$$\Rightarrow \qquad T(SS^{-1}(x)) = \alpha SS^{-1}(x)$$

$$\Rightarrow \qquad (S^{-1}TS)(S^{-1}(x)) = \alpha S^{-1}(x).$$

Let $S^{-1}(x) = y$ then $y \neq 0$ as $x \neq 0 \Rightarrow (S^{-1}TS)(y) = \alpha y \Rightarrow \alpha$ is an eigenvalue of $S^{-1}TS \Rightarrow S^{-1}TS$ and T have the same eigenvalues.

(*v*) If x is an element in the given space of T then $T(x) = \alpha x \Rightarrow (T - \alpha I)(x) = 0 \Rightarrow x \in \ker(T - \alpha I)$ and if $x \in \ker(T - \alpha I)$ and if $x \in \ker(T - \alpha I)$ then $T(x) = \alpha x$. Hence the eigenspace of T corresponding to α is $\ker(T - \alpha I)$.

Now using the above result we can say that T is singular iff $\alpha = 0$ or eigenspace of T is $\ker T$. Also it is trivial to see that if T is non-singular then 0 cannot be an eigenvalue of T and if α is an eigenvalue of T then α^{-1} is an eigenvalue of T^{-1}.

Theorem 7.1.3. If $\alpha_1, \ldots, \alpha_k$ are distinct characteristic roots of T corresponding to characteristic vectors $x_1, x_2, \ldots, x_k$ in V then $\{x_1, x_2, \ldots, x_k\}$ is linearly independent.

Proof. $T(x_i) = \alpha_i x_i$ and $x_i \neq 0$

Let $k = 2$ and $\beta_1 x_1 + \beta_2 x_2 = 0$

$\Rightarrow T(\beta_1 x_1 + \beta_2 x_2) = 0$

$\Rightarrow \beta_1\alpha_1 x_1 + \beta_2\alpha_2 x_2 = 0$ and $\beta_1\alpha_1 x_1 + \beta_2\alpha_2 x_2 = 0$

$\Rightarrow \beta_1(\alpha_1 - \alpha_2)x_2 = 0$ and $\beta_2(\alpha_1 - \alpha_2) = 0$ but $\alpha_1 \neq \alpha_2$

$\Rightarrow \beta_2 = 0$ and so $\beta_1 = 0$

$\Rightarrow \{x_1, x_2\}$ is linearly independent.

Let $\{x_1, x_2, \ldots, x_t\}$ be linearly independent for $t < k$ and

$$\beta_1 x_1 + \beta_2 x_2 + \ldots + \beta_t x_t + \beta_{t+1} x_{t+1} = 0. \qquad \ldots(1)$$

Now operating T and using $T(x_i) = \alpha_i x_i$, we get

$$\beta_1\alpha_1 x_1 + \beta_2\alpha_2 x_2 + \ldots + \beta_t \alpha_t x_t + \beta_{t+1}\alpha_{t+1} x_{t+1} = 0. \qquad \ldots(2)$$

Now multiplying (1) by α_{t+1} and subtracting from (2), we get

$$\beta_1(\alpha_1 - \alpha_{t+1})x_1 + \beta_2(\alpha_2 - \alpha_{t+1})x_2 + \ldots + \beta_t(\alpha_t - \alpha_{t+1})x_t = 0$$

but $\{x_1, x_2, \ldots, x_t\}$ be linearly independent.

$\Rightarrow \beta_1(\alpha_1 - \alpha_{t+1}) = 0,\ \beta_2(\alpha_2 - \alpha_{t+1}) = 0, \ldots, \beta_t(\alpha_t - \alpha_{t+1})x_t = 0.$

Since $\alpha_i \neq a_j \Rightarrow \beta_1 = \beta_2 = \ldots = \beta_t = 0 \Rightarrow \beta_{t+1} = 0$

$\Rightarrow \{x_1, x_2, \ldots, x_{t+1}\}$ is linearly independent.

$\Rightarrow \{x_1, x_2, \ldots, x_k\}$ is linearly independent for all k corresponding to distinct eigenvalues α_i.

Now it is very natural to think the converse of Theorem 7.1.3, but the same is not true, which we shall indicate in the following Example 2.

1. Let $T : \mathbb{R}^2 \to \mathbb{R}^2$ such that $T(x, y) = (3x + y, 2x + 4y)$, then $T \in L(\mathbb{R}^2, \mathbb{R}^2)$.

   Let $\qquad T(x, y) = \alpha(x, y)$

   $\Rightarrow \qquad \alpha x = 3x + y,\ \alpha y = 2x + 4y$

$$\Rightarrow \quad (\alpha - 3)x - y = 0 \text{ and } - 2x + (\alpha - 4)y = 0$$

$$\Rightarrow \quad \begin{vmatrix} \alpha - 3 & -1 \\ -2 & \alpha - 4 \end{vmatrix} = 0$$

$$\Rightarrow \quad (\alpha - 3)(\alpha - 4) - 2 = 0$$

$$\Rightarrow \quad \alpha^2 - 7\alpha + 12 - 2 = 0$$

$$\Rightarrow \quad (\alpha - 5)(\alpha - 2) = 0$$

$$\Rightarrow \quad \alpha = 5, \alpha = 2.$$

Now taking $\alpha = 5$, we have

$$5(x, y) = (3x + y, 2x + 4y$$

$$\Rightarrow \quad 5x = 3x + y, 5y = 2x + 4y$$

$$\Rightarrow \quad 2x - y = 0, y - 2x = 0$$

$$\Rightarrow \quad y = 2x.$$

$\Rightarrow$ eigenvectors are $(x, 2x)$ such that $x \in \mathbb{R}, x \neq 0$.

2. Let $T : \mathbb{R}^3 \to \mathbb{R}^3$ such that $T(x, y, z) = (0, y, x + z)$ then $T \in L(\ ^3, \ ^3)$.

Let $\quad T(x, y, z) = \alpha(x, y, z)$

$$\Rightarrow \quad \alpha x = 0$$

$$\alpha y = y$$

$$\alpha z = x + z$$

$$\Rightarrow \quad \begin{vmatrix} \alpha & 0 & 0 \\ 0 & \alpha - 1 & 0 \\ -1 & 0 & \alpha - 1 \end{vmatrix} = 0$$

$\Rightarrow \alpha = 0, \alpha = 1, \alpha = 1$

$\Rightarrow$ eigenvalues of T are 0, 1, 1.

For $\alpha = 0$, we get $(x, 0, -x)$ as eigenvectors of T, where $x \in \mathbb{R}$ and $x \neq 0$.

Now for $\alpha = 1$, we get $x = 0$. Hence

$(0, y, z) = (0, y, 0) + (0, 0, z)$ for $y \neq 0, z \neq 0$ as eigenvectors of T.

Now taking $x = 1, y = 1, z = 1$, we have

$$\begin{vmatrix} 1 & 0 & 0 \\ 0 & 1 & 0 \\ -1 & 0 & 1 \end{vmatrix} = 1$$

$\Rightarrow \{(1, 0, -1), (0, 1, 0), (0, 0, 1)\}$ is linearly independent and these are eigenvectors corresponding to $\lambda_1 = 0, \lambda_2 = 1, \lambda_3 = 1$ through λ_2 and λ_3 are not distinct.

Hence we can say that corresponding to eigenvalues not all distinct we can get eigenvectors which are linear independent. Therefore, the converse of Theorem 7.1.3 is not true in general.

3. Let $T : \mathbb{R}^3 \to \mathbb{R}^3$ such that $T(x, y, z) = (x, y + 2x, z)$ then $T \in L(\mathbb{R}^3, \mathbb{R}^3)$.

   Let $T(x, y, z) = \alpha(x, y, z)$

   $\Rightarrow \alpha x = z,\ \alpha y = y + 2,\ \alpha z = z$

   $$\Rightarrow \begin{vmatrix} \alpha & 0 & -1 \\ 0 & \alpha - 1 & -2 \\ 0 & 0 & \alpha - 1 \end{vmatrix} = 0, \text{ for nonzero solution}$$

   $\Rightarrow \alpha = 0, \alpha = 1, \alpha = 1.$

   For $\alpha = 0$, $z = 0$, $y = 0 \Rightarrow$ eigenvectors are $(x, 0, 0)$, $x \in \mathbb{R}$, $x \neq 0$. Now for $\alpha = 1$ we have $z = x$, $2z = 0 \Rightarrow x = z = 0$. Hence, $(0, y, 0)$, $y \in \mathbb{R}$, $y \neq 0$ are given vectors of T corresponding to $\lambda_1 = \lambda_2 = 1$.

4. Let $T : \mathbb{R}^3 \to \mathbb{R}^2$ such that $T(x, y) = (x + 2y, 2x - 2y)$ then $T \in L(\mathbb{R}^2, \mathbb{R}^2)$.

   $$\begin{aligned} \text{Let} \quad T(x, y) &= \alpha(x, y) \\ \Rightarrow \quad x + 2y &= \alpha x \\ 2x - 2y &= \alpha y \\ \Rightarrow \quad (1 - \alpha)x + 2y &= 0 \\ 2x + (-2 - \alpha)y &= 0. \end{aligned}$$

   For nonzero solution, we have

   $$\begin{aligned} -(1 - \alpha)(2 + \alpha) - 4 &= 0 \\ \Rightarrow \quad [2 - 2\alpha + \alpha - \alpha^2] + 4 &= 0 \\ \Rightarrow \quad -\alpha^2 - \alpha + 2 + 4 &= 0 \\ \Rightarrow \quad \alpha^2 + \alpha - 6 &= 0 \\ \Rightarrow \quad (\alpha + 3)(\alpha - 2) &= 0 \\ \Rightarrow \quad \alpha &= -3, \alpha = 2 \end{aligned}$$

   $\Rightarrow$ eigenvalues of T are -3 and 2.

   Now putting $\alpha = -3$, we get $y = -2x \Rightarrow (x, -2x)$, $x \in \mathbb{R}$, $x \neq 0$ are the eigenvectors of T corresponding to $\alpha = -3$. Now putting $\alpha = 2$, we get $x = 2y \Rightarrow (2y, y)$, $y \in \mathbb{R}$, $y \neq 0$ are the given vectors of T corresponding to $\alpha = 2$.

5. Let $T : \mathbb{R}^3 \to \mathbb{R}^3$ such that $T(x, y, z) = (x + y - z, -x + 3y - z, -x + 2y)$ then $T \in L(\mathbb{R}^3, \mathbb{R}^3)$.

   Let $\quad T(x, y, z) = \alpha(x, y, z)$ for $(x, y, z) \neq 0$.

$\Rightarrow (\alpha - 1)x - y + z = 0,\ x + (\alpha - 3)y + z = 0,\ x - 2y + \alpha z = 0$

$$\Rightarrow \begin{vmatrix} \alpha - 1 & -1 & 1 \\ 1 & \alpha - 3 & 1 \\ 1 & -2 & \alpha \end{vmatrix} = 0, \text{ for nonzero solution}$$

$\Rightarrow (\alpha - 1)\,[\alpha(\alpha - 3) + 2] + 1(\alpha - 1) + 1(-2 - \alpha + 3) = 0$

$\Rightarrow (\alpha - 1)\,\alpha(\alpha - 3) + 2(\alpha - 1) + (\alpha - 1) + (1 - \alpha) = 0$

$\Rightarrow (\alpha - 1)\,[\alpha^2 - 3\alpha + 2 + 1 - 1] = 0$

$\Rightarrow \alpha = 1, 1, 2$

$\Rightarrow$ eigenvalues of T are 1, 1, 2.

Now taking $\alpha = 2$, we have

$$\begin{aligned} x - y + z &= 0 \\ x - y + z &= 0 \\ x - 2y + 2z &= 0 \end{aligned}$$

$$\Rightarrow \quad \frac{x}{-2+2} = \frac{y}{1-2} = \frac{z}{-2+1}$$

$\Rightarrow (0, -r, -r)$ are eigenvectors for any $r \in \mathbb{R}, r \neq 0$ corresponding to $\alpha = 2$.

Now taking $\alpha = 1$, we have

$$\begin{aligned} -y + z &= 0 \\ x - 2y + z &= 0 \end{aligned}$$

$$\Rightarrow \quad y = z,\ x = y$$

$\Rightarrow (x, x, x)$ are eigenvectors for any $x \in \mathbb{R}, x \neq 0$.

As we have seen in Chapter 4, linear transformations $T \in L(V, V)$ are closely related to square matrices of size dim V and $L(V, V)$ is isomorphic to algebra of square matrices of order n, where $n = \dim V$. Therefore, now we discuss the eigenvalues and eigenvectors of square matrices.

2 CHARACTERISTIC POLYNOMIAL

Definition 7.2.1. Let

$$A = \begin{bmatrix} a_{11} & \cdots & a_{1n} \\ \vdots & \vdots & \vdots \\ a_{n1} & \cdots & a_{nn} \end{bmatrix}$$

be a square matrix of order n then

$$\det(\alpha I - A) = \begin{vmatrix} \alpha - a_{11} & -a_{12} & \cdots & -a_{1n} \\ -a_{21} & \alpha - a_{22} & \cdots & -a_{2n} \\ \vdots & \vdots & \vdots & \vdots \\ -a_{n1} & -a_{n2} & \cdots & \alpha - a_{nn} \end{vmatrix}$$

is called characteristic polynomial of A and $\det(\alpha I - A) = 0$ is called characteristic equation of A.

It is trivial equation of A. It is trivial to see that $\det(\alpha I - A) = f(\alpha)$ is polynomial of degree n and if we expand, we get

$f(\alpha) = (\alpha - a_{11})(\alpha - a_{22}) \ldots (\alpha - a_{nn})$ + terms containing atmost $n - 2$ factors of the form $(\alpha - a_{jj})$

$$= \alpha^n - (a_{11} + \ldots + a_{nn})\alpha^{n-1} + \ldots + \beta,$$

Thus we see that coefficient of $\alpha^{n-1} = -\operatorname{tr}(A)$.

Further if we write

$$\det(\alpha I - A) = f(\alpha)$$
$$= (\alpha - \alpha_1)(\alpha - \alpha_2) \ldots (\alpha - \alpha_n),$$

then putting $\alpha = 0$, we get

$$\det(-A) = (-1)^n \alpha_1 \alpha_2 \ldots \alpha_n$$

and $$\det(-A) = (-1)^n \det(A).$$

The solutions of $\det(\alpha I - A) = 0$ are called eigenvalues of A. Thus $\det(A) = (-1)^n$ product of eigenvalues of A.

Now before some results we discuss some examples:

1. Let $A = \begin{bmatrix} 4 & 2 \\ 3 & 5 \end{bmatrix}$, then

$$|\lambda I - A| = \begin{vmatrix} \lambda - 4 & -2 \\ -3 & \lambda - 5 \end{vmatrix}$$
$$= (\lambda - 4)(\lambda - 5)\ 2\ 6$$
$$= \lambda^2 - 9\lambda + 20 - 6$$
$$= \lambda^2 - 9\lambda + 14$$

is characteristic equation of A and $\lambda^2 - 9\lambda + 14 = 0$ gives $\lambda = 7$ and $\lambda = 2$ as given values. $\det A = 14 = 7.2$, coefficient of $\lambda = -9 = -\operatorname{tr}(A)$.

Now putting $\lambda = -7$, we have

$$\begin{bmatrix} 3 & -2 \\ -3 & 2 \end{bmatrix}\begin{bmatrix} x \\ y \end{bmatrix} = 0$$

$$\Rightarrow \qquad 3x - 2y = 0$$

$$\Rightarrow \qquad y = \frac{3}{2}x$$

$\Rightarrow$ eigenvectors of A corresponding to $\lambda = 7$ are $\left(x, \frac{3}{2}x\right)$, $x \in \mathbb{R}$, $x \neq 0$.

Also putting $\lambda = 2$, we get

$$\begin{bmatrix} -2 & -2 \\ -3 & -3 \end{bmatrix}\begin{bmatrix} x \\ y \end{bmatrix} = \begin{bmatrix} 0 \\ 0 \end{bmatrix}$$

$$\Rightarrow \qquad 2x - 2y = 0$$

$$\Rightarrow \qquad y = -x$$

$\Rightarrow$ eigenvectors of A corresponding to $\lambda = 2$ are $(x, -x)$, $x \in \mathbb{R}$, $x \neq 0$.

2. Let $A = \begin{bmatrix} 1 & 1 & 0 \\ 1 & 1 & 0 \\ 0 & 0 & 1 \end{bmatrix}$, then

$$\begin{aligned} |\lambda I - A| &= (\lambda - 1)^3 + 1(-\lambda + 1) \\ &= (\lambda - 1)\,[\lambda^2 - 2\lambda + 1 - 1] \\ &= \lambda(\lambda - 1)(\lambda - 2) \end{aligned}$$

$\Rightarrow$ eigenvalues of A are $\lambda = 0, 1, 2$.

$$\begin{aligned} \det(A) &= 1(1 - 0) - 1(1) \\ &= 0 = 0 \cdot 1 \cdot 2. \end{aligned}$$

For $\lambda = 0$, we have

$$\begin{bmatrix} -1 & -1 & 0 \\ -1 & -1 & 0 \\ 0 & 0 & -1 \end{bmatrix}\begin{bmatrix} x \\ y \\ z \end{bmatrix} = \begin{bmatrix} 0 \\ 0 \\ 0 \end{bmatrix}$$

$$\Rightarrow \qquad -x - y = 0, -z = 0$$

$\Rightarrow$ eigenvectors of A corresponding to $\lambda = 0$ are $(x, -x, 0)$, $x \in \mathbb{R}$, $x \neq 0$.

For $\lambda = 1$, we have

$$\begin{bmatrix} 0 & -1 & 0 \\ -1 & 0 & 0 \\ 0 & 0 & 0 \end{bmatrix}\begin{bmatrix} x \\ y \\ z \end{bmatrix} = \begin{bmatrix} 0 \\ 0 \\ 0 \end{bmatrix}$$

$$\Rightarrow \qquad -y = 0, -x = 0$$

$\Rightarrow$ eigenvectors of A corresponding to $\lambda = 1$ are $(0, 0, z)$, $x \in \mathbb{R}$, $z \neq 0$.

For $\lambda = 2$, we have

$$\begin{bmatrix} 1 & -1 & 0 \\ -1 & 1 & 0 \\ 0 & 0 & 1 \end{bmatrix} \begin{bmatrix} x \\ y \\ z \end{bmatrix} = \begin{bmatrix} 0 \\ 0 \\ 0 \end{bmatrix}$$

$$\Rightarrow \qquad x - y = 0, -x + y = 0, z = 0$$

$\Rightarrow$ eigenvalues of A corresponding to $\lambda = 2$ are $(x, x, 0)$, $x \in \mathbb{R}$, $x \neq 0$.

3. Let $A = \begin{bmatrix} 1 & 0 & 0 \\ -3 & 2 & 0 \\ -1 & -2 & 4 \end{bmatrix}$, then

$$|\lambda I - A| = (\lambda - 1)(\lambda - 2)(\lambda - 4)$$

$\Rightarrow \lambda = 1, 2, 4$ are the eigenvalues of A.

For $\lambda = 1$, we have

$$\begin{bmatrix} 1 & 0 & 0 \\ 3 & 1 & 0 \\ 1 & 2 & -3 \end{bmatrix} \begin{bmatrix} x \\ y \\ z \end{bmatrix} = \begin{bmatrix} 0 \\ 0 \\ 0 \end{bmatrix}$$

$$\Rightarrow \qquad 3x - y = 0, x + 2y - 3z = 0$$

$$\Rightarrow \qquad y = 3x, z = \frac{7}{3}x$$

$\Rightarrow$ eigenvectors of A corresponding to $\lambda = 1$ are $\left(x, 3x, \frac{7}{3}x\right)$, $x \in \mathbb{R}, x \neq 0$.

For $\lambda = 2$, we have

$$\begin{bmatrix} 1 & 0 & 0 \\ 3 & 0 & 0 \\ 1 & 2 & -2 \end{bmatrix} \begin{bmatrix} x \\ y \\ z \end{bmatrix} = \begin{bmatrix} 0 \\ 0 \\ 0 \end{bmatrix}$$

$$\Rightarrow \qquad x = 0, x + 2y - 2z = 0$$

$$\Rightarrow \qquad z = y$$

$\Rightarrow (0, y, y)$, $y \in \mathbb{R}$, $y \neq 0$ are the eigenvectors of A corresponding to $\lambda = 2$.

For $\lambda = 4$, we have

$$\begin{bmatrix} 3 & 0 & 0 \\ 3 & 2 & 0 \\ 1 & 2 & 0 \end{bmatrix}\begin{bmatrix} x \\ y \\ z \end{bmatrix} = \begin{bmatrix} 0 \\ 0 \\ 0 \end{bmatrix}$$

$\Rightarrow$ $3x = 0, 3x + 2y = 0, x + 2y = 0$

$\Rightarrow$ $x = 0, y = 0$

$\Rightarrow (0, 0, z), z \in \mathbb{R}, z \neq 0$ are the eigenvectors of A corresponding to $\lambda = 4$.

Theorem 7.2.2. The following holds:

(*i*) Similar matrices have the same characteristic polynomials.

(*ii*) For any square matrix A, A and A^t have the same characteristic polynomial.

(*iii*) The main diagonal entries of a triangular matrix A are the eigenvalues of A.

Proof. (*i*) Let A be similar to B then $B = P^{-1} A$

$$\Rightarrow \quad | P^{-1}AP - \alpha I | = | P^{-1} AP - \alpha \cdot P^{-1}P |$$

$$= | P^{-1} | | A - \alpha I | | P | = | A - \alpha I |$$

$\Rightarrow A$ and B have the same characteristic polynomial so they will also have the same eigenvalues.

(*ii*) $(A - \alpha I)^t = A^t - \alpha I^t = A^t - \alpha I$. Since $| A^t | = | A |$

$$\Rightarrow \quad | A - \alpha I | = | (A - \alpha I)^t | = | A^t - \alpha I |$$

$\Rightarrow A$ and A^t have the same characteristic polynomial.

(*iii*) Left as an exercise.

Definition 7.2.3. A square matrix A satisfies a polynomial

$$f(x) = x^n + a_1x^{n-1} + \ldots + a_{n-1}x + a_n = 0 \text{ if } f(A) = 0$$

i.e., $A^n + a_1A^{n-1} + \ldots + a_{n-1}A + a_nI = 0$

$$\text{If } A = \begin{bmatrix} 1 & 1 \\ 0 & 1 \end{bmatrix}, \text{ then } A^2 = \begin{bmatrix} 1 & 1 \\ 0 & 2 \end{bmatrix}$$

$\Rightarrow$ if $f(x) = x^2 - 2x + 1$ then $f(A) = 0$ so A satisfies $f(x) = 0$. Now if $g(x) = x^2 + 2x - 1$ then A is not a root of $g(x) = 0$.

Theorem 7.2.4. (Cayley-Hamilton Theorem) Every square matrix satisfies its characteristic equation.

Proof. Let A be a square matrix of order n and

$$f(\alpha) = \det(\alpha I - A) = \alpha^n + c_{n-1}\alpha^{n-1} + \ldots + c_1\alpha + c_0$$

be the characteristic polynomial of A. Let $B(\alpha) = \text{adj } (\alpha I - A)$, then the entries of $B(\alpha)$ are cofactors of the matrix $\alpha I - A$ and so they are polynomials in α of degree atmost $n - 1 \Rightarrow B(\alpha) = \text{adj } (\alpha I - A) = B_{n-1} \alpha^{n-1} + \ldots + B_1\alpha + B_0$, where B_i's are square matrices of order n with real entries. Since

$$(\alpha I - A) \text{ adj } (\alpha I - A) = \det (\alpha I - A)I$$
$$= (\text{adj } (\alpha I - A)) (\alpha I - A)$$

$$\Rightarrow \quad B(\alpha) (\alpha I - A) = |\, \alpha I - A \,|\, I$$

$$\Rightarrow (B_{n-1} \alpha^{n-1} + \ldots + B_1\alpha + B_0) (\alpha I - A)$$
$$= (\alpha^n + c_{n-1}\alpha^{n-1} + \ldots + c_1\alpha + c_0)I$$

Now comparing the coefficients of powers of α, we get

$$\begin{aligned} B_{n-1} &= I \\ B_{n-2} - AB_{n-1} &= C_{n-1}I \\ \vdots \quad &\vdots \quad \vdots \\ B_0 - AB_1 &= C_1 I \\ -AB_0 &= C_0 I \end{aligned}$$

Now multiplying the above equations by $A^n, A^{n-1}, \ldots, A, I$, respectively, and adding, we get

$$A^n + c_{n-1} A^{n-1} + \ldots + c_1 A + c_0 I = 0$$

$\Rightarrow f(A) = 0 \Rightarrow A$ satisfies its characteristic equation.

Now we give an example. Let $A = \begin{bmatrix} 2 & 1 \\ -1 & 3 \end{bmatrix}$, then the characteristic

equation of A is $\begin{vmatrix} \lambda - 2 & -1 \\ 1 & -\lambda - 3 \end{vmatrix} = 0$

$$\Rightarrow \quad (\lambda - 2)(\lambda - 3) + 1 = 0$$

$$\Rightarrow \quad \lambda^2 - 5\lambda + 7 = 0.$$

Now $$A^2 = \begin{bmatrix} 3 & 5 \\ -5 & 8 \end{bmatrix}$$

$$\Rightarrow \quad A^2 - 5A + 7I = \begin{bmatrix} 3 & 5 \\ -5 & 8 \end{bmatrix} + \begin{bmatrix} -10 & -5 \\ 5 & -15 \end{bmatrix} + \begin{bmatrix} 7 & 0 \\ 0 & 7 \end{bmatrix}$$

$$= \begin{bmatrix} 0 & 0 \\ 0 & 0 \end{bmatrix}$$

$\Rightarrow A$ satisfies its characteristic equation.

3 DIAGONALIZING MATRIX

In this section we find a non-singular matrix P for a matrix A such that $P^{-1}AP$ is a diagonal matrix.

Definition 7.3.1. A non-singular matrix P is called *diagonalizing matrix* for square matrix A if $P^{-1}AP$ is diagonal and the matrix A is called diagonalizable.

If $A = \begin{bmatrix} 3 & 1 \\ 2 & 4 \end{bmatrix}$ then $P = \begin{bmatrix} 1 & 1 \\ 2 & -1 \end{bmatrix}$ is a diagonalizing matrix for A. Since

$P^{-1} = -\frac{1}{3}\begin{bmatrix} -1 & -1 \\ -2 & 1 \end{bmatrix}$, so

$$\begin{aligned} P^{-1}AP &= -\frac{1}{3}\begin{bmatrix} -1 & -1 \\ -2 & 1 \end{bmatrix}\begin{bmatrix} 3 & 1 \\ 2 & 4 \end{bmatrix}\begin{bmatrix} 1 & 1 \\ 2 & -1 \end{bmatrix} \\ &= -\frac{1}{3}\begin{bmatrix} -1 & -1 \\ -2 & 1 \end{bmatrix}\begin{bmatrix} 5 & 2 \\ 10 & -2 \end{bmatrix} \\ &= -\frac{1}{3}\begin{bmatrix} -15 & 0 \\ 0 & -6 \end{bmatrix} = \begin{bmatrix} 5 & 0 \\ 0 & 2 \end{bmatrix} \end{aligned}$$

It may be noted that the diagonalizing matrix is not unique because if we take $Q = \begin{bmatrix} 3 & 2 \\ 6 & -2 \end{bmatrix}$, then $Q^{-1}AQ = \begin{bmatrix} 5 & 0 \\ 0 & 2 \end{bmatrix}$.

However, every matrix cannot be diagonalized. If we take $A = \begin{bmatrix} 1 & 1 \\ 0 & 1 \end{bmatrix}$ then there does not exist any matrix P with real entries such that $P^{-1}AP$ is a diagonal matrix.

Now we prove the following characterization for diagonalizable matrix.

Theorem 7.3.2. A square matrix A of order n is diagonalizable iff A has n linearly independent eigenvectors.

Proof. Let A be diagonalizable then for a non-singular matrix P

$$P^{-1}AP = D = \text{diag}\,(\lambda_1, \ldots, \lambda_n)$$

$$\Rightarrow \qquad AP = PD.$$

Let $x_j = (x_{1j}, x_{2j}, \ldots, x_{nj}), j = 1, 2, \ldots, n$ be the jth column of P, as P is non-singular $x_j \neq 0$. Hence equating j^{th} column of both sides we get $Ax_j = \lambda_j x_j$. Since P is non-singular, so $\{x_1, x_2, \ldots, x_n\}$ is linearly independent. As $x_j \neq 0$, λ_j's are eigenvalues of A and $x_1, x_2, \ldots, x_n$ are given vectors of A.

Conversely, suppose A has n-linearly independent vectors $x_1, \ldots, x_n$ and $\lambda_1, \ldots, \lambda_n$ be the eigenvalues of A corresponding to $x_1, x_2, \ldots, x_n$. Choose $P = [x_1, \ldots, x_n]$ as the matrix whose j^{th} column is x_j, then P is non-singular as columns are linearly independent. As $Ax_j = \lambda_j x_j$, we get $AP = PD$, where

$$D = \text{diag}\,(\lambda_1, \ldots, \lambda_n)$$

$$\Rightarrow \quad P^{-1}AP = D$$

$\Rightarrow A$ is diagonalizable.

It may be noted that diagonal entries of D are eigenvalues of A and columns of P are the eigenvectors of A.

Corollary 7.3.3. A square matrix A is diagonalizable if all its eigenvalues are real and distinct.

Proof. If the eigenvalues $\lambda_1, \ldots, \lambda_n$ are distinct, then by Theorem 7.1.3, the corresponding eigenvectors $x_1, \ldots, x_n$ are linearly independent. Hence, by Theorem 7.3.2 we get the result.

We now give some examples.

1. Let $A = \begin{bmatrix} 3 & -5 \\ 2 & -4 \end{bmatrix}$, then $|\alpha I - A| = 0$ gives

$$\Rightarrow \quad (\alpha - 3)(\alpha + 4) + 10 = 0$$

$$\Rightarrow \quad \alpha^2 + \alpha - 2 = 0$$

$$\Rightarrow \quad \alpha = -2, 1$$

For $\alpha = -2$, we have

$$\begin{bmatrix} -5 & 5 \\ -2 & 2 \end{bmatrix}\begin{bmatrix} x \\ y \end{bmatrix} = \begin{bmatrix} 0 \\ 0 \end{bmatrix}$$

$$\Rightarrow \quad x = y$$

$\Rightarrow$ eigenvectors are (x, x), $x \in \mathbb{R}$, $x \neq 0$.

For $\alpha = 1$, we have

$$\begin{bmatrix} -2 & 5 \\ -2 & 5 \end{bmatrix}\begin{bmatrix} x \\ y \end{bmatrix} = \begin{bmatrix} 0 \\ 0 \end{bmatrix}$$

$$\Rightarrow \quad -2x + 5y = 0$$

$$\Rightarrow \quad x = \frac{5}{2}y$$

$\Rightarrow$ eigenvectors are $\left(\frac{5}{2}y, y\right)$, $y \in \mathbb{R}$, $y \neq 0$.

Hence, a diagonalizing matrix $P = \begin{bmatrix} 1 & 5 \\ 1 & 2 \end{bmatrix}$

$$\Rightarrow \qquad P^{-1}AP = -\frac{1}{3}\begin{bmatrix} 2 & -5 \\ -1 & 1 \end{bmatrix}\begin{bmatrix} 3 & -5 \\ 2 & -4 \end{bmatrix}\begin{bmatrix} 1 & 5 \\ 1 & 2 \end{bmatrix}$$

$$= -\frac{1}{3}\begin{bmatrix} 2 & -5 \\ -1 & 1 \end{bmatrix}\begin{bmatrix} -2 & 5 \\ -2 & 2 \end{bmatrix}$$

$$= -\frac{1}{3}\begin{bmatrix} 6 & 0 \\ 0 & -3 \end{bmatrix} = \begin{bmatrix} -2 & 0 \\ 0 & 1 \end{bmatrix}.$$

2. Let $A = \begin{bmatrix} 1 & 0 & 0 \\ -3 & 2 & 0 \\ -1 & -2 & 4 \end{bmatrix}$, then eigenvalues are 1, 2, 4 and

$$P = \begin{bmatrix} 3 & 0 & 0 \\ 9 & 1 & 0 \\ 7 & 7 & 1 \end{bmatrix}.$$

It is easy to verify that $P^{-1}AP = \text{diag}(1, 2, 4)$.

3. Let $A = \begin{bmatrix} 2 & 1 & 0 \\ 0 & 1 & -1 \\ 0 & 2 & 4 \end{bmatrix}$, then

$$|\alpha I - A| = \begin{vmatrix} \alpha - 2 & -1 & 0 \\ 0 & \alpha - 1 & 1 \\ 0 & -2 & \alpha - 4 \end{vmatrix}$$

$= (\alpha - 2)\,[(\alpha - 1)(\alpha - 4) + 2]$

$= (\alpha - 2)\,[\alpha^2 - 5\alpha + 4 + 2]$

$= (\alpha - 2)^2 (\alpha - 3)$

$\Rightarrow$ eigenvalues of A are 2, 3, 2.

For $\alpha = 2$, we have

$$\begin{bmatrix} 0 & -1 & 0 \\ 0 & 1 & 1 \\ 0 & -2 & -2 \end{bmatrix}\begin{bmatrix} x \\ y \\ z \end{bmatrix} = \begin{bmatrix} 0 \\ 0 \\ 0 \end{bmatrix}$$

$\Rightarrow \qquad\qquad\qquad\qquad y = 0, z = 0$

$\Rightarrow (x, 0, 0), x \in \mathbb{R}, x \neq 0$ are the eigenvectors of A corresponding to $\alpha = 2, 2$.

For $\alpha = 2$, we have

$$\begin{bmatrix} 0 & -1 & 0 \\ 0 & 1 & 1 \\ 0 & -2 & -2 \end{bmatrix}\begin{bmatrix} x \\ y \\ z \end{bmatrix} = \begin{bmatrix} 0 \\ 0 \\ 0 \end{bmatrix}$$

$$\Rightarrow \begin{cases} x - y = 0 \Rightarrow x = y \\ 2y + z = 0 \Rightarrow z = -2y \end{cases}$$

$\Rightarrow (x, x, -2x), x \in \mathbb{R}, x \neq 0$ are the eigenvectors of A corresponding to $\alpha = 3$.

Since there are only two linearly independent vectors $\Rightarrow A$ is not diagonalizable.

4. Let $A = \begin{bmatrix} 0 & 0 & 0 \\ 0 & 1 & 0 \\ 1 & 0 & 1 \end{bmatrix}$ then 0, 1, 1 are eigenvalues of A, though they are not distinct, $P = \begin{bmatrix} 1 & 0 & 0 \\ 0 & 1 & 0 \\ -1 & 0 & 1 \end{bmatrix}$ is a diagonalizing matrix for A.

Therefore, A is a diagonalizable matrix.

Definition 7.3.4. Let A be a square matrix of order n, then a monic polynomial $m(x)$ is called minimal polynomial for A if $m(A) = 0$ and $m(x)$ is of least degree with this property.

The Cayley-Hamilton theorem shows there is always the characteristic polynomial $f(x)$ of A such that $f(A) = 0$.

Theorem 7.3.5. If A is a square matrix of order n and $m(x)$ is a monic polynomial of smallest degree such that $m(A) = 0$, then for any polynomial $g(x)$, $g(A) = 0$ iff $m(x)$ is a factor of $g(x)$.

Proof. Let $g(A) = 0$ and $g(x) = m(x)\, p(x) + r(x)$, where $r(x) = 0$ or deg $(r(x)) <$ deg $(m(x))$. Let $r(x) \neq 0$ then $g(A) = m(A)\, p(A) + r(A) \Rightarrow r(A) = 0$ which contradicts the minimality of $m(x) \Rightarrow r(x) = 0 \Rightarrow m(x)$ is a factor of $g(x)$.

Trivially, if $g(x) = m(x)\, q(x)$, then $g(A) = 0$ as $m(A) = 0$.

Corollary 7.3.6. Minimal polynomial of A always divides the characteristic polynomial of A.

Proof. Since $f(A) = 0$, where $f(x)$ is the characteristic polynomial of A. Hence, by Theorem 7.3.4, $m(x)$ divides $f(x)$ or is a factor of $f(x)$. It is trivial to see that minimal polynomial of A is unique (due to Theorem 7.3.5).

Since the minimal polynomial $m(x)$ divides the characteristic polynomial $f(x)$ of A, so the irreducible factors of $m(x)$ must include those of $f(x)$. Hence, $m(x)$ and $f(x)$ must have the same irreducible factors. This needs proof but is left as an exercise. Therefore, we can say that if α is a eigenvalue of A then α must be a root of the minimal polynomial $m(x)$ of A.

Now consider $A = I_2$ then $f(x) = x^2 - 2x + 1$ is the characteristic polynomial of A but $m(x) = x - 1$. If we take

$$A = \begin{bmatrix} 3 & 1 & 0 & 0 & 0 \\ 0 & 3 & 0 & 0 & 0 \\ 0 & 0 & 3 & 1 & 0 \\ 0 & 0 & 0 & 3 & 1 \\ 0 & 0 & 0 & 0 & 3 \end{bmatrix}$$

then trivially the characteristic polynomial $f(x)$ of A is $(x - 3)^5$. We can verify that $(A - 3I)^3 = 0 \Rightarrow m(x) = (x - 3)^3$.

We end this chapter with some more examples.

1. Let $T : P_2(x) \to P_2(x)$ such that

   $T(a + bx + cx)^2 = (2a + b) + bx + (2a + 2b + 2c)x^2$

   Now are compute eigenvalues and eigenvector of T. Let $\{1, x, x^2\}$ be the standard basis of $P_2(x)$ then

   $$m(T) = \begin{bmatrix} 2 & 1 & 0 \\ 0 & 1 & 0 \\ 2 & 2 & 1 \end{bmatrix} = A \text{ (say)}.$$

   Characteristic equation of A is

   $$\begin{vmatrix} \lambda - 2 & -1 & 0 \\ 0 & \lambda - 1 & 0 \\ -2 & -2 & \lambda - 1 \end{vmatrix} = 0$$

   $$\Rightarrow \qquad \lambda = 1, 1, 2.$$

   Any polynomial in $P_2(x)$ may be written as (a, b, c). Hence, for $\lambda = 1$, we have

$$\begin{bmatrix} -1 & -1 & 0 \\ 0 & 0 & 0 \\ -2 & -2 & 0 \end{bmatrix} \begin{bmatrix} a \\ b \\ c \end{bmatrix} = \begin{bmatrix} 0 \\ 0 \\ 0 \end{bmatrix}$$

$\Rightarrow \quad -a - b = 0$

$\Rightarrow \quad b = -a$

$\Rightarrow (a, -a, c), a \neq 0, c \neq 0$ are the eigenvectors of $A \Rightarrow 1 - x, x^2$ are eigenvectors of A.

For $\lambda = 2$, we have

$$\begin{bmatrix} 0 & -1 & 0 \\ 0 & 1 & 0 \\ -2 & -2 & 0 \end{bmatrix} \begin{bmatrix} a \\ b \\ c \end{bmatrix} = \begin{bmatrix} 0 \\ 0 \\ 0 \end{bmatrix}$$

$\Rightarrow b = 0$ and $c = 2a \Rightarrow 1 + 2x^2$ is eigenvector corresponding to $\lambda = 2$.

2. If $T \in L(\mathbb{R}^3, \mathbb{R}^3)$ and eigenvalue 5 with an eigenvector $(-1, 1, 4)$, then we calculate eigenvalues of $T^2 - 5T + 2I$.

   $(T^2 - 5T + 2I)(-1, 1, 4) = T^2(-1, 1, 4) - 5T(-1, 1, 4) + 2(-1, 1, 4)$

   $= (-25, 25, 100) - (-25, 25, 100) + 2(-1, 1, 4)$

   $= 2(-1, 1, 4)$

   $\Rightarrow 2$ is the eigenvalue of $T^2 - 5T + 2I$.

3. If $T \in L(\mathbb{R}^3, \mathbb{R}^3)$ such that $T(2, 1) = (5, 2)$ and $T(1, 2) = (7, 10)$ then we compute the eigenvalues of T and the corresponding eigenvectors.

   It is easy to see that $T(x, y) = (x + 3y, -2x + 6y)$ as $\{(1, 2), (2, 1)\}$ is a basis of $\mathbb{R}^2 \Rightarrow$ under the standard basis of $\mathbb{R}^2$, the matrix of T,

   $$m(T) = \begin{bmatrix} 1 & 3 \\ -2 & 6 \end{bmatrix} = A \text{ (say)}.$$

   Characteristic equation of A is

   $$\begin{vmatrix} \lambda - 1 & -3 \\ 2 & \lambda - 6 \end{vmatrix} = 0$$

   $\Rightarrow \quad \lambda^2 - 7\lambda + 12 = 0$

$$\Rightarrow \qquad \lambda = 4, 3.$$

For $\lambda = 4$, we have

$$\begin{bmatrix} 3 & -3 \\ 2 & -2 \end{bmatrix}\begin{bmatrix} x \\ y \end{bmatrix} = \begin{bmatrix} 0 \\ 0 \end{bmatrix}$$

$$\Rightarrow \qquad x = y$$

$\Rightarrow (x, x) \in \mathbb{R}, x \neq 0$ are characteristic vectors of A.

For $\lambda = 3$, we have

$$\begin{bmatrix} 2 & -3 \\ 2 & -3 \end{bmatrix}\begin{bmatrix} x \\ y \end{bmatrix} = \begin{bmatrix} 0 \\ 0 \end{bmatrix}$$

$$\Rightarrow \qquad 2x = 3y$$

$\Rightarrow \left(x, \frac{2}{3}x\right), x \in \mathbb{R}, x \neq 0$ are characteristic vectors of A

$\Rightarrow$ a diagonalizing matrix of A is $P = \begin{bmatrix} 1 & 3 \\ 1 & 2 \end{bmatrix}$.

4. If $T \in L(\mathbb{R}^2, \mathbb{R}^2)$ such that

$$m(T) = \begin{bmatrix} 4 & -5 \\ -4 & 3 \end{bmatrix}$$

with respect to the basis $\{(1, 2), (0, 1)\}$ then we compute the eigenvalues and eigenvectors are a diagonalizing matrix of A.

$$T(1, 2) = 4(1, 2) - 4(0, 1) = (4, 4)$$

$$T(0, 1) = -5(1, 2) + 3(0, 1) = (-5, -7)$$

$$\Rightarrow \qquad T(1, 0) = T(1, 2) - 2T(0, 1) = (14, 18)$$

$$\Rightarrow \qquad m(T) = \begin{bmatrix} 14 & -5 \\ 18 & -7 \end{bmatrix}$$

with respect to the standard basis of $\mathbb{R}^2$. Hence, the characteristic equation of A is

$$\begin{vmatrix} \lambda - 14 & 5 \\ -18 & \lambda + 7 \end{vmatrix} = 0$$

$$\Rightarrow \qquad \lambda^2 - 7\lambda - 8 = 0$$

$$\Rightarrow \qquad \lambda = 8, -1.$$

For $\lambda = 8$, we have

$$\begin{bmatrix} -6 & 5 \\ -18 & 15 \end{bmatrix}\begin{bmatrix} x \\ y \end{bmatrix} = \begin{bmatrix} 0 \\ 0 \end{bmatrix}$$

$$\Rightarrow \qquad -6x + 5y = 0$$

$$\Rightarrow \qquad y = \frac{6}{5}x$$

$\Rightarrow$ eigenvectors are $\left(x, \frac{6}{5}x\right)$, $x \in \mathbb{R}$, $x \neq 0$.

For $\lambda = -1$, we have

$$\begin{bmatrix} -15 & 5 \\ -18 & 6 \end{bmatrix}\begin{bmatrix} x \\ y \end{bmatrix} = \begin{bmatrix} 0 \\ 0 \end{bmatrix}$$

$$\Rightarrow \qquad y = 3x$$

$\Rightarrow$ eigenvectors are $(x, 3x)$, $x \in \mathbb{R}$, $x \neq 0$.

$\Rightarrow P = \begin{bmatrix} 5 & 1 \\ 6 & 3 \end{bmatrix}$ and $P^{-1}AP = \text{diag}\,(8, -1)$.

5. If $T \in L(P_2(x), P_2(x))$ such that

$$m(T) = \begin{bmatrix} 2 & 1 & 0 \\ 0 & 2 & 0 \\ 2 & 3 & 1 \end{bmatrix} = A$$

with respect to the basis $\{x^2, x - 2, x + 1\}$, then we calculate the eigenvalues of T.

$$1 = -\frac{1}{3}(x-2) + \frac{1}{3}(x+1)$$

$$\Rightarrow \qquad T(1) = \frac{1}{3}[-x^2 - 5x + 1 + x + 1]$$

$$= \frac{1}{3}[-x^2 - 4x + 2]$$

$$x = \frac{1}{3}(x-2) + \frac{2}{3}(x+1)$$

$$\Rightarrow \qquad T(x) = \frac{1}{3}[x^2 + 5x - 1 + 2x + 2]$$

$$= \frac{1}{3}[x^2 + 7x + 1]$$

$$T(x^2) = 2x^2 + 2(x + 1)$$
$$= 2x^2 + 2x + 2$$

$\Rightarrow$ the matrix of T, $m(T) = B = \frac{1}{3}\begin{bmatrix} 2 & 1 & 6 \\ -4 & 7 & 6 \\ -1 & 1 & 6 \end{bmatrix}$ with respect to the standard basis of $P_2(x)$.

Since the matrices under the change of bases are similar, so A and B are similar. Also, as the similar matrices have the same eigenvalues, A and B have the same eigenvalues. Trivially the eigenvalues of A are 2, 2, 1; so 2, 2, 1 are the eigenvalues of B.

The characteristic equation of B is

$$\frac{1}{27}\begin{vmatrix} 3\lambda - 2 & -1 & -6 \\ 4 & 3\lambda - 7 & -6 \\ 1 & -1 & 3\lambda - 6 \end{vmatrix} = 0$$

For $\lambda = 1$, we have

$$\frac{1}{27}\begin{bmatrix} 1 & -1 & -6 \\ 4 & -4 & -6 \\ 1 & -1 & -3 \end{bmatrix}\begin{bmatrix} a \\ b \\ c \end{bmatrix} = \begin{bmatrix} 0 \\ 0 \\ 0 \end{bmatrix}$$

$\Rightarrow$
$$a - b - 6c = 0 \quad \ldots(1)$$
$$4a - 4b - 6c = 0 \quad \ldots(2)$$
$$a - b - 3c = 0 \quad \ldots(3)$$

From (2) and (3) we get $c = 0 \Rightarrow a = b \Rightarrow (a, a, 0)$, $a \neq 0$ are eigenvectors of $T \Rightarrow 1 + x$, $2 + 2x$, are eigenvectors of T.

For $\lambda = 2$, we have

$$\frac{1}{27}\begin{bmatrix} 4 & -1 & -6 \\ 4 & -1 & -6 \\ 1 & -1 & 0 \end{bmatrix}\begin{bmatrix} a \\ b \\ c \end{bmatrix} = \begin{bmatrix} 0 \\ 0 \\ 0 \end{bmatrix}$$

$\Rightarrow$
$$4a - b - 6c = 0 \text{ and } a = b$$

$\Rightarrow 3a = 6c \Rightarrow a = 2c \Rightarrow (2c, 2c, c)$, $c \neq 0$ are eigenvectors of $T \Rightarrow$ $2 + 2x + x^2$ is an eigenvector of T.

EXERCISE

1. Find if the eigenvalues, eigenvectors and diagnoalising matrix exist for the following matrices:

(i) $A = \begin{bmatrix} -4 & -2 \\ 3 & 1 \end{bmatrix}$ (ii) $B = \begin{bmatrix} 1 & 0 & 3 \\ 0 & -2 & 4 \\ 0 & 0 & 5 \end{bmatrix}$

(iii) $C = \begin{bmatrix} 3 & 3 & 3 \\ 3 & 3 & 3 \\ -3 & -3 & -3 \end{bmatrix}$ (iv) $D = \begin{bmatrix} 3 & 0 & 0 \\ -2 & 3 & -2 \\ 2 & 0 & 5 \end{bmatrix}$

(v) $E = \begin{bmatrix} 3 & 2 \\ 6 & 4 \end{bmatrix}$ (vi) $G = \begin{bmatrix} 2 & -2 & 3 \\ 0 & 3 & -2 \\ 0 & -1 & 2 \end{bmatrix}$

(vii) $H = \begin{bmatrix} 0 & -2 & 1 \\ 1 & 3 & -1 \\ 0 & 0 & 1 \end{bmatrix}$

2. Show that if A is a diagonalizable square matrix then A^t and A^n, where n is a positive integer, are also diagonalizable.

3. If $B = P^{-1} AP$, where A is square matrix of order n and x is eigenvector of A corresponding to the eigenvalue α, show that $P^{-1}x$ is eigenvector of B corresponding to α.

4. If $T \in L(\mathbb{R}^2, \mathbb{R}^2)$ has matrix representation $\begin{bmatrix} -4 & -2 \\ 3 & 1 \end{bmatrix}$ with respect to the basis $\{(1, 3), (3, 1)\}$, then find the eigenvalues and eigenvectors of T.

5. If $T \in L(\mathbb{R}^2, \mathbb{R}^2)$ such that $T(x, y) = (2x + y, x - y)$, then find eigenvalues and eigenvectors of T.

6. (*a*) If $T : P_2(x) \to P_2(x)$ such that

$$T(a + bx + cx^2) = (3a - 2b + c) + 2bx + (a - 2b + 3c)x^2$$

then find eigenvalues and eigenvectors of T.

(*b*) If $S : P_2(x)$ ϑ $P_2(x)$ such that

$$S(a + bx + cx^2) = (a + 3b + 2c) + 2bx + (b + 2c)x^2$$

then find the values and eigenvectors of S.

7. If $T \in L(\mathbb{R}^2, \mathbb{R}^2)$ has the matrix $\begin{bmatrix} 5 & -2 \\ -2 & 2 \end{bmatrix}$ with respect to the basis $\{(3, 3), (1, -1)\}$, then find the eigenvalues and eigenvectors of T.

8. Find a 2×2 nondiagonal matrix whose eigenvalues are 2 and -3 and the corresponding eigenvectors are $(-1, 2)$ and $(1, 1)$.

9. Find a 3×3 nondiagonal matrix whose eigenvalues are -2, -2, 3 and the corresponding eigenvectors are $(1, 0, 1)$, $(0, 1, 1)$, $(1, 1, 1)$.

10. If α and β are eigenvalues of a square matrix A such that $\alpha \neq \beta$ and U_1 and U_2 are corresponding eigenspaces, then show that $U_1 \cap U_2 = 0$.

11. Give an example to show that A and A^t may not have the same eigenvectors.

12. If $T \in L(V, V)$ and α is an eigenvalue of T, then show that $f(\alpha)$ is an eigenvalue of $f(T)$, where $f(x) \in F[x]$.

13. If $T \in L(V, V)$ such that $ST = TS$ and α is an eigenvalue of T and W its eigenspace, then show that $S(W) \subseteq W$.

14. If $T : \mathbb{R}^3 \to \mathbb{R}^3$ such that $T(x, y, z) = (x - y, 2x + 3y + 2z, x + y + 2z)$, then find the eigenvalues and eigenvectors of T.

15. Show that a scalar α is an eigenvalue of a square matrix A iff α is a root of the characteristic polynomial of A.

8

Inner Product Spaces

In the earlier chapters on vectors spaces, we generally considered arbitrary field F and binary operation and scalar operation on vector spaces. Now we restrict the field F as either field of reals or complex and define a new concept on vector space in terms of the length of the element, a generalization of concept of length of a vectors in $\mathbb{R}^2$ and then study the consequences on abstract vector spaces.

1 INNER PRODUCTS

Let $\overline{OP}$ and $\overline{OQ}$ be two vectors in $\mathbb{R}^2$ and θ be the angle between $\overline{OP}$ and $\overline{OQ}$.

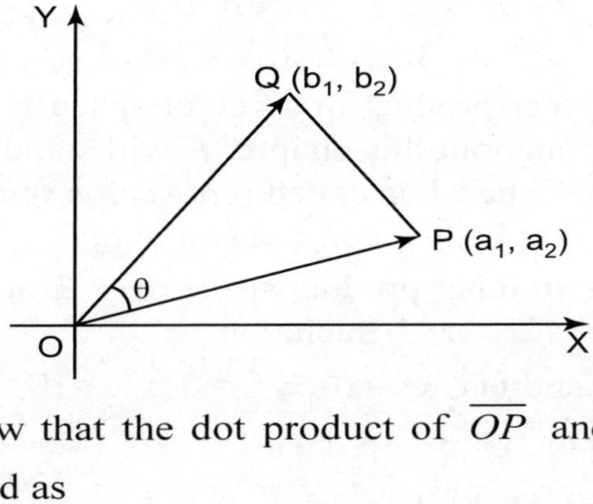

Then we know that the dot product of $\overline{OP}$ and $\overline{OQ}$ denoted by $\overline{OP} \cdot \overline{OQ}$ is defined as

$$\overline{OP} \cdot \overline{OQ} = |\overline{OP}| \cdot |\overline{OQ}| \cos\theta \qquad \ldots(1)$$

In ΔOPQ, we have

$$\overline{QP} = \overline{OP} - \overline{OQ}.$$

Applying cosine formula in ΔOPQ, we get

$$|\overline{QP}|^2 = |\overline{OP}|^2 + |\overline{OQ}|^2 - 2|\overline{OP}||\overline{OQ}|\cos\theta$$

$$\Rightarrow \quad 2|\overline{OP}||\overline{OQ}|\cos\theta = |\overline{OP}|^2 + |\overline{OQ}|^2 - |\overline{QP}|^2$$

Now if we choose the coordinate of P and Q as (a_1, a_2) and (b_1, b_2) respectively, then we get

$$2|\overline{OP}||\overline{OQ}|\cos\theta = (a_1^2 + a_2^2) + (b_1^2 + b_2^2) - (a_1 - b_1)^2 - (a_2 - b_2)^2$$

$$\Rightarrow 2|\overline{OP}||\overline{OQ}|\cos\theta = 2a_1b_1 + 2a_2b_2$$

$$\Rightarrow |\overline{OP}||\overline{OQ}|\cos\theta = a_1b_1 + a_2b_2.$$

Using equation (1), we have

$$\overline{OP}\cdot\overline{OQ} = a_1b_1 + a_2b_2.$$

Now taking x and y to be any two points in $\mathbb{R}^2$, we define

$$(x, y) = x_1y_1 + x_2y_2$$

where $x = (x_1, x_2)$ and $y = (y_1, y_2)$, (x, y) is known as inner product of x and y.

We can observe that it satisfies the following properties:

(*i*) $\langle x, x\rangle \geq 0$ and $\langle x, x\rangle = 0 \Leftrightarrow x = 0, \forall x \in \mathbb{R}^2$

(*ii*) $\langle x, y\rangle = \langle y, x\rangle, \forall x, y \in \mathbb{R}^2$

(*iii*) $\langle \alpha x, y\rangle = \alpha\langle x, y\rangle, \forall x, y \in \mathbb{R}^2, \alpha \in \mathbb{R}$

(*iv*) $\langle x + y, z\rangle = \langle x, z\rangle + \langle y, z\rangle, \forall x, y \in \mathbb{R}^2$

Now we define inner product in a vector space V over the field of reals or complex. Throughout this chapter F will stand for $\mathbb{R}$ or $\mathbb{C}$. If V is a vector space over $\mathbb{R}$ then V is called *real vector space*. If V is vector space over $\mathbb{C}$, then V is called *complex vector space.*

Definition 8.1.1. V is an inner product space over $\mathbb{R}$ or $\mathbb{C}$ if there exists a function $f : V \times V \to F(\mathbb{R}$ or $\mathbb{C})$ such that

(*i*) $f(x, x) \geq 0$ and $f(x, x) = 0 \Leftrightarrow x = 0 \; \forall x \in V$

(*ii*) $f(x, y) = \overline{f(y, x)} \; \forall x, y \in V$

(*iii*) $f(\alpha x, y) = \alpha f(x, y) \; \forall x, y \in V, \alpha \in F$

(*iv*) $f(x + z, y) = f(x, y) + f(z, y), \; \forall x, y \in V, \alpha \in F$

Most often this function f is denoted by $\langle,\rangle$. Now if we see the matters discussed earlier we can see that $\langle,\rangle$ is a generalization of dot product in $\mathbb{R}^2$.

Examples 8.1.2

1. For any $x, y \in \mathbb{R}^n$, $\langle x, y\rangle = \sum_{i=1}^{n} x_iy_i$ defined on $\mathbb{R}^n$ is an inner product and hence $\mathbb{R}^n$ is an inner product space and inner product is known as standard inner product.

2. For $x = (x_1, x_2), y = (y_1, y_2) \in \mathbb{R}^2$, $x, y\rangle = y_1(x_1 + 2x_1) + y_2(2x_1 + 5x_2)$ is an inner product on $\mathbb{R}^2$.
3. In $\mathbb{C}^n$, for $x = (x_1, x_2, ..., x_n)$, $y = (y_1, y_2, ..., y_n)$ we define $x, y\rangle = \sum_{i=1}^{n} x_i \overline{y}_i$, then it is an inner product on $\mathbb{R}^n$, and hence $\mathbb{R}^n$ is an inner product space and it is also known as standard inner product on $\mathbb{R}^n$.
4. Let V be the set of all continuous complex (or real) valued functions on the closed unit interval [0, 1]. If $f, g \in V$ define

$$\langle f, g\rangle = \int_0^1 f(t)\,\overline{g(t)}\,dt.$$

The following verification shows that the above is an inner product on V.

(*i*) $$(f, f) = \int_0^1 f(t)\,\overline{f(t)}\,dt = \int_0^1 |f(t)|\,dt \geq 0$$

and $$f, f\rangle = 0$$

$$\Rightarrow \quad \int_0^1 |f(t)|\,dt = 0$$

$$\Leftrightarrow \quad f(t) = 0 \ \forall\, t \in [0, 1] (\because f \text{ is continuous})$$

$$\Leftrightarrow \quad f(t) = 0$$

(*ii*) $$\overline{\langle g, f\rangle} = \overline{\int_0^1 g(t)\,\overline{f(t)}\,dt}$$

$$= \int_0^1 \overline{g(t)\,\overline{f(t)}}\,dt$$

$$= \int_0^1 g(t)\,\overline{g(t)}\,dt = f, g\rangle.$$

(*iii*) For $h \in V$, $\alpha, \beta \in F$ ($\mathbb{R}$ or $\mathbb{C}$)

$$\langle \alpha f + \beta g, h\rangle = \int_0^1 (\alpha f + \beta f)(t)\,\overline{h(t)}\,dt$$

$$= \int_0^1 (\alpha f(t) + \beta f(t))\,\overline{h(t)}\,dt$$

$$= \alpha \int_0^1 f(t)\,\overline{h(t)}\,dt + \beta \int_0^1 g(t)\,\overline{h(t)}\,dt$$

$$= \alpha f, h\rangle + \beta\alpha\; g, h\rangle$$

Thus Example 2 defines as inner product on V and hence V is an inner product space. We any note from Examples 1 and 2 that the inner product is not unique.

Theorem 8.1.3. In an inner product space V, for all $u, v, w, x \in V, \alpha, \beta, \gamma, \delta \in F$.

$$(i)\ \langle u, \beta v\rangle = \langle \overline{\beta}\, u, v\rangle$$

$$(ii)\ \langle \alpha u+\beta v, \gamma w+\delta x\rangle = \alpha\overline{\gamma}\langle u, w\rangle + \beta\overline{\gamma}\langle v, w\rangle + \beta\overline{\delta}\langle v, x\rangle + \alpha\overline{\delta}\langle u, x\rangle$$

$$(iii)\ \langle 0, v\rangle = \langle u, 0\rangle = 0$$

$$(iv)\ \langle u, v\rangle = 0\ \forall\ u \in V$$

$$\Rightarrow u = 0 \text{ and } \langle u, v\rangle = 0\ \forall\ u \in V$$

$$\Rightarrow v = 0$$

$$(v)\ \langle u, v + w\rangle = \langle u, v\rangle + \langle u, w\rangle$$

Proof. If can be trivially verified by definition.

Remark 1: Let $V_{\mathbb{R}}$ be n-dimensional real inner product space. Then we know that $V_{\mathbb{R}}$ is isomorphic to $\mathbb{R}^n$ *i.e.*, $V_{\mathbb{R}} \cong \mathbb{R}^n$.

Since $\mathbb{R}^n$ is called n-dimensional Euclidean space, therefore $V_{\mathbb{R}}$ is also called Euclidean space.

Remark 2: Let $V_{\mathbb{C}}$ be n-dimensional complex inner product space. Then we know that $V_{\mathbb{C}}$ is isomorphic to $\mathbb{R}^n$ *i.e.*, $V_{\mathbb{C}} \cong \mathbb{R}^n$.

Definition 8.1.4 (Normal Space)

Let V_F be the real or complex vector space. Now we define a function $g : V \to \mathbb{R}$, which satisfies the following properties:

$$(i)\ g(x) \geq 0 \text{ and } g(x) = 0 \text{ iff } x = 0\ \forall\ x \in V$$

$$(ii)\ g(\alpha x) = |\ \alpha\ |\ g(x)\ \forall\ x \in V, \alpha \in F$$

$$(iii)\ g(x + y) \leq g(x) + g(y)\ \forall\ x, y \in V$$

then g is called norm on V and V together will g is called normed space. We shall denote g by $\|\cdot\|$. Thus $\|\cdot\| : V \to \mathbb{R}$ is a function.

Definition 8.1.5. If V is an inner product space, then $\| x \| = \sqrt{(x, x)}$ for $x \in V$ is called length of x.

Theorem 8.1.6. (Cauchy-Schwartz Inequality) Let V be an inner product space. If $x, y \in$ d $\Rightarrow V$, then $|\ (x, y)\ | \leq \| x \|\ \| y \|$.

Proof. If $x = 0$, then LHS $= |\ 0, y\ | = 0$ and RHS $= \| 0 \|\ \| y \| = 0$. So

$$\langle x, y\rangle = \| x \|\ \| y \|.$$

Similarly for $y = 0$, $\langle x, y\rangle = \| x \|\ \| y \|$.

Now let $x \neq 0, y \neq 0$, then $\| x \| \neq 0$, else we shall get $\| x \| = 0 \Rightarrow \langle x, x\rangle = 0 \Rightarrow x = 0$, a contradiction to the assumption.

For any real number λ,

$$0 \leq \langle x - \lambda y, x - \lambda y\rangle$$

$$= \langle x, x\rangle - \lambda\langle y, x\rangle - \overline{\lambda}\langle x, y\rangle + \overline{\lambda}\langle y, y\rangle$$

$$= \langle x, x\rangle - \lambda\langle y, x\rangle - \overline{\lambda}\{\langle x, y\rangle - \overline{\lambda}\langle y, y\rangle\}$$

Choosing λ such that $\lambda = \dfrac{\langle x, y\rangle}{\langle y, y\rangle}$, then

$$0 \leq \| x \|^2 - \frac{\langle x, y\rangle}{\langle y, y\rangle}\langle y, x\rangle \; - 0$$

$$\Rightarrow \quad 0 \leq \| x \|^2 - \frac{\langle x, y\rangle \overline{\langle x, y\rangle}}{\| y \|^2}$$

$$\Rightarrow \quad |\langle x, y\rangle|^2 \leq \| x \|^2 \, \| y \|^2$$

$$\Rightarrow \quad |\langle x, y\rangle| \leq \| x \| \, \| y \|.$$

Remark 1 : In the case of $\mathbb{R}^2$ with the dot product (which is an inner product) Cauchy-Schwartz inequality can be real as

$$\langle x, y\rangle = \| x \| \, \| y \| \cos\theta$$

$$\Rightarrow \quad |\langle x, y\rangle| = | \, \| x \| \, \| y \| \cos\theta \, |$$

$$= \| x \| \, \| y \| \, |\cos\theta|$$

$$|\langle x, y\rangle| = \| x \| \, \| y \| \qquad (\because \cos\theta \leq 1)$$

Remark 2: If we consider ^{n} with the dot product, the Cauchy-Schwartz inequality reduces to

$$\left| \sum_{i=1}^{n} x_i y_i \right| \leq \left(\sum_i x_i^2\right)^{1/2} \left(\sum_i y_i^2\right)^{1/2} \quad \forall\, x_i, y_i \in \mathbb{R}.$$

More generally, if $V = F^n$ ($F = \mathbb{R}$ or $\mathbb{C}$) with

$$\langle x, y\rangle = x_1\overline{y_1} + x_2\overline{y_2} + \ldots + x_1\overline{y_n}$$

where $x = (x_1, x_2, \ldots, x_n)$ and $y = (y_1, y_2, \ldots, y_n)$ then the Cauchy-Schwartz inequality implies that

$$| x_1\overline{y_1} + x_2\overline{y_2} + \ldots + x_1\overline{y_n} |^2 \leq \; (| x_1 |^2 + \ldots + | x_n |^2)\,(| y_1 |^2 + \ldots + | y_n |^2).$$

Theorem 8.1.7. Every inner product space is a normed space with respect to the norm

$$\| x \| = x, x^{\;1/2} \qquad \ldots(1)$$

Proof. (*i*) $\quad \| x \| = \langle x, x\rangle^{1/2}$

$$\Rightarrow \quad \| x \| = \langle x, x\rangle \geq 0$$

$$\Rightarrow \quad \| x \| \geq 0.$$

If $\| x \| = 0$, then $\| x \|^2 = 0$

$$\Rightarrow \qquad \langle x, x\rangle = 0$$

$$\Rightarrow \qquad x = 0$$

If $x = 0$, then $\langle x, x\rangle = 0$

$$\Rightarrow \qquad \langle x, x\rangle^{1/2} = 0$$

$$\Rightarrow \qquad \| x \| = 0.$$

Thus $\| x \| = 0 \Leftrightarrow \| x \| = 0$.

(*ii*) $\| \alpha x \|^2 = \langle \alpha x, \alpha x\rangle$

$$= \alpha\bar{\alpha}\langle x, x\rangle$$

$$= | \alpha |^2 \| x \|^2 = (| \alpha | \| x \|)^2$$

$$\Rightarrow \qquad \| \alpha x \| = | \alpha | \| x \|.$$

(*iii*) $\| x + y \|^2 = \langle x + y, x + y\rangle$

$$= \langle x, x\rangle + \langle x, y\rangle + \langle y, x\rangle + \langle y, y\rangle$$

$$= \langle x, x\rangle + \langle x, y\rangle + \overline{\langle x, y\rangle} + \langle y, y\rangle$$

$$= \langle x, x\rangle + 2 \text{ Re } \langle x, y\rangle + \langle y, y\rangle$$

$$\leq \| x \|^2 + 2 | \ x, y\rangle | + \| y \|^2$$

$$\leq \| x \|^2 + 2 \| x \| \| y \| + \| y \|^2$$

(by Cauchy-Schwartz inequality)

$$\leq (\| x \| + \| y \|)^2$$

$$\Rightarrow \qquad \| x + y \|^2 \leq (\| x \| + \| y \|)^2$$

$$\Rightarrow \qquad \| x + y \| \leq \| x \| + \| y \|.$$

From above it is clear that all three properties of norm are satisfied.

Therefore, every inner product space is normed space w.r.t. the norm defined by equation (1).

Remark. Every normed space is a metric space w.r.t. the metric defined by $d(x, y) = \| x - y \|$.

Theorem 8.1.8. Every inner product space is a metric space.

Proof. Let V be an inner product space. For any $u, v \in V$ putting

$$d(u, v) = \| u - v \|, \qquad \ldots(1)$$

then it satisfies the following:

(*i*) $d(u, v) = \| u - v \| \geq 0$ ($\because$ norm is a always non-negative)

i.e., $d(u, v) \geq 0$. Further

$$d(u, v) = 0$$

$$\Leftrightarrow \qquad \| u - v \| = 0$$

$$\Leftrightarrow \qquad (u - v, u - v) = 0$$

$$\Leftrightarrow \qquad u - v = 0 \qquad \text{(by definition of inner product)}$$

$$\Rightarrow \qquad u = v.$$

(*ii*) $d(u, v) = \| u - v \|$

$$= (u - v, u - v)^{1/2}$$

$$= \langle -(v - u), -(v - u)\rangle^{1/2}$$

$$= \{(-1)\overline{(-1)}\langle v - u, v - u\rangle\}^{1/2}$$

$$= \langle v - u, v - u\rangle^{1/2} = \| v - u \|$$

$$= d(v, u).$$

(*iii*) $\| u + v \|^2 = \langle u + v, u + v\rangle$

$$= \langle u, u\rangle + \langle v, u\rangle + \langle u, v\rangle + \langle v, v\rangle$$

$$= \| u \|^2 + \overline{\langle u, v\rangle} + \langle u, v\rangle + \| v \|^2$$

$$\le \| u \|^2 + 2 \text{ Re } (u, v\rangle) + \| v \|^2 \quad (\because \ u, v\rangle \in \mathbb{C})$$

$$\le \| u \|^2 + 2 \mid \langle u, v\rangle \mid + \| v \|^2$$

$$(\because \text{Re } (\alpha) \le |\alpha|, \text{ where } \alpha \in \mathbb{C})$$

$$\le \| u \|^2 + 2 \| u \| \| v \| + \| v \|^2$$

(by Cauchy-Schwartz inequality)

$$\| u + v \|^2 \le (\| u \| + \| v \|)^2$$

$$\Rightarrow \qquad \| u + v \| \le \| u \| + \| v \| \qquad \ldots(2)$$

Now for $u, v, w \in V$,

$$d(u, v) + d(v, w) = \| u - v \| + \| v - w \|$$

$$\ge \| u - v + v - w \| \qquad \text{[by (2)]}$$

$$= \| u - w \| = d(u, w) \qquad \text{[by (1)]}$$

i.e., $d(u, v) + d(v, w) \ge d(u, w)$

Hence V is a metric space.

We conclude this section with some more examples.

1. Let $A, B \in M_2(\mathbb{R})$, such that

$$A = \begin{bmatrix} a & b \\ c & d \end{bmatrix}, B = \begin{bmatrix} \alpha & \beta \\ \gamma & \lambda \end{bmatrix}.$$

If we define $(A, B) = a\alpha + b\beta + c\gamma + d\gamma$, then $(A, A) = a^2 + b^2 + c^2 + d^2 \ge 0$ and equals 0 iff $a = b = c = d$ *i.e.*, iff

$$A = \begin{bmatrix} 0 & 0 \\ 0 & 0 \end{bmatrix}.$$

Trivially $\langle A, B\rangle = \langle B, A\rangle$ as entries are real. Also

$$\langle \lambda A + \mu A', B\rangle = \left\langle \begin{bmatrix} \alpha a + \mu a' & \lambda b + \mu b' \\ \lambda c + \mu c' & \lambda d + \mu d' \end{bmatrix}, B \right\rangle$$

$$= \alpha(\alpha a + \mu a') + \beta(\lambda b + \mu b') + \gamma(\lambda c + \mu c') + \lambda(\lambda d + \mu d')$$
$$= \lambda(\alpha a + \beta b + \gamma c + \lambda d) + \mu(\alpha a' + \beta b' + \gamma c' + \lambda d')$$
$$= \lambda\langle A, B\rangle + \mu\langle A', B\rangle.$$

Thus $\langle , \rangle$ is an inner product on $M_2(\mathbb{R})$. One can easily see that it is not inner product on $M_2(\mathbb{C})$. If

$$A = \begin{bmatrix} 1 & 3 \\ -1 & 2 \end{bmatrix}, \text{ then length of } A = \| A \| = \sqrt{1+9+4+1} = \sqrt{15}.$$

2. Let $\alpha_1, \alpha_2, \beta_1, \beta_2 \in \mathbb{C}$. Define

$$\langle(\alpha_1 + \alpha_2), (\beta_1 + \beta_2)\rangle$$
$$= 5\overline{\alpha_1}\beta_1 + i\overline{\alpha_1}\beta_2 - i\overline{\alpha_2}\beta_1 + 2\overline{\alpha_2}\beta_2$$

then $\langle(\alpha_1 + \alpha_2), (\beta_1 + \beta_2)\rangle$

$$= 5\overline{\alpha_1}\alpha_1 + i\overline{\alpha_1}\alpha_2 - i\overline{\alpha_2}\alpha_1 + 2\overline{\alpha_2}\alpha_2$$
$$= 5|\alpha_1|^2 + 2|\alpha_2|^2 + i(i2\,\text{Im}\,(\overline{\alpha_1}\alpha_2))$$
$$= 5|\alpha_1|^2 + 2|\alpha_2|^2 - 2\,\text{Im}\,(\overline{\alpha_1}\alpha_2)$$
$$= 4|\alpha_1|^2 + |\alpha_2|^2 + |\overline{\alpha_1}|^2 + |\alpha_2|^2 - 2\,\text{Im}\,(\overline{\alpha_1}\alpha_2)$$
$$\geq 4|\alpha_1|^2 + |\alpha_2|^2 + |\overline{\alpha_1}|^2 + |\alpha_2|^2 - 2|\overline{\alpha_1}\alpha_2|$$
$$= 4|\alpha_1|^2 + |\alpha_2|^2 + (|\overline{\alpha_1}| - |\alpha_2|)^2 \geq 0$$

and equals zero iff $\alpha_1 = \alpha_2 = 0$. Also

$$\overline{\langle(\alpha_1, \alpha_2), (\beta_1, \beta_2)\rangle} = 5\overline{\beta_1}\alpha_1 - i\alpha_1\overline{\beta_1} + i\alpha_2\overline{\beta_2} + 2\alpha_2\overline{\beta_2}$$
$$= \langle(\beta_1, \beta_2), (\alpha_1, \alpha_2)\rangle.$$

Similarly, the third condition can be verified. Hence, $\langle , \rangle$ is an inner product on ${}^2\mathbb{R}$.

If $z = (1, i)$, then

$$\| z \| = \sqrt{5 + i \cdot i - i(-i) + 2(-i)(i)}$$
$$= \sqrt{5 - 1 - 1 + 2} = \sqrt{5}$$

$\Rightarrow$ length of $z = \sqrt{5}$.

3. If V is an inner product space, then for any $x, y \in V$

$$|\langle x, y\rangle| = \|x\|\,\|y\|$$

if $\{x, y\}$ is linearly dependent.

If any one of x or y is 0, then nothing to do. Hence, suppose $x \neq 0$, $y \neq 0$. Let

$$|\, x, y\rangle| = \|x\|\,\|y\| \text{ and } \lambda = \frac{\langle x, y\rangle}{\langle y, y\rangle}$$

Now,

$$\langle x - \lambda y, x - \lambda y\rangle = \langle x, x\rangle - \bar{\lambda}\langle x, y\rangle - \lambda\langle y, x\rangle + \lambda\bar{\lambda}\langle y, y\rangle$$

$$= \|x\|^2 - \frac{\langle x, y\rangle\langle y, x\rangle}{\langle y, y\rangle} - \bar{\lambda}\left[\langle x, y\rangle - \frac{\langle x, y\rangle}{\langle y, y\rangle}\langle y, y\rangle\right]$$

$$= \|x\|^2 - \frac{|\langle x, y\rangle|^2}{\|y\|^2}$$

$$= \frac{\|x\|^2\,\|y\|^2 - |\langle x, y\rangle|^2}{\|y\|^2} = 0.$$

$$\Rightarrow \quad x - \lambda y = 0$$

$$\Rightarrow \quad x = \lambda y$$

$\Rightarrow \{x, y\}$ is linearly dependent.

Conversely, let $\{x, y\}$ be linearly dependent, then $x = \alpha y$ for some scalar $\alpha \Rightarrow \|x\|\,\|y\| = \|\alpha\|\,\|y\|^2 = |\alpha|\ \langle y, y\rangle = |\langle \alpha y, y\rangle| = |\langle x, x\rangle|$.

4. Let $(x_1, x_2), (y_1, y_2) \in \mathbb{R}^2$. Define

$$\langle (x_1, x_2), (y_1, y_2)\rangle = 2x_1y_1 + x_1y_2 + x_2y_1 + x_2y_2.$$

Now,

$$\langle (x_1, x_2), (x_1, x_2)\rangle = 2{x_1}^2 + x_1x_2 + x_1x_2 + {x_2}^2$$
$$= {x_1}^2 + (x_1 + x_2)^2 \geq 0$$

and equals 0 iff $x_1 = 0$,

$$x_1 + x_2 = 0$$

$\Rightarrow (x_1, x_2) = 0$. Trivially

$$\overline{\langle (x_1, x_2), (y_1, y_2)\rangle} = \langle (y_1, y_2), (x_1, x_2)\rangle \text{ as } x_1, x_2, y_1, y_2 \in \mathbb{R}.$$

$\langle \alpha(x_1, x_2) + \beta(x'_1, x'_2), (y_1, y_2)\rangle$

$=\langle(\alpha x_1 + \beta x'_1, \alpha x_2 + \beta x'_2), (y_1, y_2)\rangle$
$= 2(\alpha x_1 + \beta x'_1)y_1 + (\alpha x_1 + \beta x'_1)y_2 + (\alpha x_2 + \beta x'_2)y_1 + (\alpha x_2 + \beta x'_2)y_2$
$= \alpha(2x_1y_1 + x_1y_2 + x_2y_1 + x_2y_2) + \beta(2x'_1y_1 + x'_1y_2 + x'_2y_1 + x'_2y_2)$
$= \alpha\langle(x_1, x_2), (y_1, y_2)\rangle + \beta\langle(x'_1, x'_2), (y_1, y_2)\rangle$

Thus $\langle,\rangle$ defined above is an inner product on $\mathbb{R}^2$. If $z = (2, 3)$, then the length of $z = \| z \| = \sqrt{\langle(2, 3), (2, 3)\rangle} = \sqrt{4 + 25} = \sqrt{29}$.

If $x = (1, -1)$, then $\| x \| = 1$. Verify if

$$\langle(x_1, x_2), (y_1, y_2)\rangle = 2x_1y_1 + x_1y_2 + x_2y_1 + x_2y_2.$$

Is $\langle,\rangle$ an inner product on $\mathbb{R}^2$?

5. Let $\alpha_1, \alpha_2, \beta_1, \beta_2 \in \mathbb{C}$, then define

$$\langle(\alpha_1, \alpha_2), (\beta_1, \beta_2)\rangle = 2\alpha_1\overline{\beta_1} + \alpha_1\overline{\beta_2} + \alpha_2\overline{\beta_1} + \alpha_2\overline{\beta_2}.$$

Now,

$$\langle(\alpha_1, \alpha_2), (\beta_1, \beta_2)\rangle = 2\alpha_1\overline{\alpha_1} + \alpha_1\overline{\alpha_2} + \alpha_2\overline{\alpha_1} + \alpha_2\overline{\alpha_2}$$

$$\alpha_1\overline{\alpha_1} + \alpha_1(\overline{\alpha_1} + \overline{\alpha_2}) + \alpha_2(\overline{\alpha_1} + \overline{\overline{\alpha_2}}) = |\alpha_1|^2 + (\alpha_1 + \alpha_2)\overline{(\alpha_1 + \alpha_2)}$$
$$= |\alpha_1|^2 + |\alpha_1 + \alpha_2|^2 \geq 0$$

and equals zero iff $\alpha_1 = \alpha_{21} = 0$.

Further $\overline{\langle(\alpha_1, \alpha_2), (\beta_1, \beta_2)\rangle} = 2\overline{\alpha_1}\beta_1 + \beta_2\overline{\alpha_1} + \beta_1\overline{\alpha_2} + \beta_2\overline{\alpha_2}$

$$= \langle(\beta_1, \beta_2), (\alpha_1, \alpha_2)\rangle.$$

It is trivial to verify the third condition.

If $z = (1, i)$, then the length of $z = \| z \|$

$= \sqrt{2i(-i) + i + (-i) + 1} = \sqrt{3}$.

6. Let $A, B \in M_2(\mathbb{R})$. Define $\langle A, B\rangle = \text{tr}\,(A^tB)$.

Let

$$A = \begin{bmatrix} a & b \\ c & d \end{bmatrix},$$

then

$$\langle A, B\rangle = \text{tr}\,(A^tA) = a^2 + b^2 + c^2 + d^2 \geq 0$$

and equals zero iff $a = b = c = d = 0$. Also

$\langle A, B\rangle = \text{tr}\,(A^tB) = \text{tr}\,((A^tB)^t) = \text{tr}\,(B^tA) = \langle B, A\rangle$.

Also

$$\alpha A, \beta C, B\rangle = \text{tr}\,((\alpha A + \beta C)^tB)$$
$$= \text{tr}\,((\alpha A^t + \beta C^t)B)$$

$= \alpha \text{tr}(A^t B) + \beta \text{tr}(C^t B)$

$= \alpha\langle A, B\rangle + \beta\langle C, B\rangle$

$\Rightarrow \langle , \rangle$ defines inner product on $M_2(\mathbb{R})$. If

$$A = \begin{bmatrix} 2 & 1 \\ 0 & 3 \end{bmatrix}, \text{ then } \| A \| = \sqrt{\langle A, A\rangle} = \sqrt{\text{tr}(A^t A)} = \sqrt{14}$$

$$A^t = \begin{bmatrix} 2 & 0 \\ 1 & 3 \end{bmatrix}$$

$$A^t A = \begin{bmatrix} 4 & 2 \\ 2 & 10 \end{bmatrix}.$$

If A is of order $n \times n$, then we can also show that $\langle A, B\rangle = \text{tr}(A^t B)$ is an inner product on $M_n(\mathbb{R})$.

2 ORTHOGONALITY AND ORTHONORMALITY

If x and y are two nonzero vectors in an inner product space V, then by Cauchy-Schwartz inequality, we have

$$-1 \leq \frac{\langle x, y\rangle}{\| x \| \| y \|} \leq 1. \qquad \ldots(1)$$

In $\mathbb{R}^2$, $\langle x, y\rangle$ represents the dot product of x and y and if θ is the angle between these vectors, then by definition of dot product, we have

$$\langle x, y\rangle = \| x \| \, \| y \| \cos \theta$$

$$\Rightarrow \qquad \cos \theta = \frac{\langle x, y\rangle}{\| x \| \| y \|}. \qquad \ldots(2)$$

2.1 Particular Cases

If $\theta = 0$, $\dfrac{\langle x, y\rangle}{\| x \| \| y \|} = 1$,

if $\theta = \pi$, $\dfrac{\langle x, y\rangle}{\| x \| \| y \|} = \cos \pi = -1$,

and if $\theta = \dfrac{\pi}{2}$, $\dfrac{\langle x, y\rangle}{\| x \| \| y \|} = \cos \dfrac{\pi}{2} = 0$.

$\Rightarrow \langle x, y\rangle = 0$, *i.e.*, if $x \perp y$, then $\langle x, y\rangle = 0$.

Hence equation (1) gives that there exists unique $\theta \in [0, \pi]$ satisfying equation (2). In general, we can define

Definition 8.2.1. If x and y are two vectors in an inner product space V, we say that x is orthogonal to y if $\langle x, y\rangle = 0$.

It coincides with what we have in $\mathbb{R}^2$, with the dot product $x \cdot y = 0$ $\Rightarrow \cos\theta = 0 \Rightarrow \theta = \frac{\pi}{2}$. We write it as $x \perp y$ to denote $x, y\rangle = 0$.

Note. If x is orthogonal to y, then y is orthogonal to x, for

$$\langle y, x\rangle = \overline{\langle x, y\rangle} = \overline{0} = 0.$$

Hence x and y are orthogonal if $\langle x, y\rangle = 0$.

Example 8.2.2

1. If $u = (a, b)$, $v = (c, d) \in \mathbb{R}^2$, then $\langle u, v\rangle = a\overline{c} + b\overline{d}$ is an inner product on $\mathbb{R}^2$. Now, the vectors $u = (3, 5)$ and $v = (5i, -3i)$ are orthogonal. Since

$$\begin{aligned}\langle u, v\rangle &= \langle(3, 5), (5i, -3i)\rangle \\ &= 3(\overline{5i}) + 5(\overline{-3i}) \qquad \text{(by definition)} \\ &= 3(-5i) + 5(3i) = -15i + 15i = 0\end{aligned}$$

2. Let V be the vector space over $\mathbb{C}$ of all continuous complex valued functions on the interval $[0, 1]$. For $f, g \in V$,

$$\langle f, g\rangle = \int_0^1 f(t)\,\overline{g(t)}\,dt$$

is an inner product.

The vectors $\sin \pi t$ and $\cos \pi t$ are orthogonal with respect to the above inner product. Since

$$\begin{aligned}\sin \pi t, \cos \pi t\rangle &= \int_0^1 \sin \pi t\, \overline{\cos \pi t}\, dt \\ &= \int_0^1 \sin \pi t \cos \pi t\, dt \\ &= \frac{1}{2}\int_0^1 \sin 2\pi t\, dt \\ &= \frac{1}{4\pi}[\cos 2\pi]_0^1 \\ &= \frac{1}{4\pi}(\cos 2\pi - \cos 0) \\ &= \frac{1}{4\pi}(1-1) = 0.\end{aligned}$$

Definition 8.2.3. If W is a subspace of an inner product space V, then the orthogonal complement of W, denoted as $W^{\perp}$ is defined by

$$W^{\perp} = \{x \in V \mid \langle x, w\rangle = 0, \forall\ w \in W\}.$$

Even for any subset X of V,

$$X^{\perp} = \{v \in V \mid \langle v, x\rangle = 0, \forall\ x \in X\}.$$

Trivially if $X \subseteq Y$, then $X^{\perp} \supseteq Y^{\perp}$.

Example 8.2.4

1. Let $V = \mathbb{R}^2$ and $W = \{(x, 0) \mid x \in \mathbb{R}\}$, then W is a subspace of V and V is equipped with the standard inner product *i.e.*, for $u = (u_1, u_2) \in \mathbb{R}^2$ and $v = (v_1, v_2) \in \mathbb{R}^2$, $\langle u, v\rangle = u_1v_1 + u_2v_2$.

   Then orthogonal complement of W *i.e.* $W^{\perp}$ is

   $$W^{\perp} = \{(0, y) \mid y \in \mathbb{R}\}.$$

2. $\mathbb{R}^2$ is an inner product space with respect to inner product, defined as

   $$\langle u, v\rangle = u_1v_1 - u_2v_1 - v_1v_2 + 4v_1v_2$$

   for any $u, v \in \mathbb{R}^2$ with $u = (u_1, u_2)$, $v = (v_1, v_2)$ and let $W = \{(a, a) \mid a \in \mathbb{R}\}$, then W is a subspace of $\mathbb{R}^2$.

   It is easy to see that the orthogonal complement of W is $W^{\perp} = \{(x, 0) \mid a \in \mathbb{R}\}$. Since $(p, q) \in W^{\perp} \Rightarrow \langle (a, a), (p, q)\rangle = 0\ \forall\ a \in \mathbb{R}$.

   Let $a \neq 0$, then

   $$\begin{aligned} ap - ap - aq + 4aq &= 0 \\ \Rightarrow \qquad 3aq &= 0 \\ \Rightarrow \qquad q &= 0\ \forall\ (a, a) \in W. \end{aligned}$$

   Also any $r \in \mathbb{R}$,

   $\langle (a, a), (r, 0)\rangle = ar - ar - 0 + 4 \cdot 0 = 0\ \forall\ (a, a) \in W.$

   But order standard inner product

   $$W^{\perp} = \{(a, -a) \mid a \in \mathbb{R}\}.$$

   So the inner product plays important role.

Lemma 8.2.5. For any subspace W of V, $W^{\perp}$ is a subspace of V, such that $W \cap W^{\perp} = \{0\}$.

Proof. Obviously $0 \in W^{\perp}$, so $W^{\perp} \neq \emptyset$. Now, let $u, v \in W^{\perp}$ and $\alpha, \beta \in F$, then

$$\langle u, w\rangle = \langle v, w\rangle = 0, \forall\ w \in W^{\perp}.$$

Now, for all $w \in W$,

$$\langle \alpha u + \beta v, w\rangle = \langle \alpha\ u, w\rangle + \beta\ u, w\rangle$$

(by property of inner product)

$$=\alpha \cdot 0 + \beta \cdot 0 = 0$$

$\Rightarrow \alpha u + \beta v \in W^{\perp}$.

As $u, v \in W^{\perp}$, $\alpha, \beta \in F$ were arbitrary, hence $W^{\perp}$ is a subspace of V.

Finally, let $x \in W \cap W^{\perp}$

$\Rightarrow x \in W$ and $x \in W^{\perp}$

$\Rightarrow \quad \langle x, x \rangle = 0$ (by definition)

$\Rightarrow \quad x = 0$ (by property of inner product)

$\Rightarrow \quad W \cap W^{\perp} = \{0\}$.

Definition 8.2.6 A subset X of V is said to be an orthogonal set if for any $\langle x, y \in X,\ x, y \rangle = 0$, whenever $x \neq y$.

2.2 Pythagoras Theorem in $\mathbb{R}^2$

We know that in case of $\mathbb{R}^2$, if ΔABC is a right angle triangle such that $\angle ABC = \frac{\pi}{2}$, then by Pythagoras theorem, we have

$$CA^2 = AB^2 + BC^2. \qquad \ldots(1)$$

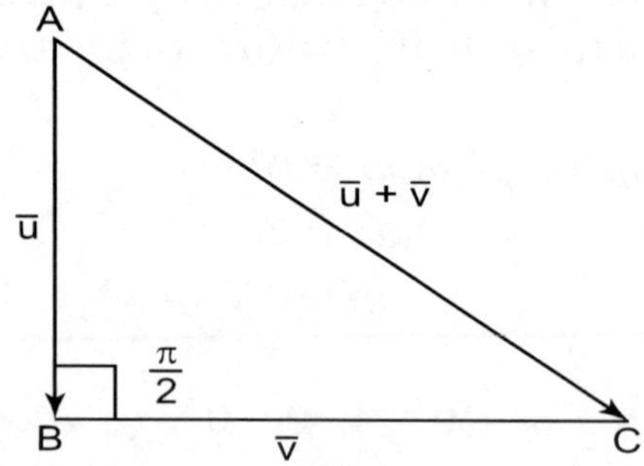

If $\overline{u}$ and $\overline{v}$ are the vectors along the directions of AB and BC, respectively, such that

$$AB = \| \overline{u} \|, \ BC = \| \overline{v} \|$$

then the resultant vector $\overline{u} + \overline{v}$ is in the direction of AC and $AC = \| \overline{u} + \overline{v} \|$.

Thus from (1), we have

$$\| \overline{u} + \overline{v} \|^2 = \| \overline{u} \|^2 + \| \overline{v} \|^2, \text{ where } \overline{u} \perp \overline{v}.$$

Now, we can generalize this concept on an inner product space, for an orthogonal set as:

Lemma 8.2.7 (Pythagoras Theorem) Let $\{v_i\}_{i=1}^{k}$ be a set of vectors such that they are pairwise orthogonal, *i.e.*, $v_i \perp v_j$, if $i \neq j$. Then

$$\left\| \sum_{i=1}^{k} v_i \right\|^2 = \sum_{i=1}^{k} \| v_i \|^2 .$$

Proof: Since

$$\begin{aligned}\left\| \sum_{i=1}^{k} v_i \right\|^2 &= \left\langle \sum_{i=1}^{k} v_i, \sum_{i=1}^{k} v_i \right\rangle \\ &= \sum_{i \leq i,\, j \leq k} \langle v_i, v_j \rangle \\ &= \sum_{i=1}^{k} \langle v_i, v_i \rangle \text{ (since } v_i, v_j\rangle = 0, \text{ for } i \neq j) \\ &= \sum_{i=1}^{k} \| v_i \|^2 .\end{aligned}$$

Remarks

1. This result is true only for Euclidean space and not for unitary space.

   Counter Example: $\mathbb{C}^2$ is an inner product space with standard inner product *i.e.*, $\langle x, y\rangle = x_1 \overline{y_1} + x_2 \overline{y_2}$, where $x = (x_1, x_2), y = (y_1, y_2) \in \mathbb{C}^2$. If we take $x = (0, i)$ and $y = (0, 1)$, then $\langle x, y\rangle = 0 \cdot \overline{0} + i \cdot \overline{1} = i \neq 0 \Rightarrow x$ is not orthogonal to y.

   But
   $$\begin{aligned}\|x + y\| &= \|(0, i) + (0, 1)\|^2 \\ &= \|(0, 1 + i)\|^2 \\ &= \langle (0, 1 + i), (0, 1 + i)\rangle \\ &= 0 \cdot \overline{0} + (1 + i)\overline{(1 + i)} = (1 + i)(1 - i) \\ &= (1 + i^2) = 1 + 1 = 2\end{aligned}$$

   $$\begin{aligned}\|x\|^2 = \langle (0, i), (0, i)\rangle &= 0 \cdot \overline{0} + i \cdot \overline{i} = 0 + i(-i) \\ &= - i^2 = 1\end{aligned}$$

   and
   $$\|y\|^2 = \langle (0, 1), (0, 1)\rangle = 0 \cdot \overline{0} + 1 \cdot \overline{1} = 1$$

   Hence $\|x + y\|^2 = \|x\|^2 + \|y\|^2$, while x is not orthogonal to y.

2. x and y are orthogonal if and only if $||x - y||^2 = ||x||^2 + ||y||^2$.
3. $||x + y|| = ||x|| + ||y||$ if and only if one is a non-negative scalar multiple of the other.
4. For any $x, y, \in V$, we have $\langle x, y \rangle$

$$= \begin{cases} \frac{1}{4}\{||x+y||^2 - ||x-y||^2 + i||x+iy||^2 - ||x-iy||^2\}, & \text{if } F = \mathbb{C} \\ \frac{1}{4}\{||x+y||^2 - ||x-y||^2\}, & \text{if } F = \mathbb{R} \end{cases}$$

Which is known as polarization identity.

5. For any two vectors $x, y \in V$.

$$||x + y||^2 + ||x - y||^2 = 2(||x||^2 + ||y||^2).$$

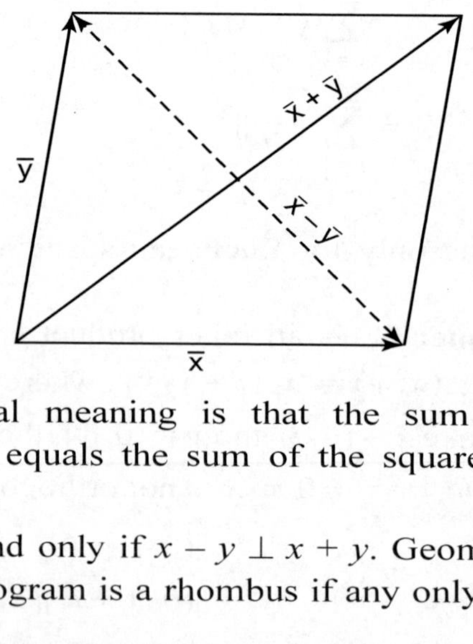

Its geometrical meaning is that the sum of the squares of the diagonals equals the sum of the squares of the sides of a parallelogram.

6. $||x|| = ||y||$ if and only if $x - y \perp x + y$. Geometrically, this means that a parallelogram is a rhombus if any only if the diagonals are perpendicular.

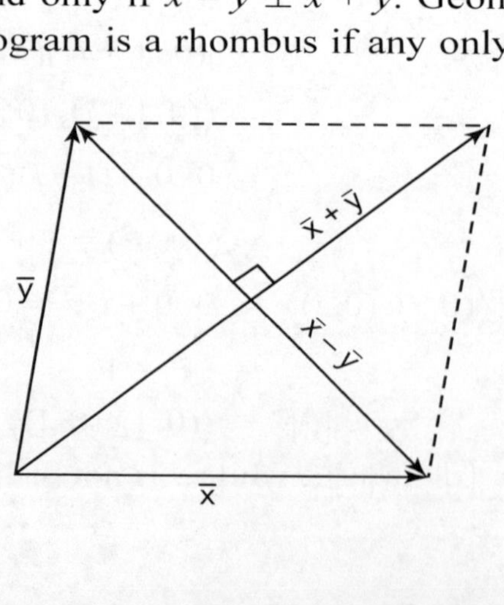

2.3 Orthogonal Projection

Let $v \in \mathbb{R}^2$ be any vector and u be any unit vector *i.e.*, $||u|| = 1$. Let θ be the angle between u and v. Then the vector $P_u(v)$ is called the projection of v on u. By trigonometry, we have

$$P_u(v) = (||v|| \cos \theta)u$$

$$P_u(v) = ||v|| \frac{\langle u, v\rangle}{||u|| \, ||v||} \quad \left(\text{since } \cos\theta = \frac{\langle u, v\rangle}{||u|| \, ||v||}\right)$$

$$= \langle u, y\rangle \qquad (\text{since } ||y|| = 1)$$

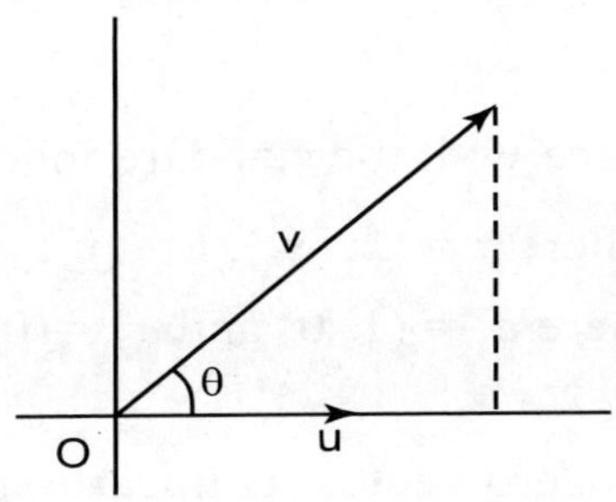

Definition 8.2.8

Let V be an inner product space. Let u be a unit vector and $v \in V$ be an arbitrary element of V. Then the orthogonal projection $P_u(v)$ of v along u is defined by

$$P_u(v) = \langle u, v\rangle u \, .$$

If u is a nonzero vector, then

$$P_u(v) = \left\langle \frac{u}{||u||}, v \right\rangle \frac{u}{||u||} = \frac{\langle u, v\rangle}{\langle u, u\rangle} u$$

Remark

$v - P_u(v) \perp u$ for any $u, v \in V_{\mathbb{R}}$.

Proof

$$\langle v - P_u(v), u\rangle = \langle v, u\rangle - \langle P_u(v), u\rangle$$

$$= \langle v, u\rangle - \left\langle \frac{\langle u, v\rangle}{\langle u, u\rangle} u, u \right\rangle$$

$$= \langle v, u\rangle - \frac{\langle u, v\rangle}{\langle u, u\rangle} \langle u, u\rangle$$

$$= \langle v, u \rangle - \overline{\langle v, u \rangle}$$

$$= \langle v, u \rangle - \langle v, u \rangle \text{ (since } \langle v, u \rangle \in \mathbb{R}) = 0$$

2.4 Orthonormality

Definition 8.2.9

A nonempty subset X of V is called an orthonormal set, if

(*i*) $\|x\| = 1, \forall\, x \in X$ and

(*ii*) $\langle x, y \rangle = 0$ for $x, y \in X$, whenever $x \neq y$.

Example

$\mathbb{C}^2$ is an inner product space with the standard inner product $\langle x, y \rangle = x_1\overline{y_1} + x_2\overline{y_2}$, where $x = (x_1, x_2), y = (y_1, y_2) \in \mathbb{C}^2$, then $\{e_1, e_2\}$ is an orthonormal set, where $e_1 = (1, 0)$ and $e_2 = (0, 1)$.

Lemma 8.2.10

Any orthogonal set of nonzero vectors is linearly independent.

Proof Let X be any non-empty orthogonal set in an inner product space with the property that $0 \notin X$. Let $\{v_1, v_2, \ldots, v_k\}$ be any subset of X, such that

$$\sum_{i=1}^{k} \alpha_i v_i = 0 \text{ where } \alpha_i \ (i = 1, 2, \ldots, k) \in F$$

$$\text{Then } \left\langle \sum_{i=1}^{k} \alpha_i v_i, v_j \right\rangle = 0 \text{ (since } \langle 0, v \rangle = 0)$$

$$\Rightarrow \sum_{i=1}^{k} \alpha_i \langle v_i, v_j \rangle = 0 \qquad \text{(by property of inner product)}$$

$$\Rightarrow \alpha_j \langle v_j, v_j \rangle = 0 \qquad \text{(since } \langle v_i, v_j \rangle = 0, i \neq j)$$

$$\Rightarrow \alpha_j = 0 \qquad \text{(since } \langle v_i, v_j \rangle \neq 0, \text{ for } v_j \neq 0$$

$\Rightarrow \alpha_j = 0\ \forall\, j$

$\Rightarrow X$ is linearly independent subset of V.

Remark Since every orthonormal set is orthogonal, therefore every orthonormal set of nonzero vectors is also linearly independent.

Corollary 8.2.11

If $\{v_1, v_2, \ldots, v_n\}$ is an orthogonal set in V and

$$v = \sum_{i=1}^{n} \alpha_i v_i \text{ , then } \alpha_i = \frac{\langle v, v_i \rangle}{\langle v_i, v_i \rangle}, \ 1 \le i \le n.$$

Proof As $v = \sum_{j=1}^{n} \alpha_j v_j$ therefore we have

$$\langle v, v_i \rangle = \left\langle \sum_{j=1}^{n} \alpha_j v_j, v_i \right\rangle = \sum_{j=1}^{n} \alpha_j \langle v_j, v_i \rangle$$

$$= \alpha_i \langle v_i, v_i \rangle \text{ (since } \langle v_j, v_i \rangle = 0, j \ne 1)$$

$$\Rightarrow \qquad \alpha_i = \frac{\langle v, v_i \rangle}{\langle v_i, v_i \rangle}, \ 1 \le i \le n.$$

Remark

If $\{v_1, v_2, \ldots, v_n\}$ is an orthonormal set, then $\langle v_i, v_i \rangle = 1$, for all i, and therefore $\alpha_i = \langle v, v_i \rangle$, $1 \le i \le n$.

2.5 Gram-Schmidt Orthonormalization Process

This is the process to construct an orthonormal basis of a finite dimensional inner product space from a given basis. First we prove the following.

Lemma 8.2.12

If $\{v_1, v_2, \ldots, v_k\}$ is an orthonormal subset of V, then for any $v \in V$, the vector $v - \sum_{i=1}^{k} \langle v, v_i \rangle v_i$ is orthogonal to each of v_i. Moreover, if $y \notin \{v_1, v_2, \ldots, v_k\}$ then for

$Z = y - \sum_{i=1}^{k} \langle y, v_i \rangle v_i$, $\| z \| \ne 0$ and for $v_{k+1} = \frac{z}{\| z \|}$, $\{v_1, v_2, \ldots, v_k, v_{k+1}\}$ is an orthogonal subset of V.

Proof: For $j \in \{1, 2, \ldots, k\}$, we have

$$\left\langle v - \sum_{i=1}^{k} \langle v, v_i \rangle v, v_j \right\rangle = \langle v, v_j \rangle - \sum_{i=1}^{k} \langle v, v_i \rangle \langle v, v_j \rangle$$

$$= \langle v, v_j \rangle - \langle v, v_j \rangle \langle v_j, v_j \rangle$$

$$\text{(since } \langle v_i, v_j \rangle = 0 \text{ for } i \ne j)$$

$$=\langle v, v_j\rangle - \langle v, v_j\rangle$$

$$(\text{since } \langle v_j, v_j\rangle = 1) = 0$$

$$\Rightarrow v - \sum_{i=1}^{k} \langle v_i, v_i\rangle\, v_i \perp v_i \text{ for each } i.$$

Now, suppose $y \notin [v_1, v_2, \ldots, v_k]$ and $||z|| = 0$, where

$$z = y - \sum_{i=1}^{k} \langle y, v_i\rangle v_i$$

$$||z|| = 0$$

$$\Rightarrow \quad z = 0$$

$$\Rightarrow \quad y - \sum_{i=1}^{k} \langle y, v_i\rangle v_i = 0 \Rightarrow y = \sum_{i=1}^{k} \langle y, v_i\rangle v_i$$

i.e., y is a linear combination of v_i's ($i = 1, 2, \ldots, k$). *i.e.*, $y \in [v_1, v_2, \ldots, v_k]$ which is a contradiction to the hypothesis. Hence $||z|| \neq 0$.

To complete the proof, we must show that

$$\langle v_{k+1}, v_i\rangle = 0, \ \forall\, i = 1, 2, \ldots, k$$

and $\langle v_{k+1}, v_{k+1}\rangle = 1$

Now,
$$\langle v_{k+1}, v_i\rangle = \left\langle \frac{z}{\| z \|}, v_i \right\rangle = \frac{1}{\| z \|}(z, v_i)$$

$$= \frac{1}{\| z \|}\left\langle y - \sum_{j=1}^{k} \langle y, v_j\rangle, v_i \right\rangle$$

$$= \frac{1}{\| z \|}\left\{\langle y, v_i\rangle - \sum_{j=1}^{k} \langle y, v_j\rangle\langle v_j, v_i\rangle\right\}$$

$$= \frac{1}{\| z \|}\{\langle y, v_i\rangle - \langle y, v_i\rangle\}$$

$$(\text{since } \langle v_j, v_i\rangle = 0 \text{ for } j \neq i)$$

$$= 0, \ \forall\, i = 1, 2, \ldots, k$$

Finally,

$$\langle v_{k+1}, v_{k1}\rangle = \left\langle \frac{z}{\| z \|}, \frac{z}{\| z \|} \right\rangle = \frac{1}{\| z \|^2}(z, z)$$

$$= \frac{1}{\| z \|^2} \times \| z \|^2 = 1 \qquad (\text{since } \langle z, z\rangle = ||z||^2)$$

Hence, the proof is complete.

Theorem 8.2.13

Every finite dimensional inner product space has an orthonormal basis.

Proof: Let V be an inner product space of dimension n. We have to find an orthonormal set $\{v_1, v_2, \dots v_n\}$ which forms a basis of V. Now, starting with the basis $\{e_1, e_2, \dots e_n\}$ of V, let us put $u_1 = e_1$.

As $e_1 \neq 0$, $||u_1|| \neq 0$. So on taking $v_1 = \dfrac{u_1}{|| u_1 ||}$, we have $v_1, v_1\rangle = 1$

As $e_2 \notin [v_1]$ there exists a vector $u_2 = e_2 - \langle e_2, v_1\rangle$ with

$||u_2|| \neq 0$ and $v_2 = \dfrac{u_2}{|| u_2 ||}$ (by Lemma 8.2.10)

is such that $\{v_1, v_2\}$ is orthonormal subset of V. Further

$$\{v_1, v_2\} \subseteq [e_1, e_2]$$

Claim: $e_3 \notin [v_1, v_2]$.

Suppose the claim is not true. Then we would have found $\alpha_1, \alpha_2 \in F$, at least one of them is nonzero such that $e_3 = \alpha_1 v_1 + \alpha_2 v_2$.

$$\Rightarrow \qquad e_3 = \alpha_1 \frac{e_1}{|| u_1 ||} + \alpha_2 \frac{u_2}{|| u_2 ||}$$

$$\left(\because \quad v_i = \frac{u_i}{|| u_i ||}, i = 1, 2 \right)$$

$$\text{i.e.,} \qquad e_3 = \alpha_1 \frac{e_1}{|| u_1 ||} + \alpha_2 \frac{\{e_2 - \langle e_2 - v_1\rangle e_1\}}{|| u_2 |||| u_1 ||}$$

which is absurd. Again by Lemma 8.2.10, there exists u_3 such that

$$u_3 = e_3 - \langle e_3 - v_1\rangle\, v_1 - \langle e_3 - v_2\rangle\, v_2 \text{ with } ||u_3|| \neq 0 \text{ and } v_3 = \frac{u_3}{|| u_3 ||}$$

is such that $\{v_1, v_2, v_3\}$ is an orthonormal subset of V. Also

$$\{v_1, v_2, v_3\} \subseteq [e_1, e_2, e_3]$$

Now, on continuing this process, we get an orthonormal set

$$\{v_1, v_2, \dots, v_i\} \subseteq [e_1, e_2, \dots e_i]$$

As before it can be shown that $e_{i+1} \notin [v_1, v_2, \dots, v_i]$. Then by Lemma 8.2.12, there exists a vector

$$u_{i+1} = e_{i+1} - \sum_{k=1}^{l} \langle e_{i+1}, v_k\rangle v_k \ \text{ with } ||v_{k+1}|| \neq 0$$

and $$u_{i+1} = \frac{u_{i+1}}{\| u_{i+1} \|}$$

is such that $\{v_1, v_2,\ldots, v_{i+1}\}$ is an orthonormal subset of V.

Since every orthonormal subset is linearly independent and dim $V = n$, this process terminates till we get an orthonormal set $\{v_1, v_2,\ldots, v_n\}$. This is also a basis of V as dim $V = n$.

Hence V has an orthonormal basis.

Example 8.2.14

Using Gram-Schmidt orthonormalization process, we find the orthonormal basis.

1. $\mathbb{R}^3$ is an inner product space over $\mathbb{R}$ with respect to the standard inner produce *i.e.*, for

$$x = \{x_1, x_2, x_3\}, y = \{y_1, y_2, y_3\} \in \mathbb{R}^3,$$
$$\langle x, y\rangle = x_1y_1 + x_2y_2 + x_3y_3$$
$$\dim \mathbb{R}^3 = 3 \text{ and } \{(2, 0, 0), (0, -1, 0), (0, 0, 3)\}$$
is a basis of $\mathbb{R}^3$.

Putting $e_1 = (2, 0, 0)$, $e_2 = (0, -1, 0)$, $e_3 = (0, 0, 3)$ we apply Gram-Schmidt orthonormalization process to obtain an orthonrmal basis of $\mathbb{R}^3$ over $\mathbb{R}$.

Taking $u_1 = e_1$ and $v_1 = \frac{v_1}{\| u_1 \|} = \frac{e_1}{\| e_1 \|}$, we have

$$v_1 = \frac{e_1}{\langle e_1, e_1\rangle^{\frac{1}{2}}} = \frac{(2, 0, 0)}{\sqrt{2\cdot 2 + 0\cdot 0 + 0\cdot 0}}$$
$$= \frac{1}{2}(2, 0, 0) = (1, 0, 0)$$

Putting $u_2 = e_2 - e_2, v_1\rangle\, v_1$, we have

$$u_2 = (0, -1, 0) - \langle (0, -1, 0), (1, 0, 0)\rangle\, (1, 0, 0)$$
$$= (0, -1, 0) - (0.1 + (-1) . 0 + 0 . 0)\, (1, 0, 0)$$
$$= (0, -1, 0) - 0. (1, 0, 0) = (0, -1, 0)$$

Then

$$v_2 = \frac{u_2}{\| u_2 \|} = \frac{(0, -1, 0)}{\sqrt{0.0 + (-1)\cdot(-1) + 0\cdot 0}} = (0, -1, 0).$$

Again putting

$$u_3 = e_3 - \langle e_3, v_1\rangle\, v_1 - \langle e_3, v_2\rangle\, v_2 \text{ we have}$$

$$\begin{aligned} u_3 &= (0, 0, 3) - \langle(0, 0, 3), (1, 0, 0)\rangle\,(1, 0, 0) \\ &\qquad - \langle(0, 0, 3), (0, -1, 0)\rangle\,(0, -1, 0) \\ &= (0, 0, 3) - (0\cdot 1 + 0\cdot 0 + 0\,.\,3)\,(1, 0, 0) \\ &\qquad - (0\cdot 0 + 0\cdot(-1, 0 + 3\cdot 0)\,(0, -1, 0) \\ &= (0, 0, 3) \end{aligned}$$

Then,

$$v_3 = \frac{u_3}{\|u_3\|} = \frac{(0, 0, 3)}{\sqrt{0.0 + 0\cdot 0 + 3\cdot 3}}$$

$$= \frac{1}{2}(0, 0, 3) = (0, 0, 1)$$

Hence, $\{(1, 0, 0), (0, -1, 0), (0, 0, 1)\}$ is an orthonormal basis of $\mathbb{R}^3$.

Let V be the vector space over $\mathbb{R}$ of the polynomials of degree < 3, then $\dim V_{\mathbb{R}} = 3$ and $\{1, x, x^2\}$ is a basis of V.

Putting $e_1 = 1, e_2 = x, e_3 = x^2$ for $f, g \in V$, define

$$\langle f, g\rangle = \int_0^1 f(x)g(x)dx$$

Apply Gram-Schmidt orthonormalization process to obtain an orthonormal basis of V over $\mathbb{R}$.

Take $u_1 = e_1$ and

$$v_1 = \frac{u_1}{\|u_1\|} = \frac{e_1}{\|e_1\|} = \frac{e_1}{\sqrt{\left(\int_0^1 e_1 e_1\, dx\right)}} = e_1 = 1$$

Putting $u_2 = e_2 - \langle e_2, v_1\rangle\, v_1 = x - \langle x, 1\rangle = x \int_0^1 x dx = \frac{1}{2}$, we have

$$v_2 = \frac{u_1}{\|u_2\|} = \frac{x - \frac{1}{2}}{\sqrt{\int_0^1 \left(x - \frac{1}{2}\right)^2 dx}} = \frac{x - \frac{1}{2}}{\sqrt{\int_0^1 \left(x^2 - x + \frac{1}{4}\right)^2 dx}} = \sqrt{3}(2x - 1)$$

Again putting $u_3 = e_3 - \langle e_3 - v_1\rangle - \langle e_3, v_2\rangle\, v_2$, we have

$$u_3 = x^2 - \int_0^1 x^2\, dx - (6x - 3)\int_0^1 (2x^3 - x^2)dx = x^2 - x + \frac{1}{6}$$

Now, $$\|u_3\|^2 = \int_0^1 \left(x^2 - x + \frac{1}{6}\right)^2 dx = \frac{1}{180}$$

$$\Rightarrow \qquad ||u_3|| = \frac{1}{6\sqrt{5}}$$

Thus, $$v_3 = \frac{u_3}{|| u_3 ||} = (6x^2 - 6x + 1)\sqrt{5}$$

Hence, $\{1, (2x - 1)\sqrt{3}, (6x^2 - 6x + 1)\sqrt{5}\}$ is an orthonormal basis of V.

Theorem 8.2.15

For each subspace W of a finite dimensional inner product space V.

$$V = W \oplus W^{\perp}.$$

Proof: Since W is a subspace of V and V is an inner product space, W is also an inner product space. Then there exists an orthonormal basis, say $\{w_1, w_2, \ldots, w_m\}$ of W, by Theorem 8.2.13.

Let $v \in V$ and $x = v - \sum_{i=1}^{m}\langle v, w_i\rangle w_i$, then trivially $\langle x, w_i\rangle = 0$ for $i = 1, 2, \ldots, m$.

Since any element of W can be written as a linear combination of $w_1, w_2, \ldots w_m$; we get $\langle x, w\rangle = 0$, $\forall\, w \in W \Rightarrow W^{\perp}$.

Hence $v \in W + W^{\perp} \Rightarrow V \subseteq W + W^{\perp} \Rightarrow V = W + W^{\perp}$.

Now, since $W \cap W^{\perp} = (0)$, therefore

$$V = W \oplus W^{\perp}.$$

Remark: If W is a subspace of V and let $\{w_i\}_{i=1}^{r}$ be an orthonormal basis of W and $\{u_j\}_{j=1}^{s}$ be an orthonormal basis of $W^{\perp}$, then $\{w_i\} \cup \{u_j\}$ is an orthornormal basis of V (since $V = W \oplus W^{\perp}$) and therefore

$$\dim V = r + s = \dim W + \dim W^{\perp}.$$

Corollary 8.2.16

If W is a subspace of a finite dimensional inner product space V, then $(W^{\perp})^{\perp} = W$.

Proof: Let $x \in W$, for any $w \in W^{\perp}$,

$$\langle x, w\rangle = 0 \Rightarrow \overline{\langle x, w\rangle} = \overline{0} \text{ i.e., } \langle w, x\rangle = 0$$

As w is arbitrary,

$$x \in (W^{\perp})^{\perp} \quad \Rightarrow \quad W \subseteq (W^{\perp})^{\perp},$$

Now, as $V = W \oplus W^{\perp}$ also $V = W^{\perp} \oplus (W^{\perp})^{\perp}$. So if $\dim V = n$, $\dim W = m$, then

$$\dim W^{\perp} = \dim V - \dim W = n - m$$

and $\dim (W^{\perp})^{\perp} = \dim V - \dim (W^{\perp}) = n - (n - m) = m$

i.e., $\dim W = \dim (W^{\perp})^{\perp}$.

Thus, $W = (W^{\perp})^{\perp}$.

2.6 Orthogonal Projection of an Inner Product Space

If W is a vector subspace of an inner product space V, then the orthogonal projection P_w of V onto W is the map $P_w(x) = w$, where $x = w + w'$ is the orthogonal decomposition of x. If $\{w_1, w_2, \ldots, w_r\}$ is an orthogonal basis of W, then $P_w(v) = \sum_{i=1}^{r} \langle v, w_i \rangle w_i$ is called projection of v on W.

Clearly, P_w is an idempotent map.

Theorem 8.2.17 (Bessel's Inequality)

Let $\{e_1, e_2, \ldots, e_n\}$ be a finite orthonormal set in an inner product space V. Then for any x in V,

$$\sum_{i=1}^{n} |\langle x, e_i \rangle|^2 \leq \| x \|^2$$

Proof (Method I) We have

$$0 \leq \left\| x - \sum_{i=1}^{n} \langle x, e_i \rangle \right\|^2$$

$$\Rightarrow \quad 0 \leq \left\langle x - \sum_{i=1}^{n} \langle x, e_i \rangle \, e_i, \, x - x - \sum_{j=1}^{n} \langle x, e_i \rangle e_j \right\rangle$$

$$= \langle x, x \rangle - x - \sum_{i=1}^{n} \langle x, e_i \rangle \langle e_i, x \rangle - \sum_{j=1}^{n} \overline{\langle x, e_j \rangle} \langle x, e_j \rangle$$

$$+ \sum_{j=1}^{n} \sum_{i=1}^{n} \langle x, e_i \rangle \overline{\langle x, e_j \rangle} \langle e_i, e_j \rangle$$

$$= \| x \|^2 - \sum_{i=1}^{n} |\langle x, e_i \rangle|^2 - \sum_{i=1}^{n} |\langle x, e_i \rangle|^2 + \sum_{i=1}^{n} |\langle x, e_i \rangle|^2$$

(Since $\langle e_i, e_j \rangle = 0$ for $i \neq j$ and $\langle e_i, e_i \rangle = 1$

$$= \| x \|^2 - \sum_{i=1}^{n} |\langle x, e_i \rangle|^2$$

$$\Rightarrow \sum_{i=1}^{n} |\langle x, e_i \rangle|^2 \leq \| x \|^2$$

Method II

Let $x \in V$, then we can write

$$x = \sum_{i=1}^{n} \langle x, e_i \rangle e_i + v, \text{ where } v = x - \sum_{i=1}^{n} \langle x, e_i \rangle e_i$$

and
$$v, e_j\rangle = \left\langle x - \sum_{i=1}^{n} \langle x, e_i \rangle e_i, e_j \right\rangle$$
$$= \langle x, e_j \rangle - \sum_{i=1}^{n} \langle x, e_i \rangle \langle e_i, e_j \rangle$$
$$= \langle x, e_j \rangle - \langle x, e_j \rangle = 0, \forall j = 1, 2, \ldots, n.$$

Now putting $w = \sum_{i=1}^{n} \langle x, e_i \rangle e_i$, we have

$$x = w + v \text{ with } w, v\rangle$$
$$= \left\langle \sum_{i=1}^{n} \langle x, e_i \rangle e_i, v \right\rangle = \sum_{i=1}^{n} \langle x, e_i \rangle \langle e_i, v \rangle = 0$$

Thus,
$$||x||^2 = \langle w + v, w + v \rangle$$
$$= \langle w, w \rangle + \langle w, v \rangle + \langle v, w \rangle + \langle v, v \rangle$$
$$= ||w||^2 + 0 + 0 + ||v||^2 \text{ (since } w \perp v)$$
$$= ||w||^2 \leq ||v||^2$$
$$\Rightarrow \quad ||w||^2 \leq ||x||^2.$$

But
$$||w||^2 = \left\langle \sum_{i=1}^{n} \langle x, e_i \rangle e_i, \sum_{i=1}^{n} \langle x, e_j \rangle e_j \right\rangle$$
$$= \sum_{i=1}^{n} \sum_{j=1}^{n} \langle x, e_i \rangle \overline{\langle x, e_j \rangle} \langle e_i, e_j \rangle$$
$$= \sum_{i=1}^{n} | \langle x, e_i \rangle |^2 \qquad \left(\text{since } \langle \mathrm{e}_i, e_j \rangle = \begin{cases} 0 & i \neq j \\ 1 & i = j \end{cases} \right)$$

$$\Rightarrow \sum_{i=1}^{n} | \langle x, e_i \rangle |^2 \leq || x ||^2$$

This completes the proof.

Remarks.

1. The geometrical interpretation of the above inequality is the sum of the squares of the components of a vector in various perpendicular

directions does not exceed the square of the length of the vector itself.

2. Bessel the geometrical interpretation of the above inequality is the sum of the squares of the components of a vector in various perpendicular directions does not exceed the square of the length of the vector itself.
3. Bessel's inequality, for $n = 1$, is essentially the Cauchy Schwartz inequality. If $n = 1$ and $e_1 = e$, then

$$\sum_{i=1}^{n} |\langle x, e_i \rangle|^2 = |\langle x, e \rangle|^2$$

Now, we shall prove Riesz Representation theorem. First we see that if V is an inner product space, then for any $v \in V$ the map $f_v : V \to F$ such that $f_v(x) = \langle x, v \rangle$ is trivially linear. Therefore, $f_v \in V^*$.

Theorem 8.2.18 (Risez Representation)

If V is finite dimensional inner product space and $f \in V^*$, then there exists unique $y \in V$, such that $f(x) = (x, y)$, $x \in V$.

Proof: Let $\{v_1, v_2, \ldots, v_n\}$ be an orthonormal basis of V. Let $y = \overline{f(v_1)}v_1 + \overline{f(v_2)}v_2 + \ldots. + \overline{f(v_n)}v_n$, then as remarked above, $f_y \in V^*$. Such that $f_y(x) = \langle x, y \rangle$. Now,

$$f_y(v_i) = \left\langle v_i, \sum_{i=1}^{n} \overline{f(v_i)}v_i \right\rangle = f(v_i) \text{ as } \langle v_i, v_j \rangle = \delta_{ij}$$

$$\Rightarrow \quad f_y(v_i) = f(v_i)\ \forall\ i$$

$$\Rightarrow \quad f_y\left(\sum_{i=1}^{n} \lambda_i v_i\right) = \sum_{i=1}^{n} \lambda_i f_y(v_i) = \sum_{i=1}^{n} \lambda_i f(v_i) = f\left(\sum_{i=1}^{n} \lambda_i v_i\right)$$

$$\Rightarrow \quad f_y = f.$$

$$\Rightarrow \quad f(x) = \langle x, y \rangle$$

Now, suppose there exists $y' \in V$, such that

$$f(x) = \langle x, y' \rangle$$

$\Rightarrow \langle x, y \rangle = \langle x, y' \rangle$ for all $x \in V$

$\Rightarrow \langle x, y - y' \rangle = 0$ for all $x \in V$

$\Rightarrow \langle y - y', y - y' \rangle = 0 \quad \Rightarrow \quad y = y'$

$\Rightarrow \exists$ a unique $y \in V$ such that $f(x) = \langle x, y \rangle$.

The section ends with solutions of some problems and examples. Throughout V stands for inner product space.

1. If U and W are subspaces of V, then
   $(U \cap W)^{\perp} = (U^{\perp} + W^{\perp})$ and $(U + W)^{\perp} = U^{\perp} \cap E^{\perp}$.
   $U, W \subseteq U + W \Rightarrow (U + W)^{\perp} \subseteq U^{\perp}$,
   $(U + W)^{\perp} \subseteq W^{\perp}$
   $\Rightarrow (U + W)^{\perp} \subseteq U^{\perp}, \cap W^{\perp}$
   Let $x \in U^{\perp}, \cap W^{\perp}$
   $\Rightarrow \langle x, u\rangle =$ and $\langle x, w\rangle = 0$ for all $u \in U, w \in W$
   $\Rightarrow \langle x, u + w\rangle = 0 \Rightarrow x \in (U + W)^{\perp}$
   $\Rightarrow (U + W)^{\perp} = U^{\perp} \cap W^{\perp}$.
   Since $U^{\perp}, W^{\perp} \subseteq V$
   $\Rightarrow (U^{\perp} + W^{\perp})^{\perp} = (U^{\perp})^{\perp} \cap (W^{\perp})^{\perp} = U \cap W \Rightarrow U^{\perp} + W^{\perp} = (U \cap W)^{\perp}$.
2. If $V = U \oplus W$, then $V = U^{\perp} \oplus W^{\perp}$, since $V^{\perp} = 0 \Rightarrow U^{\perp} \cap W^{\perp} = 0$.
   Also $U \cap W = 0 \Rightarrow 0^{\perp} = V = U^{\perp} + W^{\perp} \Rightarrow V = U^{\perp} \oplus W^{\perp}$.
3. If $U = [(1, -1, 2)]$, $W = [(1, 0, 3), (-1, 2, -1)]$ and $S = \{(x, y, z) | x + 2y - 4z = 0\}$, then we find $U^{\perp}$, $W^{\perp}$ and $S^{\perp}$ in ^{3}.
   Trivially $S^{\perp} = [(1, 2, -4)]$ as S is the equation of the plane passing through the origin and direction ratios of the normal are proportional to 1, 2, – 4.
   $U^{\perp} = \{(a, b, c\} \mid a - b + 2c = 0\}$
   due to the standard inner product.
   Now, take $a = 0$, then $b = 2c$, taking $b = 0$, we get $a = -2c$
   $\Rightarrow U^{\perp} = [(0, 2, 1), (-2, 0, 1)]$
   Now, let $(a, b, c) \in W^{\perp}$
   $\Rightarrow a + 3c = 0, -a + 2b - c = 0$
   $\Rightarrow b = -c \Rightarrow W^{\perp} = [(-3, -1, 1)]$.
4. If W is a subspace of V, then for any $x \in V$,
   $$||x||^2 = ||w||^2 + ||w_1||^2$$
   for some $w \in W$ and $w_1 \in W^{\perp}$.
   Since $V = W \oplus W^{\perp} \Rightarrow x = w + w_1, w \in W, w_1 \in W^{\perp}$
   $$\Rightarrow \quad ||x|| = \langle x, x\rangle = \langle w + w_1, w + w_1\rangle$$
   $$= \langle w, w\rangle + \langle w_1, w_1\rangle \text{ as } w, w_1\rangle = 0$$
   $$\Rightarrow \quad ||x||^2 = ||w||^2 + ||w_1||^2.$$

5. If V is a real or complex vector space, then $|| \cdot || : V \to \mathbb{R}$ is a norm iff

   (*a*) $||x|| \neq 0$ if $x \neq 0$

   (*b*) $||\lambda x + \mu y|| \leq |\lambda|\ ||x|| + |\mu|\ ||y||$; $x, y \in V$ and λ, μ are scalars.

   If $|| \cdot ||$ is a norm function then trivially conditions (*a*) and (*b*) are satisfied. Now, suppose (*a*) and (*b*) are satisfied. Choosing $\mu = 0$, we get $||\lambda x|| \leq |\lambda|\ ||x||$ for all scalar λ and $x \in V$. Let $\lambda \neq 0 \Rightarrow \left\|\frac{1}{\lambda}x\right\| \leq \frac{1}{|\lambda|}||x||\ \forall\ x \in V$. Since $\lambda x \in V \Rightarrow ||x|| \leq \frac{1}{|\lambda|}||\lambda x||$

   $\Rightarrow |y|\ ||x|| = ||\lambda x||\ \forall\ x \in V$.

   Now taking $x = 0$, we get $|\lambda|\ ||0|| = ||0||$. Now, if $||0|| \neq 0$

   $\Rightarrow |\lambda| = 1$ which is not always true.

   $\Rightarrow ||0|| = 0$. Further if $||x|| = 0$ and if $x \neq 0$, then by (*a*) $||x|| \neq 0$

   $\Rightarrow x = 0$. Thus $||x|| = 0$ iff $x = 0$. Also $||0|| = ||x - x|| \leq 2||x||$

   $\Rightarrow ||x|| \leq 0$. Also choosing $\lambda = \mu = 1$, we get $||x + y|| \leq ||x|| + ||y||$.

   $\Rightarrow || \cdot ||$ as given above is a norm on V.

6. If W is subspace of V, then $f \in A(W)$ iff $\sigma(f) \in W^{\perp}$, where $\sigma(f) = \overline{f(e_1)}\, e_1 + \overline{f(e_2)}\, e_2 + \ldots + \overline{f(e_n)}\, e_n$ for $\{e_1, e_2, \ldots, e_n\}$ as orthnormal basis of V.

Let $\{e_1, e_2, \ldots, e_r\}$ be an orthonormal basis of W, then $\{e_1, e_2, \ldots, e_r\}$ can be extended to an orthonormal basis $\{e_1, e_2, \ldots, e_r, e_{r+1}, \ldots, e_n\}$ of V. Now, for any $w \in W$,

$$w = \sum_{i=1}^{r} \lambda_i e_i$$

Let $f \in A(W) \Rightarrow f(w) = 0 \Rightarrow f(e_i) = 0$ for $i = 1, 2, \ldots, r$.

$$\Rightarrow \qquad \langle \sigma(f), w \rangle = \left\langle \sum_{i=r+1}^{r} \overline{f(e_i)}\, e_i, \sum_{i=1}^{r} \lambda_i e_i \right\rangle = 0$$

$$\Rightarrow \qquad \sigma(f) \in W^{\perp}.$$

Now, if $\sigma(f) \in W^{\perp}$

$$\Rightarrow \qquad \left\langle \sum_{i=r+1}^{r} \overline{f(e_i)}\, e_i, \sum_{i=1}^{r} \lambda_i e_i \right\rangle = 0$$

$$\Rightarrow \qquad \left\langle \sum_{i=1}^{r} \overline{f(e_i)}\, e_i, e_j \right\rangle = 0 \text{ for } j = 1, 2, \ldots, r$$

$$\Rightarrow \qquad \overline{f(e_j)}\langle e_j, e_j\rangle = 0 \text{ for } j = 1, 2, \ldots, r$$

$$\Rightarrow \qquad f(e_j) = 0 \text{ for } j = 1, 2, \ldots, r$$

$$\Rightarrow \qquad f(w) = 0$$

$$\Rightarrow \qquad f \in A(W).$$

7. If $W = [(1, -1, 2, 3), (2, 1, -4)]$, then under standard inner product in $\mathbb{R}^4$, we find $W^\perp$.

   Let $(a, b, c, d) \in W^\perp \Rightarrow a - b + 2c + 3d = 0$ and $2a + b - c + 4d = 0$. Choosing $b = 0$, we have $\frac{\sigma}{11} = \frac{c}{2} = \frac{d}{-5} = r$. Similarly, choosing $c = 0$, we have

$$\frac{a}{-7} = \frac{b}{2} = \frac{d}{3} = s$$

$$\Rightarrow \qquad W^\perp = [(11, 0, 2, -5), (-7, 2, 0, 3)].$$

8. If $W = [(1, i, 1 - i), (i - 1, 0)]$ is a subspace of $\mathbb{C}^3$, then we find $W^\perp$ under standard inner product. Let $(a, b, c) \in W^\perp$

   $\Rightarrow a - ib + c\,(1 + i) = 0$ and $-ai - b = 0 \Rightarrow b = -ai$

   $\Rightarrow (1 + i)c = -a + bi = -a + a = 0 \Rightarrow c = 0$

   $\Rightarrow (1, -i, 0)$ spans $W^\perp \Rightarrow W^\perp = [(1, -i, 0)]$.

9. If $(\alpha_1, \alpha_2), (\beta_1, \beta_2) \in \mathbb{C}^2$ and $\langle(\alpha_1, \alpha_2), (\beta_1, \beta_2)\rangle = 2\alpha_1\overline{\beta_1} + \alpha_1\overline{\beta_2} + \alpha_2\overline{\beta_1} + \alpha_2\overline{\beta_2}$, then we know that $\langle\ ,\ \rangle$ is an inner product on $\mathbb{C}^2$. Let

   $W = [(i - 1, i)]$ and $(a, b) \in W^\perp, (a, b) \neq 0$

   $\Rightarrow 2a\,(-1 - i) + a(-i) + b\,(-i - 1) + b(-i) = 0$

   $\Rightarrow a\,[-2 - 2i - i] + b\,(-2i - 1) = 0$

   $\Rightarrow a\,(3i + 2) + b(2i + 1) = 0$

   $\Rightarrow a\,(3i + 2)\,(2i - 1) + b(-5) = 0$

$$\Rightarrow \qquad b = \frac{8 - i}{-5}a$$

   $\Rightarrow W^\perp = \left[\left(a, \frac{8 - i}{-5}\right)\right]$ choose any a and get the answer

   $= [(-5, 8 - i)]$.

10. If $A, B \in M_2(\mathbb{R})$, then $\langle A, B\rangle = tr\,(A^t B)$ defined as inner product on $M_2(\mathbb{R})$. The set

$$\left\{ v_1 = \begin{bmatrix} 1 & 0 \\ 0 & 0 \end{bmatrix}, v_2 = \begin{bmatrix} 1 & 1 \\ 0 & 0 \end{bmatrix}, v_3 = \begin{bmatrix} 1 & 1 \\ 1 & 0 \end{bmatrix}, v_4 = \begin{bmatrix} 1 & 1 \\ 1 & 1 \end{bmatrix} \right\}$$

is a basis of $M_2(\mathbb{R})$. Using Gram-Schmidt process, we calculate an orthonormal basis of $M_2(\mathbb{R})$. Trivially $\|v_1\| = 1$. Let

$$u_2 = v_2 - \langle v_2, v_1 \rangle v_1$$

$$= \begin{bmatrix} 1 & 1 \\ 0 & 0 \end{bmatrix} - 1 \begin{bmatrix} 1 & 0 \\ 0 & 0 \end{bmatrix} = \begin{bmatrix} 0 & 1 \\ 0 & 0 \end{bmatrix}$$

then $\langle u_2, u_2 \rangle = 1$. Let

$$u_3 = v_3 - \langle v_3, u_2 \rangle u_2 - \langle v_3, v_2 \rangle v_1$$

$$= \begin{bmatrix} 1 & 1 \\ 1 & 0 \end{bmatrix} - 1 \begin{bmatrix} 0 & 1 \\ 0 & 0 \end{bmatrix} - 1 \begin{bmatrix} 1 & 0 \\ 0 & 0 \end{bmatrix}$$

$$= \begin{bmatrix} 0 & 0 \\ 1 & 0 \end{bmatrix}$$

Trivially $\langle u_3, u_3 \rangle = 1$. Let

$$u_4 = v_4 - \langle v_4, u_3 \rangle u_3 - \langle v_4, u_2 \rangle u_2 - \langle v_4, v_1 \rangle v_1$$

$$= \begin{bmatrix} 1 & 1 \\ 1 & 1 \end{bmatrix} - 1 \begin{bmatrix} 0 & 0 \\ 1 & 0 \end{bmatrix} - 1 \begin{bmatrix} 0 & 1 \\ 0 & 0 \end{bmatrix} - 1 \begin{bmatrix} 1 & 0 \\ 0 & 0 \end{bmatrix}$$

$$= \begin{bmatrix} 0 & 0 \\ 0 & 1 \end{bmatrix}$$

Trivially $\langle u_4, u_3 \rangle = 1$.

$$\Rightarrow \left\{ \begin{bmatrix} 1 & 0 \\ 0 & 0 \end{bmatrix}, \begin{bmatrix} 0 & 1 \\ 0 & 0 \end{bmatrix}, \begin{bmatrix} 0 & 0 \\ 1 & 0 \end{bmatrix}, \begin{bmatrix} 0 & 0 \\ 0 & 1 \end{bmatrix} \right\}$$

is the required orthonormal basis of $M_2(\mathbb{R})$.

11. Consider $A = \{(1, -1, 0), (1, 1, 0), (1, 0, -1)\}$ as a basis $\mathbb{R}^3$. Under the standard inner product on $\mathbb{R}^3$, we find an orthonormal basis (using the Gram-Schmidt process).

Let $v_1 = (1, -1, 0)$, $v_2 = (1, 1, 0)$, $v_3 = (1, 0, -1)$,

$$u_1 = \frac{v_1}{\| v_1 \|} = \frac{1}{\sqrt{2}} (1, -1, 0)$$

Define $$w_3 = v_3 - \langle v_3, u_1 \rangle u_1 - \langle v_3, u_2 \rangle u_2$$

$$= (1, 0, 1) - \frac{1}{2}(1, -1, 0) - \frac{1}{2}(1, 1, 0)$$

$$= (0, 0, -1)$$

Let $u_3 = \dfrac{w_3}{\| w_3 \|} = (0, 0, -1)$

$$\Rightarrow \left\{\frac{1}{\sqrt{2}}(1, -1, 0), \frac{1}{\sqrt{2}}(1, 1, 0), (0, 0, -1)\right\}$$

is the orthonormal basis.

12. If $a + bx + cx^2, \alpha + \beta x + \gamma x^2 \in P_2(x)$ and $\langle a + bx + cx^2, \alpha + \beta x + \gamma x^2\rangle = a\alpha + b\beta + c\gamma$, then $\langle\ ,\ \rangle$ defines an inner product on $P_2(x)$. If $\{1 - x, 1 + x, 1 - x_2\}$ is considered a basis of $P_2(x)$, then as done in Example 11,

$$\left\{\frac{1-x}{\sqrt{2}}, \frac{1+x}{\sqrt{2}}, -x^2\right\}$$

is orthonormal basis of $P_2(x)$.

13. Let $A = \{(2, 2, 1), (0, 4, 1), (8, 3, 5)\} \subseteq \mathbb{R}^3$, then A is a basis of $\mathbb{R}^3$. We find the orthonormal basis corresponding to A.

Let $v_1 = (2, 2, 1), (0, 4, 1), v_3 = (8, 3, 5)$ and $u_1 = \dfrac{v_1}{\| v_1 \|} = \dfrac{1}{3}(2, 2, 1)$

Define

$$w_3 = v_3 - \langle v_3, u_1\rangle\, u_1 = (0, 4, 1) - (2, 2, 1)$$
$$= (-2, 2, 0).$$

Let $u_2 = \dfrac{w_2}{\| w_2 \|} = \dfrac{1}{2\sqrt{2}}(-2, 2, 0) = \dfrac{1}{\sqrt{2}}(-1, 1, 0)$

Define

$$w_3 = v_3 - \langle v_3, u_1\rangle\, u_1 - \langle v_3, u_2\rangle\, u_2$$
$$= (8, 3, 5) - 3(2, 2, 1) - \left(\frac{5}{2}, \frac{-5}{2}, 0\right)$$
$$= \left(-\frac{1}{2}, -\frac{1}{2}, 2\right) = \frac{1}{2}(-1, -1, 4)$$

$$\Rightarrow \qquad u_3 = \frac{1}{3\sqrt{2}}(-1, -1, 4)$$

$$\Rightarrow \{u_1, u_2, u_3\} = \left\{\frac{1}{3}(2, 2, 1), \frac{1}{\sqrt{2}}(-1, 1, 0), \frac{1}{3\sqrt{2}}(-1, -1, 4)\right\}$$

is the orthonormal basis.

3 ISOMETRIES

As we know that if V is an inner product space then V is a normed space under the norm.

$$||x|| = \sqrt{\langle x, x\rangle}, x \in V .$$

Further we have seen that a metric can be defined in an inner product space V to make it a metric space, for instance $d(x, y) = ||x - y||$, $x, y \in V$ (see Theorem 8.1.6). In case of isomorphism we know that the linear map preserves the operations of the vector space and further it is one-one and onto.

The question is: Is it possible to find some linear maps which also preserve the inner product or which do not change the distance or which do not change the length of the vector in inner product space? The purpose of this section is to discuss this problem and its related theories.

Let us consider $T : \mathbb{R}^2 \to \mathbb{R}^2$ such that $T(x, y) = (y, x)$ under the standard inner product $\mathbb{R}^2$ is an inner product space. Let $z \in \mathbb{R}^2$ such that $z = (x, y)$, then $||z|| = \sqrt{x^2 + y^2}$. Also

$$T(z) = (y, x) \Rightarrow ||T(z)|| = \sqrt{y^2 + x^2} \Rightarrow ||z|| = ||T(z)||$$

$\Rightarrow T$ pressure the length of the vector. However, if we take $T_1 : \mathbb{R}^2 \to \mathbb{R}^2$ such that $T_1(x, y) = (x, 0)$, then $||(x, y)|| \neq ||T_1(x, y)||$.

$$d(z_1, z_2) = ||z_1 - z_2||, z_1, z_2 \in \mathbb{R}^2$$

$$z_1 = (x_1, y_1), z_2 = (x_2, y_2), \text{ we have}$$

$$d(z_1, z_2) = \sqrt{(x_1 - x_2)^2 + (y_1 - y_2)^2}$$

and

$$d(Tz_1, Tz_2) = ||(y_1, x_1) - (y_2, x_2)|| = \sqrt{(y_1 - y_2)^2 + (x_1 - x_2)^2}$$

$$\Rightarrow \quad d(z_1, z_2) = d(Tz_1, Tz_2)$$

$\Rightarrow T$ does not change the distance. It preserves the distance. Hence we can define.

Definition 8.3.1

If V is an inner product space and $T \in L(V, V)$, then T is called isometry if for all $x, y \in V$.

$$d(x, y) = d(T(x), T(y)).$$

From the above definition, it means that

$$\|x - y\| = \|T(x) - T(y)\| = \|T(x - \psi)\| \text{ for all } x, y \in V.$$

$$\Rightarrow \quad \|x\| = \|T(x)\| \text{ taking } y = 0.$$

Further, if $\|x\| = \|T(x)\|$ for all $x \in V$

$$\Rightarrow \quad \|x - y\| = \|T(x - \psi)\| \text{ for all } y \in V, x - y \in V.$$

$$= |T(x) - T(y)\|$$

$\Rightarrow T \in L(V, V)$ is isometry iff $\|T(x)\| = \|x\|$ for all $x \in V$

If $x \in$ ket T, then $T(x) = 0 \Rightarrow \|x\| = 0 \Rightarrow x = 0 \Rightarrow T$ is one-one. Since V is finite dimensional

$\Rightarrow$ T is non-singular. Hence, isometries are always non-singular. An isometrty on Euclidean space is called *orthogonal* while on *unitary space*, it is known as *unitary*.

Now, we show that an isometry also preserves the inner product.

Theorem 8.3.2 If $T \in L(V, V)$, then T is isometry iff

$$\langle T(x), T(y)\rangle = \langle x, y\rangle \text{ for all } x, y \in V.$$

Proof Let T be an isometry. It is trivial to see that

$$\|x + y\|^2 = \|x\|^2 + \|y\|^2 + \; x, y\rangle + \; y, x\rangle.$$

If V is Euclidean i.e. V is real inner product space that

$$\langle x, y\rangle = \overline{\langle y, x\rangle} = \langle y, x\rangle.$$

$$\Rightarrow \langle x, y\rangle. = \frac{\|x + y\|^2 - \|x\|^2 - \|y\|^2}{2} \text{ for all } x, y \in V.$$

Now, replacing x and y by $T(x)$ and $T(y)$, we get

$$\langle T(x), T(y)\rangle = \frac{\|T(x) + (y)\|^2 - \|T(x)\|^2 - \|T(y)\|^2}{2}$$

$$= \frac{\|T(x + y)\|^2 - \|x\|^2 - \|y\|^2}{2} \text{ as } T \text{ is isometry}$$

$$= \frac{\|x + y\|^2 - \|x\|^2 - \|y\|^2}{2} \text{ as } T \text{ is isometry}$$

$$= \langle x, y\rangle.$$

Now, suppose V is unitary then

$$\langle x, y\rangle = \overline{\langle y, x\rangle}$$

$$\Rightarrow \quad \langle x, y\rangle + \overline{\langle x, y\rangle} = ||x + y||^2 - ||x||^2 - ||y||^2 \ \forall\, x, y \in V \qquad (1)$$

Also

$$\begin{aligned} ||x + iy||^2 &= \langle x + iy, x + iy\rangle \\ &= x, x\rangle - \langle x, y\rangle + i\langle y, x\rangle - i^2\langle y, y\rangle \\ &= ||x||^2 - i\langle x, y\rangle + i\overline{\langle x, y\rangle} + ||y||^2 \end{aligned}$$

$$\Rightarrow \quad -i(x, y) + i\overline{\langle x, y\rangle} = ||x + iy||^2 - ||x||^2 - ||y||^2$$

$$\Rightarrow \quad \langle x, y\rangle - \overline{\langle x, y\rangle} = i\,[||x + iy||^2 - ||x||^2 - ||y||^2] \qquad (2)$$

$$\Rightarrow \quad \langle x, y\rangle = \frac{1}{2}\,[||x + y||^2 + i||x + iy||^2 - (1 + i)(||x||^2 + ||y||^2)]$$

For all $x, y \in V$. Now replace x and y by $T(x)$ and $T(y)$ and using $T(iy) = iT(y)$, we get

$$\langle T(x), T(y)\rangle = \frac{1}{2}\,[||T(x + y)||^2 + i||T(x + iy)||^2 - (1 + i)(||T(x)||^2 + ||T(y)||^2)].$$

Since T is an isometry

$$\Rightarrow \langle T(x), T(y)\rangle = \frac{1}{2}\,[||x + y||^2 + i||x + iy||^2 - (1 + i)(||x||^2 + ||y||^2)] = \langle x, y\rangle.$$

$\Rightarrow T$ preserves the inner product. Conversely, if

$$\langle T(x), T(y)\rangle = \langle x, y\rangle \text{ for all } x, y \in V$$

$\Rightarrow \langle T(x), T(x)\rangle = \langle(x, x\rangle \Rightarrow ||T(x)||^2 = ||x||^2 \Rightarrow ||T(x)|| = ||x||.$

$\Rightarrow T$ is an isometry.

Now, we characterize isometry in terms of orthonormal basis and prove the following:

Theorem 8.3.3: It $T \in L(V, V)$, then T is isometry iff T maps orthonormal basis into orthonormal basis.

Proof: Let $\{x_1, x_2, \ldots, x_n\}$

$$\begin{aligned} \langle T(x_i), T(x_j)\rangle = \langle x_i, x_j\rangle &= 1 \text{ if } i = j \\ &= 0 \text{ if } i \neq j \end{aligned}$$

$\Rightarrow \{T(x_1), T(x_2), \ldots, T(x_n)\}$ is an orthonormal set. Since an orthonormal set is linearly independent

$\Rightarrow \{T(x_1), T(x_2), \ldots, T(x_n)\}$ is linearly independent

$\Rightarrow \{T(x_1), T(x_2), \ldots, T(x_n)\}$ is an orthonormal basis of V.

Conversely, suppose T maps orthonormal basis to orthonormal basis. Let $x \in V \Rightarrow x = \sum_{i-1}^{n} \alpha_i x_i$, then

$$\langle x, x\rangle = \left\langle \sum_{i=1}^{n} \alpha_i x_i, \sum_{i=1}^{n} \alpha_i x_i \right\rangle = \sum_{i=1}^{n} |\alpha_i|^2 \text{ (using } x_i, x_j\rangle = \delta_{ij}).$$

Also

$$\langle T(x), T(x)\rangle = \left\langle \sum_{i=1}^{n} \alpha_i T(x_i), \sum_{i=1}^{n} \alpha_i T(x_i) \right\rangle = \sum_{i=1}^{n} |\alpha_i|^2$$

$$\text{(using } \langle T(x_i), T(x_j)\rangle = \delta_{ij})$$

$\Rightarrow \quad ||T(x)|| = ||x||$

$\Rightarrow$ is an isometry.

Corollary 8.3.4 If $T \in L(V, V)$ and T is isometry and W is a subspace of V, then

$$||T(x)|| = ||w|| + ||w\ ||,$$

Where $w \in W$, $w' \in W$.

Theorem 8.3.5 The set of isometries is a group.

Proof Let V be an inner product space and $S, T \in L(V, V)$

$$||(S \text{ o } T)\ (x)|| = ||S(T(x))|| = ||T(x)|| = ||x||$$

$\Rightarrow S \text{ o } T$ is an isometry. Now, for any isometry T there exists $T_1 \in L(V, V)$

$$T \text{ o } T_1 = I = T_1 \text{ o } T.$$

We show that T_1 is also an isometry. Since $||T(x)|| = ||x||$ for all $x \in V$

$\Rightarrow \quad ||T(T_1(x))|| = ||T_1(x)||$

$\Rightarrow \quad ||I(x)|| = ||T_1(x)||$

$\Rightarrow \quad ||T_1(x)|| = ||x||.$

$\Rightarrow T_1$ is isometry. Hence the set of isometries form a group.

It may be noted that the sum of two isometries need not be an isometry. For example, if $T : \mathbb{R}^2 \to \mathbb{R}^2$ such that $T(x, y) = (-y, -x)$ and $S : \mathbb{R}^2 \to \mathbb{R}^2$ such that $S(x, y) = (y, x)$, then S and T are isometries but $S + T = 0$ which is obviously not an isometry. It is trivial to see that if T is an isometry then αT is an isometry iff $|\alpha| = 1$.

Now, using Riesz Representation Theorem 8.2.18, we prove the following:

Theorem 8.3.6 If $T \in L(V, V)$ is an isometry on the inner product space V, then there exists a unique isometry $S \in L(V, V)$ such that

$$S \circ T = T \circ S = I.$$

Proof Let $y \in V$, then for every $x \in V$, $\langle T(x), y\rangle \in$ ($\mathbb{R}$ or $\mathbb{C}$). Now, define $f : V \to \mathbb{R}$ or $\mathbb{C}$ such that

$$f(x) = \langle T(x), y\rangle.$$

Then f is well defined and

$$\begin{aligned} f(\alpha x + \beta z) &= \langle(\alpha x + \beta z), y\rangle \\ &= \alpha\langle T(x), y\rangle + b\langle T(z), y\rangle \\ &= \alpha f(x) + \beta f(z) \end{aligned}$$

$\Rightarrow f \in V^*$. Hence by, Riesz Representation theorem there exists a unique element $y' \in V$, such that

$$f(x) = \langle x, y'\rangle.$$

Now, define $S : B \to V$ such that $S(y) = y'$. Then

$$\langle T(x), y\rangle = \langle x, S(y)\rangle \text{ for } x, y \in V.$$

Now,

$$\begin{aligned} \langle T(x), \alpha y_1 + \beta y_2\rangle &= \overline{\alpha}\langle T(x), y_1\rangle + \overline{\beta}\langle T(x), y_2\rangle; y_1, y_2 \in V \\ &= \overline{\alpha}\langle x, S(y_1)\rangle + \overline{\beta}\langle x, S(y_2)\rangle \\ &= \overline{\alpha}\langle x, S(y_1)\rangle + \overline{\beta}\langle x, S(y_2)\rangle \quad (1) \end{aligned}$$

and

$$\langle(x), \alpha y_1 + \beta y_2\rangle = \langle x, S(\alpha y_1 + \beta y_2)\rangle. \quad (2)$$

Since (1) and (2) are true for all $x \in V$

$$\Rightarrow \quad S(\alpha y_1 + \beta y_2) = \alpha S(y_1) + \beta S(y_2) \Rightarrow S \in L(V, V)$$

Now, suppose there exists $S' \in L(V, V)$

$$\langle(x), y\rangle = \langle x, S'(y)\rangle$$

then

$$\langle x, (S - S')(y)\rangle = 0 \ \forall\, x, y \in V \Rightarrow S - S' = 0 \Rightarrow S = S'.$$

Now, we show that S is one-one. Let $S(y) = 0$, then

$$\langle(x), y\rangle = \langle x, S(y)\rangle = 0 \text{ for all } x \in V.$$

Since T is an isometry, T^{-1} exists

$$\Rightarrow \quad \langle T(T^{-1}(y)), y\rangle = 0$$

$$\Rightarrow \quad \langle y, y\rangle = 0 \Rightarrow y = 0$$

$\Rightarrow S$ is one-one. Since V is finite dimensional $\Rightarrow S$ is non-singular. As

$$\langle T(x), y\rangle = \langle x, S(y)\rangle \text{ for all } x, y \in V$$

$$\Rightarrow \quad \langle(x), T(y)\rangle = \langle x, S(T(y))\rangle$$

$$\Rightarrow \quad \langle x, y\rangle = \langle x, (S \circ T)(y)\rangle \text{ (by Theorem 8.3.2)}$$

$\Rightarrow\quad \langle x, (S \circ T - I)y\rangle = 0$ for all $x, y \in V$

$\Rightarrow\quad \langle (S \circ T - I)(y), (y), (S \circ T - I)(y)\rangle = 0$

$\Rightarrow\quad S \circ T - I = 0 \Rightarrow S \circ T = I.$

Similarly, $T \circ S = I \Rightarrow S = T^{-1}$. By Theorem 8.3.5, we see that S is an isometry.

Hence, in case of isometry, we have

$$\langle (x), y\rangle = \langle x, T^{-1}(y)\rangle.$$

4 ADJOINT AND NORMAL TRANSFORMATIONS

Now, as seen from Theorem 8.3.6, $T(x), y\rangle = x, T^{-1}(y)\rangle$ is true for all x, $y \in V$ and isometry T. Here we shall discuss the general case for any $T \in L(V, V)$.

Definition 8.4.1 Let V be an inner product space and $T \in L(V, V)$, then *a* map $T^* : V \to V$ given as $T(x), y\rangle = x, T^*(y)\rangle$ is called the adjoint of T. The existence of such a unique map is guaranteed by Theorem 8.3.6.

As done in Theorem 8.3.6, it is trivial to check that $T^* \in L(V, V)$. Also if T is an isometry, then $T^* = T^{-1}$ and vice versa.

Before dealing with some results it will be more proper to discuss some examples.

However, for $T \in L(V, V)$, to find T^* will be more easier if we find the relation between $m(T)$ and $m(T^*)$. If V is an inner product space and $T \in L(V, V)$ choose an orthonormal basis $\{e_1, e_2, \ldots, e_n\}$ of V. Let $m(T) = (a_{ij})$ and $m(T^*) = (b_{ij})$ be the $n \times n$ matrices representing T and T^*, respectively, with respect to the above basis. Now,

$$T(e_i) = \sum_{j=1}^{n} a_{ji} e_j \text{ and } T^*(e_i) = \sum_{j=1}^{n} b_{ji} e_j .$$

Then

$$\langle (e_i), e_j\rangle = \left\langle \sum_{j=1}^{n} a_{ji} e_j , e_j \right\rangle = a_{ji} \text{ (using } e_i, e_j\rangle = \delta_{ij}).$$

Similarly

$$\langle T^*(e_i), e_j\rangle = b_{ji}$$

$$\Rightarrow\qquad b_{ji} = \overline{\langle e_j, T^*(e_i)\rangle} = \overline{\langle T(e_j), e_i\rangle} = \overline{a_{ij}}$$

$$\Rightarrow\qquad m(T^*) = \left(\overline{m(T)}\right)'.$$

Now, we give two examples, one each for Euclidean and unitary space.

1. Let $T : \mathbb{R}^3 \to \mathbb{R}^3$ such that $T(x, y, z) = (x + 2y, x - z, x + 3y - 2z)$. Then with respect to the standard basis

$$m(T) = \begin{bmatrix} 1 & 2 & 0 \\ 1 & 0 & -1 \\ 1 & 3 & -2 \end{bmatrix}$$

$$\Rightarrow \qquad m(T^*) = \begin{bmatrix} 1 & 1 & 1 \\ 2 & 0 & 3 \\ 0 & -1 & -2 \end{bmatrix}$$

$$\Rightarrow \qquad T^*(x, y, z) = (x + y + z, 2x + 3z, -y - 2z).$$

2. Let $T : \mathbb{C}^3 \to \mathbb{C}^3$ such that $T(x, y, z) = (x + iy + (i + 1)z, ix - y, 2ix + 3y + (2i - 1)z)$. Then choosing the standard basis $\{(1, 0, 0), (1, 0, 0), (0, 0, 1)\}$, we have

$$m(T) = \begin{bmatrix} 1 & i & i+1 \\ i & -1 & 0 \\ 2i & 3 & 2i-1 \end{bmatrix}$$

$$\Rightarrow \qquad m(T^*) = \begin{bmatrix} 1 & -i & -2i \\ -i & -1 & 3 \\ 1-i & 0 & -1-2i \end{bmatrix}$$

$$\Rightarrow \qquad T^*(x, y, z) = (x - iy - 2iz, -ix - y + 3z, (1 - i)x - (1 + 2i)z).$$

3. Let $T : \mathbb{C}^3 \to \mathbb{C}^3$ such that $T(x, y, z) = (5x + (3 + 3i)y + (3 - 2i)z, (2 - 3i)x + 2y + (1 + i)z, (3 + 2i)x + (1 + i)y + 7z)$. Then $T \in L(\mathbb{C}^3, \mathbb{C}^3)$ and with respect to the standard basis, we have

$$m(T) = \begin{bmatrix} 5 & 2+3i & 3-2i \\ 3-3i & 2 & 1+i \\ 3+2i & 1-i & 7 \end{bmatrix}$$

$$\Rightarrow \qquad m(T^*) = \begin{bmatrix} 5 & 2+3i & 3-2i \\ 2-3i & 2 & 1+i \\ 3+2i & 1-i & 7 \end{bmatrix}$$

$$\Rightarrow \qquad T^*(x, y, z) = T(x, y, z) \text{ i.e } T^* = T.$$

Now, we prove the following elementary result.

Theorem 8.4.2 Let $S, T \in L(V, V)$, then the following hold:

(*i*) $O^* = O$, where O is zero transformation

(*ii*) $I^* = I$

(*iii*) $(S + T)^* = S^* + T^*$

(*iv*) $(\alpha T)^* = \bar{\alpha} T^*$, α is *a* scalar in R or C

(*v*) $(ST)^* = T^*S^*$

(*vi*) $(T^*)^* = T$

(*vii*) If T is non-singular, then T^* is also non-singular

Proof (*i*) and (*ii*) are trivial

$$\begin{aligned}\langle (S + T)(x), y\rangle &= \langle S(x), y\rangle + \ T(x), y\rangle \\ &= \langle x, S^* (y)\rangle + \langle x, T^*(y)\rangle \\ &= \langle x, S^* (y) + T^* (y)\rangle \\ &= \langle x, (S^* + T^*)(y)\rangle \qquad (1)\end{aligned}$$

Also

$$\langle (S + T)(x), y\rangle = \langle x, (S^* + T^*)(y)\rangle \qquad (2)$$

Since (1) and (2) are ture for all $x, y \in V \Rightarrow (S + T)^* = S^* + T^*$,

$$\begin{aligned}\langle (\alpha T)(x), y\rangle &= \langle \alpha T(x), y\rangle = \alpha \langle T(x), y\rangle = \alpha\langle x, T^* (y)\rangle \\ &= x, \langle \bar{\alpha} T^* (y)\rangle\end{aligned}$$

and

$$\langle (\alpha T)(x), y\rangle = \langle x, (\alpha T)^* y\rangle \Rightarrow (\alpha T)^* = \bar{\alpha} T^*.$$

$\Rightarrow$ (*iii*) and (*iv*) are satisfied.

(*v*) $\quad \langle (ST)(x), y\rangle = S(T(x)), y\rangle = \ T(x), S^*(y)\rangle = \ x, T^*S^* (y)\rangle$

and

$$\langle (ST)(x), y\rangle = \langle x, (ST)^* (y)\rangle \Rightarrow (ST)^* = T^*S^*.$$

(*vi*) $\quad \langle T^*(x), y\rangle = \langle x, (T^*) (y)\rangle$. Also

$$\langle T^*(x), y\rangle = \overline{\langle y, T^*(x)\rangle} = \overline{\langle T(y), x\rangle} = \langle x, T(y)\rangle.$$

Hence $(T^*)^* = T$.

(*vii*) If T is non-singular, then $TT^{-1} = I$

$\Rightarrow (T^{-1})^* T^* = I^* = I$ [from (vi) and (ii)].

$\Rightarrow (T^*)^{-1} = (T^{-1})^*$

$\Rightarrow T^*$ is non-singular.

Now, using the results proved in Section 3 and the remark of this section we can state the following:

Theorem 8.4.3 If $T \in L(V, V)$ then the following are equivalent:

(*i*) $T^*T = I$, (*ii*) $\langle T(x), T(y)\rangle = \langle x, y\rangle$ and (*iii*) $\|T(x)\| = \|x\|$ for all $x, y \in V$.

Definition 8.4.4 Let $T \in L(V, V)$ then T is called self-adjoint if $T^* = T$ and if $T^* = -T$, then T is called skew-adjoint.

If we see Example 3, we find T self-adjoint. Consider $T : \mathbb{C}^2 \to \mathbb{C}^2$ such that $T(x, y) = (2ix + (4 + i)y, (-4 + i)x + 3iy)$, then

$$m(T) = \begin{bmatrix} 2i & 4+i \\ -4+i & 3i \end{bmatrix}$$

$$\Rightarrow \quad m(T^*) = \begin{bmatrix} -2i & -4-i \\ 4-i & -3i \end{bmatrix} = -\begin{bmatrix} 2i & 4+i \\ -4+i & 3i \end{bmatrix}$$

$$\Rightarrow \quad T^* = -T$$

$\Rightarrow$ T is skew adjoint.

It may be noted that Hermitian and skew-Hermitian matrices give rise to self-adjoint and skew-adjoint transformation and vice versa.

Theorem 8.4.5 If $T \in L(V, V)$ is self-adjoint, then the following hold:

(*i*) $\langle (x), x\rangle$ is a rea for all $x \in V$.

(*ii*) All characteristic roots of T are real.

(*iii*) The characteristic subspaces corresponding to distinct characteristic roots are orthogonal.

Proof:

(*i*) $\langle T(x), x\rangle = \langle x, T^*(x)\rangle = \langle x, T(x)\rangle = \overline{\langle T(x), x\rangle} \Rightarrow \langle (x), x\rangle$ is real.

(*ii*) Let α be *a* characteristic root of $T \Rightarrow T(x) = \alpha x$ for some $x \neq 0$

$$\langle T(x), x\rangle = \langle \alpha x, x\rangle = \alpha\, x, x\rangle,$$

since $\langle T(x), x\rangle$ and $\langle x, x\rangle$ is α is real.

(*ii*) Let U_1 and U_2 be the characteristic subspaces corresponding to characteristic roots $\alpha \neq \beta$, then for any nonzero $x \in U_1$ and $y \in U_2$

$$T(x) = \alpha x,\ T(y) = \beta y$$

$$\langle T(x), y\rangle = \langle x, T(y)\rangle \text{ as } T \text{ is self-adjoint}$$

$$\Rightarrow \quad \langle \alpha x, y\rangle = \langle x, \beta y\rangle$$

$$\Rightarrow \quad \langle \alpha\, x, y\rangle = \overline{\beta}\,\langle x, y\rangle \text{ but by (ii) } \overline{\beta} = \beta$$

$$\Rightarrow \quad (\alpha - \beta)\langle x, y\rangle = 0 \text{ but } \sigma \neq \beta$$

$$\Rightarrow \quad \langle x, y\rangle = 0, \Rightarrow \neq U_1 \perp U_2.$$

Theorem 8.4.6 If $T \in L(V, V)$ is skew-adjoint, then the following hold:

(*i*) $\langle T(x), x\rangle$ is purely imaginary for all $x \in V$.

(*ii*) All characteristic roots of T are purely imaginary.

(*iii*) The characteristic subspaces corresponding to distinct characteristic roots are orthogonal.

Since the proof is on similar lines to that of Theorem 8.4.5, only a sketch will be given.

Proof:

(*i*) $\langle T(x), x\rangle = \overline{\langle T(x), x\rangle}$

$\Rightarrow \langle T(x), x\rangle + \overline{\langle T(x), x\rangle} = 0 \Rightarrow \langle T(x), x\rangle$ is purely imaginary.

(*ii*) If $T(x) = \alpha x$, then $\langle T(x), x\rangle = \alpha\langle x, x\rangle$, since $\langle x, x\rangle$ is real and $\langle T(x), x\rangle$ is purely imaginary $\Rightarrow \alpha$ is purely imaginary.

(*iii*) $T(x) = \alpha x$, $T(y) = \beta y$, then

$$\langle \alpha x, y\rangle = \langle x, \beta y\rangle$$

$\Rightarrow (\alpha - \overline{\beta})\langle x, y\rangle = 0$, since α and β are purely imaginary and distinct

$\Rightarrow \langle x, y\rangle = 0, \Rightarrow U_1 \perp U_2$.

Now, we shall show that if $T^* = T$ and $T \neq 0$, then $T^n \neq 0$ for any n i.e. T cannot be nilpotent.

Theorem 8.4.7: If $T \in L(V, V)$ such that $T \neq 0$ and $T^* = T$, then T can never be nilpotent.

Proof: Let T be nilpotent $\Rightarrow \exists\, k$ such that $T^k = 0$ but $T^{k-1} \neq 0$, then $k > 1$. As $T^{k-1} \neq 0$, we can find $x \in V$, such that

$$T^{k-1}(x) \neq 0$$

$$\Rightarrow \quad \langle T^{k-1}(x), T^{k-1}(x)\rangle = \langle x, (T^{k-1})^* T^{k-1}(x)\rangle$$
$$= \langle x, T^{2k-2}(x)\rangle = 0 \text{ as } T^{2k-2} = 0$$

$\Rightarrow T^{k-1}(x) = 0$ which is *a* contradiction $\Rightarrow T$ cannot be nilpotent.

Theorem 8.4.8: If $T \in L(V, V)$, then $T = P + Q$, where $P, Q \in L(V, V)$ and $P^* = P$, $Q^* = -Q$ and decomposition is unique.

Proof: Let

$$P = \frac{T + T^*}{2} \text{ and } Q = \frac{T - T^*}{2},$$

Then P and Q satisfy the given assertion.

Definition 8.4.9 Let $T \in L(V, V)$ then T is called normal if $TT^* = T^*T$.

Let $T : \mathbb{R}^2 \to \mathbb{R}^2$ such that $T(x, y) = (ix, (1 + i)y)$, then

$$m(T) = \begin{bmatrix} i & 0 \\ 0 & 1+i \end{bmatrix}, \; m(T^*) = \begin{bmatrix} -i & 0 \\ 0 & 1-i \end{bmatrix},$$

then $m(T)m(T^*) = m(T^*)m(T) \Rightarrow T$ is normal.

Now, let us consider $T : R^2 \to R^2$ such that $T(x, y) = (ix + 2y, (1 + i)y)$, then

$$m(T) = \begin{bmatrix} i & 2 \\ 0 & 1+i \end{bmatrix} \text{ and } m(T^*) = \begin{bmatrix} -i & 2 \\ 0 & 1-i \end{bmatrix},$$

then $m(T)m\,(T^*) \neq m(T^*)\,m(T) \Rightarrow T$ is not a normal transformation.

Theorem 8.4.10 If the $\in L(V, V)$ is normal then the following hold:

(*i*) $||T(x)|| = ||T^*(x)||$ for all $x \in V$.

(*ii*) ker T = ker T^*.

(*iii*) If α is a characteristic root of T, then $\overline{\alpha}$ is a characteristic root of T^*.

(*iv*) If $T(x) = \alpha x$ and $T(y) = $ by and $a \neq b$, then $\langle x, y\rangle = 0$.

Proof:

(*i*) $||T(x)||^2 = \langle T(x), T(x)\rangle = \langle x, T^*T(x)\rangle$

$= \langle T^*(x), T^*(x)\rangle$ as $TT^* = T^*T$ and $T^{**} = T$

$= ||T^*(x)||^2$

$\Rightarrow \qquad ||T(x)|| = ||T^*(x)||$

(*ii*) Let $x \in \ker T \Rightarrow T(x) = 0$. Now,

$$\begin{aligned} \langle T^*(x), T^*(x)\rangle &= \langle x, TT^*(x)\rangle \\ &= \langle x, T^*T(x)\rangle = x, 0\rangle = 0 \end{aligned}$$

$\Rightarrow \qquad T^*(x) = 0$

$\Rightarrow \qquad x \in \ker T^* \Rightarrow \ker T \subseteq \ker T^*.$

Let $y \in \ker T^* \Rightarrow T^*(y) = 0$. As,

$$\begin{aligned} \langle T(y), T(y)\rangle &= \langle y, T^*T(y)\rangle \\ &= \langle y, TT^*(y)\rangle = \langle y, 0\rangle = 0 \end{aligned}$$

$\Rightarrow \qquad T(y) = 0$

$\Rightarrow \qquad y \in \ker T \Rightarrow \ker T = \ker T^*.$

(*iii*) Let $T(x) = \alpha x$, $x \in V$, $x \neq 0$. Now,

$$\begin{aligned} (T - \alpha I)(T - \alpha I)^* &= (T - \alpha I)(T^* - \overline{\alpha} I) \\ &= TT^* - \overline{\alpha} T - \alpha T^* + a\overline{\alpha} \\ &= T^*T - \overline{\alpha} T^* - \overline{\alpha} T + \overline{\alpha}\alpha \\ &= T^*(T - \alpha I) - \overline{\alpha}(T + \alpha) \\ &= (T^* - \overline{\alpha} I)(T - \alpha I) \end{aligned}$$

$$= (T - \alpha I)^*(T - \alpha I)$$

$\Rightarrow (T - \alpha I)$ is a normal transformation.

Now, $T(x) = \alpha x \Rightarrow (T - \alpha I)(x) = 0 \Rightarrow x \in \ker (T - \alpha I)$. Hence, by (ii)

$x \in \ker (T - \alpha I)^* \Rightarrow (T - \alpha I)^*(x) = 0 \Rightarrow T^*(x)\ \bar{\alpha} x$

$\Rightarrow \bar{\alpha}$ is *a* characteristic root of T.

(iv) $\langle T(x), y\rangle = \langle \alpha x, y\rangle = \langle x, \bar{\beta} y\rangle$ by (iii)

$$= \beta\,(x, y)$$

$$\Rightarrow \quad (a - b)\langle x, y\rangle = 0 \text{ but } \alpha \neq \beta$$

$$\Rightarrow \quad \langle x, y\rangle = 0 \quad \Rightarrow \quad x \perp y.$$

Now, we end this section with solutions of some problems and examples.

1. If $T \in L(V, V)$ and W is a subspace of V invariant under T such that $T = T^*$, then $W^\perp$ is also invariant under T.

   Let $x \in W^\perp \Rightarrow \langle x, w\rangle = 0$ for all $w \in W$.

   As $\langle T(x), w\rangle = \langle x, T^{-1}(w)\rangle$ and $T^{-1}(W) = W$

   $\Rightarrow \langle T(x), w\rangle = 0$ as $T(w) \in W$

   $\Rightarrow \langle T(x), w\rangle = 0$ for all $w \in W$

   $\Rightarrow T(x) \in W^\perp$.

2. If W is invariant under $T \in L(V, V)$ such that T is an isometry then $W^\perp$ is also T-invariant.

   Let $x \in W^\perp \Rightarrow\ x, w\rangle = 0$ for all $w \in W$.

   As $\langle T(x), w\rangle = \langle x, T^{-1}(w)\rangle$ and $T^{-1}(W) = W$

   $\Rightarrow \langle T(x), w\rangle = 0$ for all $w \in W$.

   $\Rightarrow T(x) \in W^\perp \Rightarrow W^\perp$ is T-invariant.

3. If $T \in L(V, V)$ is self-adjoint and $T^k(x) = 0\ k \geq 1$, then $T(x) = 0$, $x \in V$.

   For $k = 1$, then result is true. Let it be true for $k = t$.

   Let $T^{t+1}(x) = 0$, then

$$\langle T^{t+1}(x), T^{t-1}(x)\rangle = \langle T^t(x), T^*(T^{t-1})(x))\rangle$$

$$\Rightarrow \quad 0 = \langle T^t(x), T^t(x)\rangle$$

$$\Rightarrow \quad T^t(x) = 0$$

$$\Rightarrow \quad T(x) = 0.$$

   In general, if $T \in L(V, V)$ such that T is normal, then $T^k(x) = 0 \Rightarrow T(x) = 0$. Let $S = T^*\,T$, then S is self-adjoint and

$$S^k = (T^*)^k T^k \text{ as } TT^* = T^*T.$$
$$\Rightarrow \quad S^k(x) = (T^*)^k T^k(x) = 0$$
$$\Rightarrow \quad S(x) = 0$$
$$\Rightarrow \quad \langle T^*T(x), x\rangle = 0$$
$$\Rightarrow \quad \langle T(x), T(x)\rangle = 0$$
$$\Rightarrow \quad T(x) = 0.$$

4. If V is Euclidean inner product space, then $L(V, V) = A \oplus B$, where A and B are sets of self-adjoint and skew-adjoint transformations: $V \to V$.

   Let $S, T \in A$

$$\Rightarrow \quad S^* = S, T^* = T$$
$$\Rightarrow \quad (S-T)^* = S^* - T^* = S - T$$

   and

$$(\lambda T)^* = \bar{\lambda} T^* = \lambda T \Rightarrow S - R, \lambda T \in A$$

   $\Rightarrow A$ is subspace of $L(V, V)$. Similarly, it is easy to see that B is a subspace of $L(V, V)$. Let $T \in L(V, V)$

$$\Rightarrow \quad T = \frac{T+T^*}{2} + \frac{T-T^*}{2} \in A + B$$
$$\Rightarrow \quad L(V, V) = A + B.$$

   Now, if $S \in A \cap V \Rightarrow S^* = S$ and $S^* = -S \Rightarrow S = 0 \Rightarrow L(V, V) = A \oplus B$.

5. Let $T : \mathbb{R}^3 \to \mathbb{R}^3$ such that $T(x, y, z) = (-y, x, 0)$, then for any $u \in \mathbb{R}^3$ $\langle T(u), u\rangle = 0$ under the standard inner product in $\mathbb{R}^3$, but $T \neq 0$. Also $T(1, 2, 3), (4, 1, 0)\rangle = -8 + 1 = -7 \neq 0 \Rightarrow \exists\, u, v \in \mathbb{R}^3$ such that $\langle T(u), v\rangle \neq 0$.

   However, if for any inner product space V, if $T(u), v\rangle = 0$ for all $u, v \in V$, then $T = 0$ as $T(u), T(u)\rangle = 0$ Þ $T(u) = 0$ for all $u \in V \Rightarrow T = 0$.

6. If $T \in L(V, V)$ and V is *a* unitary space, then T is self-adjoint iff $\langle T(x), x\rangle$ is real for all $x \in V$.

   If T is self-adjoint, then Theorem 8.4.5, $\langle T(x), x\rangle$ is real.

   Now, let $\langle T(x), x\rangle$ be real

$$\Rightarrow \quad \langle T(x), x\rangle = \langle x, T^*(x)\rangle = \langle T^*(x), x\rangle$$
$$\Rightarrow \quad \langle T - T^*(x), x\rangle = 0 \text{ for all } x \in V. \qquad (1)$$

   Let

$$S = T - T^*$$
$$\Rightarrow \quad \langle S(x+y), x+y\rangle = 0 \text{ by } (1)$$

$\Rightarrow \quad \langle S(y), x\rangle + S(x), y\rangle = 0 \quad (2)$

Using $S(x), x\rangle = 0$ for all $x \in V$. Since (2) is true for all $x, y \in V$ and iy is also an element of V

$$\Rightarrow \quad \langle S(iy), x\rangle + \langle S(x), iy\rangle = 0$$
$$\Rightarrow \quad i\langle S(y), x\rangle - i\langle S(x), y\rangle = 0$$
$$\Rightarrow \quad \langle S(y), x\rangle - \langle S(x), y\rangle = 0$$

Adding (2) and (3), we get $\langle S(y), x\rangle = 0$ for all $x, y \in V \Rightarrow S = 0$ as done in Example 5.

$$\Rightarrow \quad T - T^* = 0$$
$$\Rightarrow \quad T = T^* \Rightarrow T \text{ is self-adjoint.}$$

7. Let $T : \mathbb{C}_2 \to \mathbb{C}^2$ given as $T(x, y) = (ix + (i - 1)y, (i + 1)x)$, then

$$m(T) = \begin{bmatrix} i & i-1 \\ i+1 & 0 \end{bmatrix}, \; m(T^*) = \begin{bmatrix} -i & 1-i \\ -i-1 & 0 \end{bmatrix}$$

$$m(T)m(T^*) = \begin{bmatrix} -i^2 - (i^2 - 1) & i - i^2 \\ -i^2 - i & 1 - i^2 \end{bmatrix} = \begin{bmatrix} 3 & i-1 \\ 1-i & 2 \end{bmatrix}$$

$$m(T^*)m(T) = \begin{bmatrix} -i^2 + 1 - i^2 & -i^2 + i \\ -i^2 - i & 1 - i^2 \end{bmatrix} = \begin{bmatrix} 3 & i+1 \\ 1-i & 2 \end{bmatrix}$$

Hence, $TT^* = T^*T \Rightarrow T$ is normal.

EXERCISE

1. Prove or disprove that the following defines an inner product:
   (*a*) $\langle x, y\rangle = x_1y_1 + x_1y_2 + x_2y_1 + 2x_2y_2$ on $\mathbb{R}^2$, where $x = (x_1, x_2)$, $y = (y_1, y_2)$.
   (*b*) $\langle x, y\rangle = x_1y_2 + x_2y_1$ on $\mathbb{R}^2$.
   (*c*) $\langle x, y\rangle = x_1y_1 + x_3y_3$ on $\mathbb{R}^3$.
   (*d*) $\langle x, y\rangle = x_1^2y_1^2 + x_2^2y_2^2 + x_3^2y_3^2$ on $\mathbb{R}^3$.
   (*e*) $\langle x, y\rangle = x_1y_1 - x_1y_2 + 3x_2y_2 - x_2y_1$ on $\mathbb{R}^2$.
2. If $\{x_1, x_2, \ldots x_r\}$ is an orthonormal set in an inner product space V, then show that

$$\sum_{i=1}^{r} |\langle x, x_i\rangle|^2 = \langle x, x\rangle \text{ iff } x \in [x_1, x_2, \ldots, x_r].$$

3. If $\{x_1, x_2, \ldots, x_n\}$ is an orthonormal basis of V, then for any x, $y \in V$

$$\langle x, y\rangle = \sum_{i=1}^{r}\langle x, x_i\rangle\langle x_i, y\rangle .$$

4. If V is an inner product space and U is a subspace of V, then show that U is also an inner product space.

5. If $T \in L(V, V)$ is a projection on U along W, then show that T^* is a projection on $W^{\perp}$ along $U^{\perp}$.

6. If $T \in L(V, V)$ is self-adjoint, then show that $T(x) = 0$ provided $T^2(x) = 0, x \in V$.

7. If $S, T \in L(V, V)$ are self-adjoint, then show that ST is self-adjoint iff $ST = TS$ and αT is self-adjoint iff α is real.

8. If any $T \in L(V, V)$, show that $T + T^*$ and $T - T^*$ are self-adjoint and skew-adjoint, respectively.

9. For $T \in L(\mathbb{C}^2, \mathbb{C}^2)$ Find T^* in the following:

(*a*) $T(x, y) = (ix + (1 - i)y, (1 + i)x - iy)$.

(*b*) $T(x, y) = (2x + iy, (i - 1)x + y)$.

(*c*) $T(x, y) = ((i + 1)x + y, x - y)$.

10. If $W = [(i, 1, i), (1, i, 1)]$, then find the orthonormal basis of W.

11. If $T \in L(V, V)$, then show that $T = S + iP$, where $S, P \in L(V, V)$ and are self-adjoint and V is unitary space.

12. Show that $T \in L(V, V)$ is an isometry iff T maps a unit vector into a unit vector.

13. If $W = [(1, 2, -1), (1, 3, 1)]$, $U = [(1, 2, 3)]$, then find an orthonormal basis of $V^{\perp}$ and $U^{\perp}$.

14. Using Gram-Schmidt process find an orthonormal basis of the following:

(a) $\{(1, 2, 1), (1, 1, 0), (0, -1, -4)\}$ in $\mathbb{R}^3$.

(b) $\{(1, 1, 1), (1, -1, 1), (0, 1, -1)\}$ in $\mathbb{R}^3$.

15. Show that

$$A = \frac{1}{2}\begin{bmatrix} 1 & i & 1+i \\ -i & 1 & -1+i \\ -1+i & 1+i & 0 \end{bmatrix}$$

Is *a* normal matrix. Find the corresponding linear transformation $T : \mathbb{C}^3 \to \mathbb{C}^3$ and T^*.

9

Bilinear and Quadratic Forms

In previous chapter we have studied linear transformations between two vector spaces. In particular we also studied linear functional where we restricted ourselves for one vector space and the other to be thc field over which the vector space has been defined. If V is a vector space over F then a linear map f ; $V \to F$ is defined to be the linear functional. Also we did inner product as a function: $V \times V \to R$ (or C) which is linear in first variable. Let us consider another vector space U_F and the Cartesian product $U \times V$ which is again a vector space over F. In this chapter we discuss the function $f : U \times V \to F$ which is linear in both the variables and their different consequences are also dealt.

1 BILINEAR FUNCTIONAL

Definition 9.1.1 Let U and V be vector spaces over F then a function $f: U \times V \to F$ such that

$$f(a_1x_1 + a_2x_2, y) = a_1 f(x_1, y_0) + a_2 f(x_2, y)$$

and

$$f(x + a_1y_1, + a_2y_2) = a_1 f(x, y_1) + a_2 f(x, y_2)$$

for all $x, x_1, x_2 \in U$ and $y_1, y_1, y_2 \in V$ and $a_1, a_2 \in F$, is called *bilinear functional*.

In general, if $x_1, x_2, \ldots, x_m \in$ U and $y_1, y_2, \ldots, y_n \in V$, then

$$f\left(\sum_{i-1}^{m} \alpha_1 x_i, \sum_{j=1}^{n} \beta_j y_j\right) = \sum_{i=1}^{m}\sum_{j=1}^{n} \alpha_i \beta_j f\,(x_i, y_j) \text{ for all } \alpha_i, \beta_j, \in F.$$

This is very easy to obtain from the definition.

If $U = V$ then the inner product $\langle\, , \rangle$ is bilinear functional iff F is the field of real numbers.

If $f : \mathbb{R}^2 \times \mathbb{R}^2 \to \mathbb{R}$ such that $F(x, y) = x_1 y_1 + x_2 y_2$ where $x = (x_1, x_2)$ and $y = (y_1, y_2)$, then it is easy to see that f is a bilinear functional on $\mathbb{R}^2$. However, if $f : U \times V \to F$ is a bilinear functional, then f is also called bilinear form on U and V.

From the definition it is easy to see that

$$f(x, 0) = f(0, y) = 0 \text{ for all } x \in U, y \in V.$$

Example 9.1.2 If $\sigma_1, \sigma_2, \ldots, \sigma_r \in U^*$ and $\phi_1, \phi_2 \ldots, \phi_r \in V^*$, then for any $x \in U$ and $y \in V, f: U \times V \to$ F defined as

$$f(x, y) = \sigma_1 (s) \phi_1 (y) + \sigma_2 (x) \phi_2 (y) + \ldots + \sigma_r (x)\phi_r (y) + \ldots + \sigma_r (x) \phi_r (y)$$

is a bilinear functional on U and V.

Example 9.1.3 Let $f: \mathbb{R}^2 \times \mathbb{R}^3 \to \mathbb{R}$ such that $f((x_1, x_2), (y_1, y_2, y_3)) = x_1 y_2$.

$$f(a(x_1, x_2) + b(x_1', x_2'), (y_1, y_2, y_3,))$$
$$= (ax_1 + bx_1') y_2 = ax1\ y_2 + bx_1' y_2$$
$$= af((x_1, x_2), (y_1, y_2, y_3)) + bf((x_1', x_2'), (y_1, y_2, y_3)).$$

Similarly, for the other variable we can verify.

$\Rightarrow f$ is *a* bilinear form on $\mathbb{R}^2$ and $\mathbb{R}^3$.

Throughout we shall take U and V to be finite dimensional vector spaces of dimensions m and n, respectively.

Definition 9.1.4 Let $\{x_1, x_2, \ldots, x_n\}$ and $\{y_1, y_2, \ldots, y_n\}$ be bases of U and V, respectively and f be a bilinear functional on U and V, then the matrix of f with respect to the above bases is $A = [a_{ij}]$ of order $m \times n$, where

$$f(x_i, y_i) = a_{ij} \text{ for } i = 1, 2, \ldots, m \text{ and } j = 1, 2, \ldots, n.$$

Let us recall that, if $\{x_1, x_2 \ldots, x_m\}$ I a basis of U and $x \in U$, then there exists unique $\alpha_1, \alpha_2, \ldots, \alpha_m \in F$ such that $x = \sum_{i=1}^{m} a_i x_i$. Further

$$X = \begin{bmatrix} \alpha_1 \\ \alpha_2 \\ \vdots \\ \alpha_m \end{bmatrix}$$

is known as coordinate matrix of x.

Theorem 9.1.5 If $A = [a_{ij}]$, then A is matrix of a bilinear form f on U and V iff

$$f(x, y) = \begin{bmatrix} \alpha_1 \\ \alpha_2 \\ \vdots \\ \alpha_m \end{bmatrix}^t A \begin{bmatrix} \\ \\ \\ \end{bmatrix} = X^t AY$$

where X and Y are coordinate matrices of x and y, respectively, with respect to bases $\{x_1, x_2, \ldots, x_m\}$ of U and $\{y_1, y_2, \ldots, y_n\}$ of $V, x \in U$, $y \in V$.

Proof Let $A = [a_{ij}]$ be the matrix of f relative to bases $\{x_1, x_2, \dots, x_m\}$ of U and $\{y_1, y_2, \dots, y_n\}$ of V. Trivially

$$f(x, y) = f\left(\sum_{i=1}^{m} \alpha_i x_i, \sum_{j=1}^{n} \beta_j y_j\right)$$

$$= \sum_{i=1}^{m} \alpha_i \left(\sum_{j=1}^{n} \beta_j\, f(x_i, y_j)\right)$$

$$= \sum_{i=1}^{m}\sum_{j=1}^{n} \alpha_i \beta_j f\,(x_i, y_j) = \sum_{i=1}^{m}\sum_{j=1}^{n} \alpha_i \beta_j \alpha_{lf}\,.$$

Now,

$$X^t AY = [a_1, a_2, \dots, \alpha_m] \begin{bmatrix} a_{11} & a_{12} & \cdots & a_{1n} \\ a_{21} & a_{22} & \cdots & a_{2n} \\ \vdots & \vdots & \vdots & \vdots \\ a_{m1} & a_{m2} & \cdots & a_{mn} \end{bmatrix} \begin{bmatrix} \beta_1 \\ \beta_2 \\ \vdots \\ \beta_n \end{bmatrix}$$

$$=[\alpha_1, \alpha_2, \dots, \alpha_m] \begin{bmatrix} \sum_{j=1}^{n} a_{1j}\beta_j \\ \sum_{j=1}^{n} a_{2j}\beta_j \\ \vdots \quad \vdots \\ \sum_{j=1}^{n} a_{mj}\beta_j \end{bmatrix}$$

$$= \sum_{i=1}^{m} \alpha_i \quad \sum_{j=1}^{n} \alpha_i \beta_j \;=\; \sum_{j=1}^{n}\sum_{j=1}^{n} \alpha_i \beta_j a_{ij}$$

$\Rightarrow f\,(x, y) = X^t\, AY$ for all $x \in U$, $y \in V$. Choose $x = x_i$, $y = y_i$, then the coordinate matrices of x and y are

$$\underset{\longrightarrow}{i^{th}} \begin{bmatrix} 0 \\ 0 \\ \vdots \\ 1 \\ 0 \\ \vdots \\ 0 \end{bmatrix}, \begin{bmatrix} 0 \\ 0 \\ \vdots \\ 1 \\ 0 \\ \vdots \\ 0 \end{bmatrix} \underset{\longleftarrow}{j^{th}}$$

Now, we give few examples.

Examples 9.1.6

1. Define $f: \mathbb{R}^2 \times \mathbb{R}^2 \to \mathbb{R}$ such that
$$f((x_1, x_2), (y_1, y_2)) = x_1 e^{y_1} + x_2 e^{y_2}$$
then it is easy to see that f is linear in the first vector variable but f is not linear in the second, so f is not a bilinear form on $\mathbb{R}^2$.
2. Define $f: \mathbb{R}^2 \times \mathbb{R}^2 \to \mathbb{R}$ such that
$$f((y_1, x_2), (y_1, y_2)) = |x_1|y_1 + |x_2|y_2$$
then it is easy to see that f is linear in the second vector variable but f is not linear in first vector variable. Therefore, f is not a bilinear form on $\mathbb{R}^2$.
3. Let $f: \mathbb{R}^2 \times \mathbb{R}^2 \to \mathbb{R}$ such that
$$f((x_1\ x_2), (y_1, y_2)) = 2x_1y_2 - 3x_2y_1 + 4x_1y_1.$$
Let $a\{(1, 2), (2, 1)\}$ be a basis of $\mathbb{R}^2$, then we calculate the matrix of f with respect to this basis. Let $e_1 = (1,2)$ and $e_2 = (2, 1)$, then if
$$A = \begin{bmatrix} a_{11} & a_{12} \\ a_{21} & a_{22} \end{bmatrix}$$
is the matrix of f, we have
$$a_{11} = f(e_1, e_1) = 2 \cdot 1 \cdot 2 - 3 \cdot 1 \cdot 2 + 4 = 2$$
$$a_{12} = f(e_1, e_2) = 2 - 12 + 8 = -2$$
$$a_{21} = f(e_2, e_1) = 2 \cdot 1 \cdot 2 - 3 \cdot 1 \cdot 1 + 4 \cdot 2 = 13$$
$$a_{22} = f(e_2, e_2) = 2 \cdot 2 \cdot 1 - 3 \cdot 1 \cdot 2 + 4 \cdot 2 \cdot 2 = 14$$
$$\Rightarrow \qquad A = \begin{bmatrix} 2 & -2 \\ 13 & 14 \end{bmatrix}$$
4. Let $f: \mathbb{R}^2 \times \mathbb{R}^3 \to \mathbb{R}$ such that
$$f((x_1, x_2), (x_1, x_2, x_3)) = 2x_1y_3 - x_2y_2 + 7x_2\,y_2.$$
Let $\{(1, 0), (0, 1)\}$ and $\{(1, 0, 1), (0, 1, -1), (1, 1, 1)\}$ be bases of $\mathbb{R}^2$ and $\mathbb{R}^3$, respectively, and
$$A = \begin{bmatrix} a_{11} & a_{12} & a_{13} \\ a_{21} & a_{22} & a_{23} \end{bmatrix}.$$
$$a_{11} = f((1, 0), (1, 0, 1)) = 2 \cdot 1 \cdot 1 - 0 + 0 + 0 = 2$$
$$a_{12} = f((1, 0), (1, 0, -1)) = 2 \cdot 1 \cdot (-1) - 0 + 1 = -1$$
$$a_{13} = f((1, 0), (1, 1, 1)) = 2 + 1 = 3$$
$$a_{21} = f((0, 1), (1, 0, 1)) = 0$$

$$a_{22} = f((1, 0), (0, 1, -1)) = -1 + 7 = 6$$
$$a_{213} = f((0, 1), (1, 1, 1)) = -1 + 7 = 6$$
$$\Rightarrow \qquad A = \begin{bmatrix} 2 & -1 & 3 \\ 6 & 6 & 6 \end{bmatrix}.$$

5. If $A = \begin{bmatrix} 2 & 5 & -3 \\ 1 & -2 & 4 \end{bmatrix}$ is the matrix of a bilinear form f on $\mathbb{R}^2$ and $\mathbb{R}^3$ with respect to the bases given in Example 4, then we find $f((2, 3), (1, -2, 4))$

The coordinate matrix of (2, 3) is trivially $\begin{bmatrix} 2 \\ 3 \end{bmatrix}$. Now (1, –2, 4)

$= 0(1, 0, 1) + (-3)(0, 1, -1) + 1(1, 1, 1)$.

$\Rightarrow$ Coordinate matrix of (1, –2, 4) is $\begin{bmatrix} 0 \\ -3 \\ 1 \end{bmatrix}$.

Hence by Theorem 9.1.3,

$$f((2, 3), (1, -2, 4)) = [2 \;\; 3] \begin{bmatrix} 2 & 5 & -3 \\ 1 & -2 & 4 \end{bmatrix} \begin{bmatrix} 0 \\ -3 \\ 1 \end{bmatrix}$$
$$= [2 \;\; 3] \begin{bmatrix} -18 \\ 10 \end{bmatrix} = 6.$$

The set of all bilinear functional on U and V will be denoted by $L(U, V; F)$ i.e.

$$L(U, V; F) = \{f : U \times V \to F \mid f \text{ is bilinear}\}$$

For any $f, g \in L(U, V; F)$, define $(f + g)(x, y) = f(x, y) + g(x, y)$ and $(\alpha f)(x, y) = \alpha f(x, y)$ for all $a \in F, x \in U, y \in V$. It is trivial to see that $L(U, V; F)$ is a vector space over F.

As done in Chapter 6, we prove the following which is very similar to that of Theorem 6.1.3. Hence, we shall give a sketchy proof.

Theorem 9.1.7: If dim $U = m$ and dim $V = n$, then dim $(U, V; F) = mn$.

Proof: Define $f_{ij} : U \times V \to F$ such that

$$f_{ij}(x_k, y_t) = \begin{cases} 1 & \text{if } i - k \text{ and } j = t \\ 0 & \text{otherwise} \end{cases}$$

trivially $f_{ij} \in L(U, V; F)$. Let

$$\sum_{\substack{i=1\\j=1}}^{m,n} \alpha_{ij} f_{ij} = 0 \text{ i.e. } \sum_{i=1}^{m}\sum_{j=1}^{n} \alpha_{ij} f_{ij} = 0$$

$$\Rightarrow \quad \sum_{i=1}^{m}\sum_{j=1}^{n} \alpha_{ij} f_{ij}(x_k, y_t) = 0$$

$\Rightarrow a_{kt} = 0$ for all $k = 1, 2, \ldots, m$ and $t = 1, 2 \ldots, n$.

$\Rightarrow \{f_{ij}\}$ is linearly independent.

Let $f \in L(U, V; F)$ and $A = [a_{ij}]$ be the matrix of f with respect to bases $\{x_1, x_2, \ldots, x_m\}$ of U and $\{y_1, y_2, \ldots, y_n\}$ of V, then we know that $f(x_k, y_t) = a_{kt}$.

Now, it is easy to see that

$$f(x_k, y_t) = \sum_{i=1}^{m}\sum_{j=1}^{n} \alpha_{ij} f_{ij}(x_k, y_t) \Rightarrow f = \sum_{i=1}^{m}\sum_{j=1}^{n} \alpha_{ij} f_{ij}$$

$\Rightarrow \{f_{ij}\}$ $i = 1, 2, \ldots, m, j = 1, 2, \ldots, n$; is a basis of $L(U, V; F)$.

Hence, dim $(L(U, V; F)) = mn$. Thus, dim $(L(V, V; F)) = n^2$. Therefore $L(V, V; F)$ is isomorphic to vector space of all $n \times n$ matrices over F.

Hence, the concepts of matrices can be carried over to $L(V, V; F)$. We know that in the field of reals $1 + 1 \neq 0$. Now onwards we consider field F such that $1 + 1 \neq 0$.

2 SYMMETRIC BILINEAR FUNCTIONALS

As discussed above we take $U = V$. Now, if $f \in L(V, V; F)$ i.e. f is a bilinear form on V and $\{x_1, x_2 \ldots, x_n\}$ is a basis of V, then $A = [a_{ij}]$ is a matrix of f with respect to $\{x_1, x_2, \ldots, x_n\}$ if $f(x_i, x_j) = a_{ij}$ and for any $x, y \in V$ if

$$x = \sum_{i=1}^{n} a_i x_i, \; y = \sum_{i=1}^{n} \beta_i x_i ,$$

then

$$f(x, y) = \begin{bmatrix} \alpha_1 \\ \alpha_2 \\ \vdots \\ \alpha_n \end{bmatrix}^t A \begin{bmatrix} \beta_1 \\ \beta_2 \\ \vdots \\ \beta_n \end{bmatrix} = X^t AY$$

where X and Y are the coordinate matrices of x and y, respectively.

Definition 9.2.1 A bilinear form f on V is called symmetric iff $f(x, y) = f(y, x)$ for all $x, y \in V$. *A* bilinear form f on V is called skew-symmetric if $f(x, y) = -f(y, x)$ for all $x, y \in V$.

If f is skew-symmetric, then $f(x, x) = -f(x, x) \Rightarrow f(x, x) = 0$ provided $1 + 1 \neq 0$.

Let us consider

$$A = \begin{bmatrix} 1 & 2 \\ 2 & 1 \end{bmatrix}$$

And $\{(0, 1), (0, 1)\}$ basis of $\mathbb{R}^2$. Let

$$f((1, 0), (1, 0)) = 1, f((1, 0), (0,1)) = 2,$$
$$f((1, 0), (1, 0)) = 2, f((0, 1), (0,1)) = 1.$$

Let $x = (x_1, x_2) \in \mathbb{R}^2$ and $y = (y_1, y_2) \in \mathbb{R}_2$, then

$$\begin{aligned} f(x, y) &= f(x_1 (1, 0) + x_2 (0, 1), y_1 (1, 0) + y_2 (0, 1)) \\ &= x_1 y_1 f((1, 0), (1, 0)) + x_1 y_2 f((1, 0), (0, 1)) \\ &\quad + x_1 y_1 f((1, 0), (1, 0)) + x_2 y_2 f((0, 1), (0, 1)) \\ &= x_1 y_1 + 2x_1 y_2 + 2x_2 y_1 + x_2 y_2 \\ &= x_1(y_1 + 2y_2) + x_2(y_2 + 2y_1). \end{aligned}$$

Now, $\quad f(y, x) = y_1(x_1 + 2x_2) + y_2(x_2 + 2x_1) = y_1 x_1 + 2y_1 x_2 + y_2 x_2$

$\Rightarrow \quad f(x, y) = f(y, x) \Rightarrow f$ is symmetric.

Therefore, in general we can prove the following elementary result.

Theorem 9.2.2: A bilinear form is symmetric iff its matrix is symmetric.

Proof: Let f be bilinear form on V and $\{x_1, x_2, \ldots, x_n\}$ be a basis. Hence, for any $x, y \in V$

$$f(x, y) = X' AY$$

where X and Y are coordinate matrices of x and y, respectively. Since $X' AY$ is a scalar, we have

$$(X' AY)' = Y' At\, X = X' AY \Rightarrow f(x, y) = \text{as } A' = A.$$

Conversely, let $A = [a_{ij}]$ be a matrix representation of symmetric form f, then

$$f(x_i, x_j) = f(x_j, x_i) \Rightarrow a_{ij} = a_{ji}$$

for all i and $j \Rightarrow A$ is symmetric.

Similarly, we can show that a bilinear form f is skew-symmetric iff the matrix of f with respect to any basis is skew-symmetric.

Let

$$L_S = \{f \in L(V, V; F) | f(x, y) = f(y, x) \text{ for all } x, y \in V\}$$

and

$L_A = \{f \in L(V, V; F) | f(x, y) = -f(y, x) \text{ for all } x, y \in V\}$.

Trivially for any $f, g \in L_S$,

$(f - g)(x, y) = f(x, y) - g(x, y) = f(y, x) - g(y, x) = (f - g)(y, x)$

$\Rightarrow f - g \in L_S$ and $(\alpha f)(x, y) = \alpha f(x, y) = \alpha f(y, x) = (\alpha f)(y, x)$

$\Rightarrow$ af ∈ LS Þ LS is a subspace of $L(V, V; F)$.

Similarly, we can show that L_A is also a subspace of $L(V, V; F)$.

Now, using Theorem 9.2.2, it is easy to see that $L_S \cong$ vector space of skew-symmetric matrices of order n.

Trivially the map will be $f \to m(f)$, where $m(f)$ is the matrix of f with respect to some fixed basis.

Hence, as done is Chapter 3, dim $L_S = \dfrac{n(n+1)}{2}$ and dim $L_A = \dfrac{n(n-1)}{2}$.

Further, if $f \in L_S \cap L_A \Rightarrow f(x, y) = f(x, y)$ for all $x, y \in V$ and $f(x, y) = -f(x, y)$ for all $x, y \in V \Rightarrow f(x, y) = 0$ for all $x, y \in f = 0$. Hence

$$\dim (L_S + L_A) = \dim L_S + \dim L_A = \frac{n(n+1)}{2} + \frac{n(n-1)}{2}$$

$$= n^2 = \dim (L(V, V; F)).$$

Therefore, the following theorem is derived.

Theorem 9.2.3: For any vector space V, $L(V, V; F) = L_S \oplus L_A$.

Note: an independent proof for Theorem 9.2.3 can be given as for any $f \in L(V, V; F)$, take

$$\eta(x, y) = \frac{f(x, y) + f(y, x)}{2}$$

and
$$\phi(x, y) = \frac{f(x, y) - f(y, x)}{2}$$

then η and ϕ are symmetric and skew-symmetric and $f = \eta + \phi$.

Now, due to Theorem 9.2.3, we can say that every bilinear form can be uniquely written as a sum of symmetric and skew-symmetric form.

3 QUADRATIC FORM

Definition 9.3.1: If V is a vector space over F, then a map $q : V \to F$ is called *quadratic form* if there is a bilinear symmetric form f such that $q(x) = f(x, x)$ for all $x \in V$.

Now, take $\{x_1, x_2, \ldots, x_n\}$ as *a* basis of V and

$$X = \begin{bmatrix} \alpha_1 \\ \alpha_2 \\ \vdots \\ \alpha_n \end{bmatrix}$$

a coordinate matrix of $x \in V$, the $q(x) = f(x, x) = X^t AX$, where A is the matrix of f. Thus,

$$q(x) = [\alpha_1, \alpha_2, \ldots, \alpha_n] \begin{bmatrix} a_{11} & a_{12} & \cdots & a_{1n} \\ a_{21} & a_{22} & \cdots & a_{2n} \\ \vdots & \vdots & \vdots & \vdots \\ a_{n1} & a_{n2} & \cdots & a_{nn} \end{bmatrix} \begin{bmatrix} \alpha_1 \\ \alpha_2 \\ \vdots \\ \alpha_n \end{bmatrix} = \sum_{i,j}^{n} a_{ij}\alpha_i\alpha_j \, .$$

If q is a quadratic form on V relative to symmetric bilinear form f on V, then

$$f(x, y) = \frac{1}{2}[q(x + y) - q(x) - q(y)]$$

$$\begin{aligned} q(x + y) - q(x) - q(y) &= f(x + y, x + y) - f(x, x) - f(y, y) \\ &= f(x, x) + f(x, y) + f(y, x) + f(y, y) - f(x, x) - f(y, y) \\ &= f(x, y) + f(y, x) \\ &= f(x, y) + f(x, y) \text{ as } f \text{ is symmetric} \\ &= 2f(x, y) \end{aligned}$$

$$\Rightarrow \qquad f(x, y) = \frac{1}{2}[q(x + y) - q(x) - q(y)].$$

This is called polar form of f.

We can easily verify that

$$f(x, y) = \frac{1}{4}[q(x + y) - q(x - y)].$$

Example 9.3.2

Let $q(x, y) = x^2 - xy + y^2$, then the symmetric matrix for q is $\begin{bmatrix} 1 & -\frac{1}{2} \\ -\frac{1}{2} & 1 \end{bmatrix}$.

If $q(x, y, z) = x^2 + xz - yz + 3y^2 - 4yz + 4z^2$, then the symmetric matrix for q is

$$A = \begin{bmatrix} 1 & 0 & \frac{1}{2} \\ 0 & 3 & -2 \\ \frac{1}{2} & -2 & 4 \end{bmatrix}$$

If $q(x, y, z) = xz + yz + x^2$, then the symmetric matrix for q is

$$A = \begin{bmatrix} 1 & 0 & \frac{1}{2} \\ 0 & 0 & \frac{1}{2} \\ \frac{1}{2} & \frac{1}{2} & 0 \end{bmatrix}.$$

It may be note that is finding the matrix $A = [a_{ij}]$ for the quadratic form $q(x_1, x_2, \ldots, x_n)$ we proceed as follows:

Put

$$a_{11} = \text{coeff. } (x^2{}_1).\ a_{12} = \frac{1}{2}(\text{coeff. } (x_1\, x_2), \ldots, a_{1n} = \frac{1}{2}(\text{coeff. } (x_1\, x_n))$$

$$a_{21} = \frac{1}{2}\ (\text{coeff. } (x_2 x_1)),\ a_{22} = \text{coeff. } (x^2{}_1), \ldots, a_{2n} = \frac{1}{2}(\text{coeff. } (x_2\, x_n))$$

$$\vdots \qquad\qquad \vdots \qquad\qquad \vdots\ \vdots\ \vdots \qquad \vdots$$

and so on. Lastly, we prove the following result.

Theorem 9.3.3: For symmetric matrix A of a quadratic form q on V if X' $AX = 0$ for all coordinate matrix X, then $A = 0$.

Proof: Choose

$$X = \overrightarrow{i^{th}} \begin{bmatrix} 0 \\ 0 \\ \vdots \\ 1 \\ 0 \\ \vdots \\ 0 \end{bmatrix},$$

Then trivially

$$[0\ 0\ \dots\ 1\ 0\ \dots\ 0]\begin{bmatrix} a_{11} & a_{12} & \cdots & a_{1n} \\ a_{21} & a_{22} & \cdots & a_{2n} \\ \vdots & \vdots & \vdots & \vdots \\ a_{n1} & a_{n2} & \cdots & a_{nn} \end{bmatrix}\begin{bmatrix} 0 \\ 0 \\ \vdots \\ 1 \\ 0 \\ \vdots \\ 0 \end{bmatrix} = a_{ii}$$

$\Rightarrow a_{ii} = 0$ for all i.

Now, choose

$$X = \begin{bmatrix} 0 \\ 0 \\ \vdots \\ 1 \\ \vdots \\ 1 \\ \vdots \\ 0 \end{bmatrix} \begin{matrix} \\ \\ \\ \leftarrow i^{\text{th}} \\ \leftarrow j^{\text{th}} \\ \\ \\ \\ \end{matrix}$$

Then as done above it is easy to see that $X^t AX = a_{ij} + a_{ji}$
$\Rightarrow a_{ij} + a_{ji} = 0$ since A is symmetric
$\Rightarrow 2a_{ij} = 0 \Rightarrow a_{ij} = 0$ for all i and j
$\Rightarrow A = 0$.

Corollary 9.3.4: Each quadratic form corresponds to unique symmetric matrix.

Proof: Let A and B be matrices for quadratic form q on V, then $X^t AX = X^t BX$ for all coordinate matrices X. Hence, $X^t(A - B)X = 0$, but $A - B$ is symmetric (trivial) $\Rightarrow A - B = 0 \Rightarrow A = B$.

EXERCISE

1. If $f : \mathbb{R}^2 \times \mathbb{R}^2 \to \mathbb{R}$, such that

$$f((x_1, x_2, x_3), (y_1, y_2)) = x_1y_2 - 4x_2y_1 + 3x_3y_1 - 5x_2y_2,$$

then compute the matrix of f with respect to:

(*a*) Standard bases of $\mathbb{R}^3$ and $\mathbb{R}^2$.

(*b*) $B = \{(1, 0, 1), (0, 1, -1), (1, 1, 1)\}$ and $B' = \{(1, 3), (2, 3)\}$ as bases of $\mathbb{R}^3$ and $\mathbb{R}^2$, respectively.

(*c*) Verify Theorem 9.1.5.

2. If $f : \mathbb{R}^3 \times \mathbb{R}^3 \to \mathbb{R}$, such that
$f((x_1, x_2, x_3), (y_1, y_2, y_3)) = x_1 y_3 - 4x_2y_1 + 7x_1y_2 + 5x_2y_2 - 2x_3y_3$, then find the matrix of f with respect to standard basis of $\mathbb{R}^3$.

3. Find the symmetric matrices of the following quadratic forms:

(*a*) $q(x, y) = x^2 - 2xy + y^2$.

(*b*) $q(x_1, x_2, x_3) = x^2{}_1 + 2x_1x_3 - 4x_3{}^2 + x_2x_3 + 5x_1x_2$.

(*c*) $q(x_1, x_2, x_3) = x^2{}_1 + 5x_1x_2 - 7x_3x_3 + 10x_1x_3 + 4x_1{}^2$.

(*d*) $q(x_1, x_2, x_3) = x^2{}_1 + x_1x_2 - x_2x_3$.

Index